Solid Mechanics with Python

Non-Cartesian Coordinates

Series in Computational Methods

Series Editor: Gui-Rong Liu *(University of Cincinnati, USA)*

Published

Vol. 6 *Solid Mechanics with Python: Non-Cartesian Coordinates*
by G R Liu

Vol. 5 *Solid Mechanics with Python: Cartesian Coordinates*
by G R Liu

Vol. 4 *Vector Calculus: Formulations, Applications and Python Codes*
by G R Liu

Vol. 3 *Mechanics of Materials: Formulations and Solutions with Python*
by G R Liu

Vol. 2 *Calculus: Formulations and Solutions with Python*
by G R Liu

Vol. 1 *Numbers and Functions: Theory, Formulation, and Python Codes*
by G R Liu

Series in Computational Methods
Volume 6

Solid Mechanics with Python

Non-Cartesian Coordinates

G. R. Liu
University of Cincinnati, USA

World Scientific

NEW JERSEY • LONDON • SINGAPORE • BEIJING • SHANGHAI • HONG KONG • TAIPEI • CHENNAI • TOKYO

Published by

World Scientific Publishing Co. Pte. Ltd.
5 Toh Tuck Link, Singapore 596224
USA office: 27 Warren Street, Suite 401-402, Hackensack, NJ 07601
UK office: 57 Shelton Street, Covent Garden, London WC2H 9HE

Library of Congress Cataloging-in-Publication Data
Names: Liu, G. R. (Gui-Rong) author
Title: Solid mechanics with Python: non-Cartesian coordinates / G. R. Liu, University of Cincinnati, USA.
Description: New Jersey : World Scientific, [2026] | Series: Series in computational methods ; vol. 6 | Includes bibliographical references and index.
Identifiers: LCCN 2026001672 | ISBN 9789819827015 hardcover | ISBN 9789819827022 ebook for institutions | ISBN 9789819827039 ebook for individuals
Subjects: LCSH: Elastic analysis (Engineering)--Computer programs | Python (Computer program language) | Mechanics, Applied--Mathematical models | Engineering mathematics--Data processing
Classification: LCC TA653 .L58 2026
LC record available at https://lccn.loc.gov/2026001672

British Library Cataloguing-in-Publication Data
A catalogue record for this book is available from the British Library.

For any available supplementary material, please visit
https://www.worldscientific.com/worldscibooks/10.1142/14697#t=suppl

Desk Editors: Nambirajan Karuppiah/Steven Patt

Typeset by Stallion Press
Email: enquiries@stallionpress.com

About the Author

Gui-Rong-Liu received his Ph.D. from Tohoku University, Japan, in 1991. He was a Postdoctoral Fellow at Northwestern University, USA, from 1991 to 1993. He was a Professor at the National University of Singapore until 2010. He is currently a Professor at the Department of Aerospace Engineering and Engineering Mechanics, University of Cincinnati, USA. He was the Founder of the Association for Computational Mechanics (Singapore) (SACM) and served as the President of SACM until 2010. He served as the President of the Asia-Pacific Association for Computational Mechanics (APACM) (2010–2013) and an Executive Council Member of the International Association for Computational Mechanics (IACM) (2005–2010; 2020–2026). He authored a large number of journal papers and books, including two bestsellers, *Mesh Free Method: Moving Beyond the Finite Element Method* and *Smoothed Particle Hydrodynamics: A Meshfree Particle Methods.* He is the Editor-in-Chief of the *International Journal of Computational Methods* and served as an Associate Editor for *IPSE* and *MANO.* He is the recipient of numerous awards, including the Singapore Defence Technology Prize, NUS Outstanding University Researcher Award, NUS Best Teacher Award, APACM Computational Mechanics Award, JSME Computational Mechanics Award, ASME Ted Belytschko Applied Mechanics Award, Zienkiewicz Medal from APACM, AJCM Computational Mechanics Award, Humboldt Research Award, SACM Medal from the Association of Computational Mechanics (Singapore), and the Master Engineering Educator Award (UC, CEAS). He has been listed as one among the world's top 1% most influential scientists (Highly Cited Researchers) by Thomson Reuters for a number of years.

Contents

Chapter 1

Introduction

1.1 Computational methods

This book is part of a series that explores the exciting field of computational methods. The goal is to provide a solid foundation in this discipline by combining theoretical principles with practical applications.

The series encompasses general methods and techniques used throughout science, technology, engineering, and mathematics (STEM) education, as well as in research across various scientific and engineering disciplines. It may be viewed as a comprehensive encyclopedia of computational methods, covering the theory, formulation, and coding techniques necessary to get started. Readers who have completed primary and lower secondary education (approximately nine years of schooling) can begin with the foundational volumes on mathematics. As they progress through the mechanics volumes, they will be equipped to tackle research and design projects that leverage computational methods.

To take advantage of rapid advancements in computer technology, code examples are extensively integrated throughout the series. This integration enables concepts, theories, and formulations to be vividly demonstrated through practical examples with clear visualizations. Python will initially serve as the primary programming language in the foundational volumes, with the possibility of introducing other languages for more advanced topics. By effectively utilizing code, readers can devote more time to understanding the core principles. The computer will handle the complex calculations, tedious formula derivations, time-consuming operations, and extensive data processing, allowing readers to focus on the broader concepts.

This approach empowers readers to explore the fascinating world of computational methods with both strong theoretical understanding and practical skills.

1.2 Why start and contribute to this book series

The editor-in-chief of this book series, Dr. G. R. Liu, has been working in areas related to computational methods for over 45 years. He developed his first FEM code for nonlinear problems in 1980 and has since published more than 600 journal papers and over a dozen monographs in this field. After decades of studying, applying, and developing computational methods, he sought a way to help others learn them more effectively, systematically, and smoothly. He concluded that creating this book series would be the best way to achieve that goal.

1.3 Solid mechanics

Volume 5 of this series focuses on solid mechanics, emphasizing the analysis of stresses and displacements in loaded, bulky, linear elastic solids defined in the Cartesian coordinate system. This volume further extends the study to non-Cartesian coordinate systems.

All physical devices and structures used in daily life are made of solids. This book introduces the subject of **Solid Mechanics** for linear elastic materials, which is fundamental to the safe and effective analysis and design of devices and structures in science and engineering. Solid mechanics explores the relationships among forces, stresses, strains, and deformations within solid materials.

This volume covers the fundamental principles of solid mechanics, with particular emphasis on the mechanical behavior of solids in two and three dimensions under various types of loading. The solids are described in non-Cartesian coordinate systems, such as polar, cylindrical, and spherical systems. Consequently, the solids studied here have geometries best represented in these coordinates, such as plates with circular holes experiencing stress concentration and circular-arch bridges.

The methodologies and techniques for solving solid mechanics problems are thoroughly explained, formulated, and demonstrated through numerous examples and illustrations. Two major approaches — the stress method and the displacement method — are introduced. Python is employed to derive formulations, solve problems symbolically or numerically, and plot results graphically.

The material in this book can be learned in various ways, including classroom teaching, online courses, or self-study. Since Python codes are provided, readers can clearly see how the theory is formulated and how solutions are obtained in terms of formulas, numerical values, and graphical results. By experimenting with the provided codes and developing their own solutions to related problems, readers can further deepen their understanding.

The book is written in **Jupyter Notebook** format, integrating theory, formulation, and coding into a single document. This provides a unified environment for reading, exercises, practice, and exploration.

This book references Ref. [1], which served as the author's textbook during his university years; Refs. [2, 3], which were frequently used as reference books during his PhD and research work; and Refs. [4,5], which were adopted as textbooks and reference materials for university teaching.

Numerous examples are presented in this volume. For instance, a complete set of analytical solutions (displacements, strains, and stresses) for circular-arch bridges is derived here for the first time. A stress concentration factor of 6 is found by the author for plates with holes subjected to combined stresses. However, the application of solid mechanics is vast, and the examples and case studies presented in this book are intended to be illustrative rather than exhaustive.

1.4 Who may read this book

This book is written for beginners interested in learning computational methods for solving problems in science, engineering, and nature. The target audience includes university students, graduate students, researchers, and professionals in any STEM discipline. Engineers and practitioners may also find it useful for building a strong foundation in the core concepts of computational methods.

Readers are encouraged to first read Volume 5 of this series, *Mechanics of Materials: Formulations and Solutions with Python* [6], or take an introductory course in mechanics of materials.

This book is written for human readers. It may not be used for training or testing any automated systems without the author's permission.

1.5 Use of the book and codes

Readers who purchase the book may contact the author at **liugr100@gmail.com** to request a free soft copy in Jupyter Notebook format, including the codes (which may be updated), for academic use after registration.

Conditions of use

The following conditions apply to the use of the book and the codes developed by the author, whether in hard copy, soft copy, or any other medium or format:

- **User responsibility**: Users are solely responsible for any risks associated with the use of the book, codes, or techniques.
- **Code purpose**: The codes are primarily intended to demonstrate concepts and may not be optimized for efficiency or robustness. Many of the codes have not been independently tested.
- **Limited use**: The book and codes are for personal use by the individual purchaser. Redistribution or copying without the author's permission is prohibited.
- **No user support**: No technical support is provided for the codes.
- **Citation**: Proper citation and acknowledgment are required when using the book, codes, ideas, or techniques.

The codes often rely on various external packages and modules. Their behavior can be affected by the versions of Python and these packages. If a code example does not run as expected, a version mismatch may be the cause. You can verify the versions within a code cell in Jupyter Notebook. For example, to check the current Python environment version, you can use the following:

```
!python -V                    #! is used to execute an external command
```

```
Python 3.9.16
```

```
!jupyter notebook --version
```

```
6.1.5
```

Troubleshooting version issues:

If you encounter a version mismatch error, you can either adapt the code to match your installed versions or install the required versions on your system. Searching the web using the exact error message often provides useful solutions or leads. Additionally, large language models may assist in identifying possible fixes. This approach is the author's preferred method for resolving code execution issues.

Since the text is integrated with code cells, a code cell may depend on the successful execution of preceding cells. If an error occurs while running a cell, try executing all earlier cells first.

Learning python:

This book does not cover the basics of Python, as many excellent resources are available online. Readers interested in a concise introduction to using Python for scientific computations may refer to Chapters 2 and 3 in Ref. [7]. Python is relatively easy to learn, and readers can use the codes and examples provided in this book to gain practical experience while studying the technical topics.

1.6 Use of external modules or dependencies

Module imports in Jupyter notebooks:

Frequently used modules in each chapter are imported in the first code cell, which should be executed before running the subsequent cells. If you encounter a name error (such as `NameError: name 'sp' is not defined`), return to the beginning of the chapter and execute the initial import cell.

```
#Often used external modules are in commonImports placed in folder grbin

import sys                              # for accessing the computer system
sys.path.append('../grbin/')              # Add in the path to your system

from commonImports import *          # Import dependencies from '../grbin/'
import grcodes as gr                     # Import the module of the author
importlib.reload(gr)                # For in case, when grcodes is modified

from continuum_mechanics import vector
from continuum_mechanics.solids import sym_grad, strain_stress
init_printing(use_unicode=True)          # For latex-like quality printing

# Digits in print-outs
np.set_printoptions(precision=4,suppress=True,
                    formatter={'float_kind': '{:.4e}'.format})
```

The above cell includes a common import file, `commonImports.py` (given as follows), which contains all the frequently used modules in this book.

This helps avoid repeatedly importing lengthy external modules and reduces redundancy:

```
# commonImports.py:
from __future__ import print_function
import numpy as np                                # for numerical computation
import sympy as sp                               # sympy module for computation
import numpy.linalg as lg                         # numpy linear algebra module
import scipy.linalg as sg                         # scipy linear algebra module
import scipy.integrate as si
from scipy.stats import ortho_group
import importlib
import itertools
import inspect
import csv
import pandas as pd
import random
from IPython.display import display, Math

import autograd.numpy as anp                            # Thinly-wrapped numpy
from autograd import grad

from grcodes import drawArrow, plotfig, printM, printx # frequently used

import math as ma
from sympy import sin, cos, exp, symbols, lambdify, init_printing, latex
from sympy import pi, Matrix, sqrt, oo, integrate, diff, Derivative
from sympy import MatrixSymbol, simplify, nsimplify, Function
from sympy import factor, expand, nsimplify, Matrix, ordered, hessian
from sympy.plotting import plot as splt
init_printing(use_unicode=True) # for latex-like quality printing format

from matplotlib.ticker import MultipleLocator
import matplotlib.pyplot as plt                         # for plotting figures
import matplotlib as mpl
```

Module reloading:

Importing the same module multiple times generally does not cause any issues. Python caches imported modules and skips re-importing them if they are already loaded. However, if you modify a module after it has been imported and want those changes to take effect, you need to reload the module. This can be done using the `importlib.reload()` function:

```
# grcodes was imported as gr, reload it when grcodes is modified
importlib.reload(gr);    # This is also included in the first code cell
```

To view the codes in the imported module, you can use the following command by uncommenting it. (Note: This may generate a long output):

```
import inspect
#source_code = inspect.getsource(gr)    # to view everything in gr module
# to view any function in gr:
source_code = inspect.getsource(gr.checkEquilibriumPolar)
print(source_code)
```

```
def checkEquilibriumPolar(σvg, R, br, bθ):
    '''Check the equilibrium equations for a given stress vector and
    body forces br, bθ in the polar coordinate system.'''
    r, θ = R
    equi1=(σvg[0].diff(r)+σvg[2].diff(θ)/r+(σvg[0]-σvg[1])/r+br)
    equi2=(σvg[2].diff(r)+σvg[1].diff(θ)/r+2*σvg[2]/r+bθ)

    return (equi1.simplify(), equi2.simplify())
```

Editing codes:

- You can use any text editor to view and modify the codes.
- Alternatively, you can copy the code into a Jupyter Notebook cell and edit it directly.

Using functions from imported modules:

To call a function from an imported module, use the module name followed by a dot (.) and then the function name. For example, if you import a module named `gr` that contains a function called `printx`, you would use it as follows:

```
x = 8
gr.printx('x')        # when gr. is used, printx() is from grcodes module
```

```
x = 8
```

1.7 Use of help()

Exploring modules and objects:

To view detailed information about a module or its objects after importing, you can use the `help()` function. This helps explore the available functionalities and attributes in a module. For example:

```
help(gr.strainStressFrDisp)
```

```
Help on function strainStressFrDisp in module grcodes:

strainStressFrDisp(urg, uθg, r, θ, E, ν)
    Compute 2D strains and stresses for given displacements urg and
    uθg in polar coordinates r and θ for material with E and ν.
    return: εvg, σvg (strain and stress in vector form)
```

Code comments:

The code cells throughout this book include comments that offer additional explanations and guidance. These comments, marked with the hash symbol `#`, are placed within the code cells to the right of the relevant lines, providing helpful notes and quick insights for readers.

Acknowledgments

Wikipedia has served as a valuable reference in preparing this book. Both **NumPy** and **SymPy** were used extensively in developing the demonstration codes. Additionally, discussions with **ChatGPT**, **Gemini**, and **DeepSeek** were instrumental in refining the code and shaping the content of this book.

References

[1] S. Timoshenko and J. N. Goodier, *Theory of Elasticity*, McGraw-Hill, New York, 1970.
[2] J. D. Achenbach, *Wave Propagation in Elastic Solids*, North Holland, Amsterdam, 1984.
[3] Y. C. Fung, *Foundations of Solid Mechanics*, Prentice-Hall of India, New Delhi, 1968.
[4] A. P. Boresi, K. Chong and J. D. Lee, *Elasticity in Engineering Mechanics*, John Wiley & Sons, New York, 2010.
[5] A. P. Boresi, R. J. Schmidt and O. M. Sidebottom, *Advanced Mechanics of Materials*, John Wiley & Sons, New York, 1985.
[6] G. R. Liu, *Mechanics of Materials: Formulations and Solutions with Python*, World Scientific, New Jersey, 2024.
[7] G. R. Liu, *Machine Learning with Python: Theory and Applications*, World Scientific, New Jersey, 2023.

Chapter 2

Formulations in Non-Cartesian Coordinates

Let us import the necessary external modules or dependencies for later use in this chapter:

```
# Place cursor in this cell, and press Ctrl+Enter to import dependencies.
import sys                                    # for accessing the computer system
sys.path.append('../grbin/')                     # Add in the path to your system

from commonImports import *             # Import dependencies from '../grbin/'
import grcodes as gr                         # Import the module of the author
#importlib.reload(gr)                     # When grcodes is modified, reload it

from continuum_mechanics import vector
from continuum_mechanics.solids import sym_grad, strain_stress
init_printing(use_unicode=True)            # For latex-like quality printing
np.set_printoptions(precision=4,suppress=True)     # Digits in print-outs
```

```
#!pip install continuum_mechanics                        # if not yet installed
```

2.1 Introduction

In the analysis and design of solid structures and devices, a wide range of solid mechanics problems in multiple dimensions must be solved to determine displacements, stresses, and strains. Many valuable solutions have been successfully developed using the **Cartesian coordinate system** [1]. However, for certain problems, it is often more convenient and insightful to formulate and solve them in **non-Cartesian coordinate systems**.

For example, problems with **axial symmetry** are naturally expressed in a **cylindrical coordinate system**, while those with **spherical**

symmetry are more suitably analyzed in a **spherical coordinate system**. In such cases, the governing equations can often be simplified — sometimes even reduced to one-dimensional forms — greatly facilitating analytical or numerical solutions.

A variety of coordinate systems serve as mathematical frameworks for modeling different types of physical problems. A particularly important family is the **orthogonal curvilinear coordinate systems**, which includes Cartesian, polar, cylindrical, and spherical coordinates, among others. These systems have been extensively studied; readers interested in further details may refer to Ref. [2] or the Wikipedia page on Curvilinear coordinates (https://en.wikipedia.org/wiki/Curvilinear_coordinate). In addition, several useful Python tools related to this topic can be found in the Continuum Mechanics (https://continuum-mechanics.readthedocs.io) library, which will also be utilized throughout this book.

When working in a non-Cartesian coordinate system, it is essential to establish a complete and consistent set of governing equations. The most practical approach is to begin with the well-known equations formulated in the Cartesian system and then systematically **transform** them into their corresponding non-Cartesian forms. This transformation process requires defining the relationships between Cartesian and non-Cartesian coordinates.

In an **orthogonal curvilinear coordinate system**, each point within a solid is defined by the intersection of three mutually perpendicular curved surfaces. These surfaces form locally orthogonal axes, each associated with a **scale factor**, typically denoted (α, β, γ) in three dimensions. For the Cartesian coordinates (x, y, z), the scale factors are simply $(1, 1, 1)$, making Cartesian space the foundational reference system. In contrast, in non-Cartesian curvilinear systems, these scale factors generally vary with position and play a central role in coordinate transformations. With appropriate transformation rules, many relationships originally derived in the Cartesian system can be extended to curvilinear coordinates — a concept that will be frequently applied in this book.

Building on this foundation, we now proceed to develop formulations and solutions in several non-Cartesian coordinate systems, focusing on the most commonly used ones: **polar**, **cylindrical**, and **spherical** coordinates. To establish the core principles, we will first present the **polar–Cartesian** relations in detail, followed by the **cylindrical–Cartesian** and **spherical–Cartesian** transformations. Where appropriate, existing Python snippets and modules will be introduced to facilitate computation.

Since it would be impractical to include every possible formula, the emphasis will be placed on the most relevant relationships and techniques for the topics addressed in this book. The **methods, derivations, and Python implementations** presented here are broadly applicable and can be readily extended to other curvilinear coordinate systems as needed.

2.2 Polar coordinate system

2.2.1 *Transformation between Cartesian and polar coordinates*

The **polar coordinate system** is one of the simplest and most commonly used curvilinear coordinate systems. It consists of two coordinates: the **radial distance** r and the **polar angle** θ. Their relationships with the Cartesian coordinates x and y are illustrated in Fig. 2.1.

Consider a point P in $\mathbb{R}^2$. It can be represented in **Cartesian coordinates** by (x, y) and in **polar coordinates** by (r, θ). Both coordinate systems share the same origin $(0, 0)$ as the reference point. The mathematical relationships between these two systems can be derived directly from geometry.

Expressing x and y in terms of r and θ,

$$x = r\cos\theta, \quad y = r\sin\theta \tag{2.1}$$

These expressions define x and y as elementary and differentiable functions of the independent variables r and θ [3].

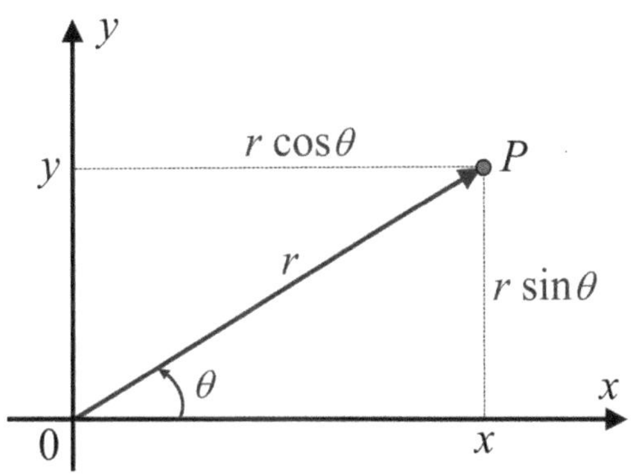

Figure 2.1. Polar coordinate system defined in relation to the Cartesian coordinate system.

Conversely, expressing r and θ in terms of x and y,

$$r = \sqrt{x^2 + y^2}, \quad \theta = \arctan \tfrac{y}{x} \tag{2.2}$$

where

- $0 \leq r < \infty$, $0 \leq \theta < 2\pi$,
- $\theta = \pi/2$ for $x = 0$ and $y > 0$, and $\theta = \frac{3\pi}{2}$ for $x = 0$ and $y < 0$.

The functions in Eq. (2.2) are also elementary and differentiable.

For convenience, the **position vector** in Cartesian coordinates can be written as [4]

$$\mathbf{X} = \begin{bmatrix} x \\ y \end{bmatrix} \tag{2.3}$$

and in polar coordinates as

$$\mathbf{R} = \begin{bmatrix} r \\ \theta \end{bmatrix} \tag{2.4}$$

A **position vector** defines the location of a point in a given coordinate system. The relationships between these vectors in different coordinate systems can be readily computed and visualized using symbolic tools such as **SymPy**.

2.2.2 *SymPy definition of position vectors and relations*

```
x, y, θ = sp.symbols('x, y, θ', real=True)  # define symbolic variables
r = sp.symbols('r', nonnegative=True)

X = sp.Matrix([x, y])         # Position vector in Cartesian coordinates
R = sp.Matrix([r, θ])             # Position vector in Polar coordinates

# Cartesian coordinates written as functions of polar coordinates:
x_rθ = r*sp.cos(θ)                    # x as a function expressed in polar
y_rθ = r*sp.sin(θ)
X_rθ = sp.Matrix([x_rθ, y_rθ])     # vector function of x and y in polar
display(Math(f" {latex(X)}={latex(X_rθ)}"))

# Polar coordinates written as functions of Cartesian coordinates:
r_xy = sp.sqrt(x**2+y**2)       # r as a function expressed in Cartesian
θ_xy  = sp.atan(y/x)
R_xy = sp.Matrix([r_xy,θ_xy])    # vector function of r & θ in Cartesian

display(Math(f" {latex(R)}={latex(R_xy)}"))
```

$$\begin{bmatrix} x \\ y \end{bmatrix} = \begin{bmatrix} r\cos(\theta) \\ r\sin(\theta) \end{bmatrix}$$

$$\begin{bmatrix} r \\ \theta \end{bmatrix} = \begin{bmatrix} \sqrt{x^2+y^2} \\ \operatorname{atan}\left(\frac{y}{x}\right) \end{bmatrix}$$

2.2.3 *Displacements transformations*

A displacement vector (or any other vector) expressed in polar coordinates, $\begin{bmatrix} u_r & u_\theta \end{bmatrix}^\top$, can be obtained by transforming the corresponding vector expressed in Cartesian coordinates, $\begin{bmatrix} u_x & u_y \end{bmatrix}^\top$, as follows:

$$\begin{bmatrix} u_r \\ u_\theta \end{bmatrix} = \underbrace{\begin{bmatrix} \cos\theta & \sin\theta \\ -\sin\theta & \cos\theta \end{bmatrix}}_{\mathbf{T}} \begin{bmatrix} u_x \\ u_y \end{bmatrix} = \begin{bmatrix} u_x\cos\theta + u_y\sin\theta \\ -u_x\sin\theta + u_y\cos\theta \end{bmatrix} \tag{2.5}$$

Here, $\mathbf{T}$ is the transformation matrix that relates vector components between the two coordinate systems. A detailed derivation of Eq. (2.5) can be found in Chapter 2 of Ref. [5].

Using Eq. (2.5), we have the following special cases:

- **When** $\theta = 0$: The point P lies on the x-axis, with $r = x$, leading to the identities

$$u_r = u_x, \quad u_\theta = u_y \tag{2.6}$$

- **When** $\theta = \pi/2$: The point P lies on the y-axis, with $r = y$, leading to the identities

$$u_r = u_y, \quad u_\theta = -u_x \tag{2.7}$$

These identities are illustrated in Fig. 2.2.

These results provide physical insight into how displacement components transform between the two coordinate systems and will be useful in subsequent derivations.

To compute the displacement vector in Cartesian coordinates from that in polar coordinates, we simply use

$$\begin{bmatrix} u_x \\ u_y \end{bmatrix} = \underbrace{\begin{bmatrix} \cos\theta & -\sin\theta \\ \sin\theta & \cos\theta \end{bmatrix}}_{\mathbf{T}^\top} \begin{bmatrix} u_r \\ u_\theta \end{bmatrix} = \begin{bmatrix} u_r\cos\theta - u_\theta\sin\theta \\ u_r\sin\theta + u_\theta\cos\theta \end{bmatrix} \tag{2.8}$$

which is the reverse transformation of Eq. (2.5).

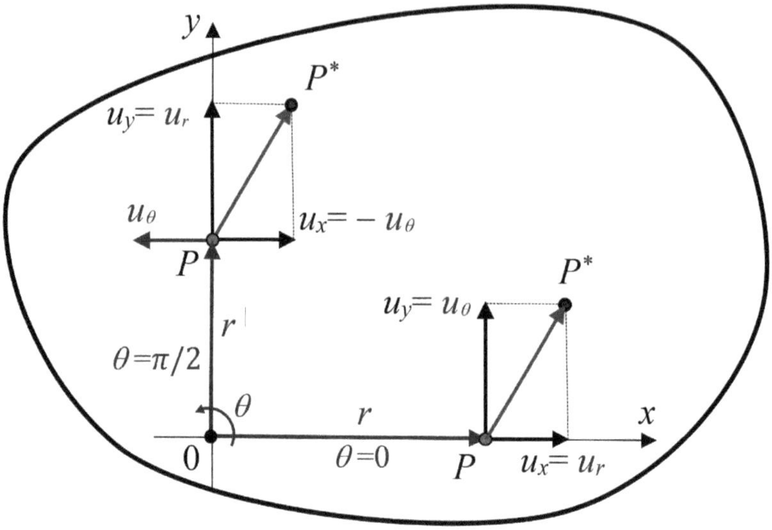

Figure 2.2. Relations between the displacements in Cartesian and polar coordinate systems at $\theta = 0$ and $\theta = \pi/2$.

This coordinate transformation applies to any **vector** (first-order tensor). It is a fundamental transformation that will be frequently used in our derivations for various field variables.

It is important to note that the transformation matrix **T** is **unitary**; therefore, its **transpose** can be used in place of its inverse [5].

In **SymPy**, these vectors and the transformation matrix can be defined as follows:

```
ux, uy = sp.symbols('u_x, u_y', real=True)
ur, uθ = sp.symbols('u_r, u_θ', real=True)

urθ = Matrix([ur, uθ])          # form the displacement vector in polar
uxy = Matrix([ux, uy])       # form the displacement vector in Cartesian

T = Matrix([[ sp.cos(θ), sp.sin(θ)],          # T: Transformation matrix
            [-sp.sin(θ), sp.cos(θ)]])                # Cartesian to Polar

display(Math(f"{latex(urθ)}={latex(T@uxy)}"))        # Cartesian to Polar
display(Math(f"{latex(uxy)}={latex(T.T@urθ)}"))      # Polar to Cartesian
```

$$\begin{bmatrix} u_r \\ u_\theta \end{bmatrix} = \begin{bmatrix} u_x \cos(\theta) + u_y \sin(\theta) \\ -u_x \sin(\theta) + u_y \cos(\theta) \end{bmatrix}$$

$$\begin{bmatrix} u_x \\ u_y \end{bmatrix} = \begin{bmatrix} u_r \cos(\theta) - u_\theta \sin(\theta) \\ u_r \sin(\theta) + u_\theta \cos(\theta) \end{bmatrix}$$

2.2.4 *Strain transformations*

Strains in the polar coordinate system can be derived using kinematic relations similar to those presented in Ref. [5] for the Cartesian coordinate system. The most straightforward approach is to apply coordinate transformations, since the strain–displacement relationships in Cartesian coordinates are already well established.

This process involves transforming **second-order tensors**, with detailed explanations provided in Section 4.8 of Ref. [5]. In this section, we will directly use the formulas introduced there.

The **strain tensor** is symmetric, and its relationship with displacements in Cartesian coordinates can be expressed as

$$\begin{bmatrix} \varepsilon_{xx} & \varepsilon_{xy} \\ \varepsilon_{yx} & \varepsilon_{yy} \end{bmatrix} = \begin{bmatrix} \frac{\partial u_x}{\partial x} & \frac{1}{2}\left(\frac{\partial u_x}{\partial y} + \frac{\partial u_y}{\partial x}\right) \\ \frac{1}{2}\left(\frac{\partial u_x}{\partial y} + \frac{\partial u_y}{\partial x}\right) & \frac{\partial u_y}{\partial y} \end{bmatrix} \tag{2.9}$$

Using the transformation matrix $\mathbf{T}$ from Eq. (2.5), the strains in polar coordinates can be computed as

$$\underbrace{\begin{bmatrix} \varepsilon_{rr} & \varepsilon_{r\theta} \\ \varepsilon_{r\theta} & \varepsilon_{\theta\theta} \end{bmatrix}}_{\boldsymbol{\varepsilon}_p} = \underbrace{\begin{bmatrix} \cos\theta & \sin\theta \\ -\sin\theta & \cos\theta \end{bmatrix}}_{\mathbf{T}} \underbrace{\begin{bmatrix} \varepsilon_{xx} & \varepsilon_{xy} \\ \varepsilon_{xy} & \varepsilon_{yy} \end{bmatrix}}_{\boldsymbol{\varepsilon}_C} \underbrace{\begin{bmatrix} \cos\theta & -\sin\theta \\ \sin\theta & \cos\theta \end{bmatrix}}_{\mathbf{T}^\top} \tag{2.10}$$

or

$$\boldsymbol{\varepsilon}_p = \mathbf{T}\boldsymbol{\varepsilon}_C\mathbf{T}^\top \tag{2.11}$$

where $\boldsymbol{\varepsilon}_p$ represents the strain tensor defined in polar coordinates and $\boldsymbol{\varepsilon}_C$ is the same strain tensor defined in Cartesian coordinates.

For the inverse transformation, we simply use the transpose of $\mathbf{T}$:

$$\boldsymbol{\varepsilon}_C = \mathbf{T}^\top\boldsymbol{\varepsilon}_p\mathbf{T} \tag{2.12}$$

Equation (2.10) can be simplified as follows:

```
εxx, εyy, εxy = symbols("ε_xx, ε_yy, ε_xy")
εrr, εθθ, εrθ = symbols("ε_rr, ε_θθ, ε_rθ")

ε_C = Matrix([[εxx, εxy],                    # ε_C: strain tensor in Cartesian
              [εxy, εyy]])

ε_p_ = Matrix([[εrr, εrθ],                     # ε_p: strain tensor in polar
               [εrθ, εθθ]])

ε_p = (T@ε_C@T.T).applyfunc(expand).applyfunc(simplify)
display(Math(f"{latex(εrr)}={latex(ε_p[0,0])}"))    # Cartesian to polar
display(Math(f"{latex(εθθ)}={latex(ε_p[0,1])}"))
display(Math(f"{latex(εrθ)}={latex(ε_p[1,1])}"))
```

$$\varepsilon_{rr} = \varepsilon_{xx}\cos^2(\theta) + \varepsilon_{xy}\sin(2\theta) + \varepsilon_{yy}\sin^2(\theta)$$

$$\varepsilon_{\theta\theta} = -\frac{\varepsilon_{xx}\sin(2\theta)}{2} + \varepsilon_{xy}\cos(2\theta) + \frac{\varepsilon_{yy}\sin(2\theta)}{2}$$

$$\varepsilon_{r\theta} = \varepsilon_{xx}\sin^2(\theta) - \varepsilon_{xy}\sin(2\theta) + \varepsilon_{yy}\cos^2(\theta)$$

We have derived the general relationship between the strains in the **polar** and **Cartesian** coordinate systems. Based on this relationship, the following special cases can be identified:

- When $\theta = 0$: For any point located on the x-axis, we have $r = x$, which leads to the following identities:

$$\begin{bmatrix} \varepsilon_{rr} & \varepsilon_{r\theta} \\ \varepsilon_{r\theta} & \varepsilon_{\theta\theta} \end{bmatrix} = \begin{bmatrix} \varepsilon_{xx} & \varepsilon_{xy} \\ \varepsilon_{xy} & \varepsilon_{yy} \end{bmatrix} \tag{2.13}$$

- When $\theta = \pi/2$: For any point on the y-axis, we have $r = y$, which leads to the following identities:

$$\begin{bmatrix} \varepsilon_{rr} & \varepsilon_{r\theta} \\ \varepsilon_{r\theta} & \varepsilon_{\theta\theta} \end{bmatrix} = \begin{bmatrix} \varepsilon_{yy} & -\varepsilon_{xy} \\ -\varepsilon_{xy} & \varepsilon_{xx} \end{bmatrix} \tag{2.14}$$

These identities will be used in subsequent derivations of the strain expressions in polar coordinates. They can be verified using the following code snippet (readers may uncomment the indicated lines to display the formulas):

```
dic_0 = {sin(θ):0, sin(2*θ):0, cos(θ):1, cos(2*θ):1}
#display(Math(f"{latex(ε_p_)}={latex(ε_p.subs(dic_0))}"))
dic_pi2 = {sin(θ):1, sin(2*θ):0, cos(θ):0, cos(2*θ):-1}
#display(Math(f"{latex(ε_p_)}={latex(ε_p.subs(dic_pi2))}"))
```

2.2.5 *Transformation of gradient of a function*

2.2.5.1 *Gradient of scalar functions*

Consider a scalar differentiable function $f(x, y)$ expressed in Cartesian coordinates x and y. The same function can also be represented in polar coordinates (r, θ) by substitution using Eq. (2.1):

$$f(x = r\cos\theta, y = r\sin\theta) = f(r, \theta).$$

The **gradient** of a scalar function $f(x, y)$ in Cartesian coordinates is a vector composed of its partial derivatives [4]:

$$\text{gradient of } f(x,y) : \nabla_{xy} f(x,y) \equiv \begin{bmatrix} \dfrac{\partial}{\partial x} \\ \dfrac{\partial}{\partial y} \end{bmatrix} f(x,y) = \begin{bmatrix} \dfrac{\partial f}{\partial x} \\ \dfrac{\partial f}{\partial y} \end{bmatrix}$$

Similarly, the gradient of a scalar function $f(x, y)$ in polar coordinates is a vector formed by its partial derivatives:

$$\text{gradient of } f(r,\theta) : \nabla_{r\theta} f(r,\theta) \equiv \begin{bmatrix} \dfrac{\partial}{\partial r} \\ \dfrac{\partial}{\partial \theta} \end{bmatrix} f(r,\theta) = \begin{bmatrix} \dfrac{\partial f}{\partial r} \\ \dfrac{\partial f}{\partial \theta} \end{bmatrix}$$

Often, it is necessary to express the gradient of $f(x, y)$ in terms of r, θ, and the gradient of $f(r, \theta)$ with respect to r and θ. A typical application is in relating strains between the two coordinate systems, since strains are derivatives of displacements in either Cartesian or polar coordinates, as illustrated in Eq. (2.9).

The relationship between the gradients of an arbitrary differentiable function $f(x, y)$ can be established using the **chain rule of differentiation** [3]:

$$\begin{aligned} \frac{\partial f}{\partial x} &= \frac{\partial f}{\partial r}\frac{\partial r}{\partial x} + \frac{\partial f}{\partial \theta}\frac{\partial \theta}{\partial x} = \frac{\partial f}{\partial r}\cos\theta - \frac{\partial f}{\partial \theta}\frac{\sin\theta}{r} \\ \frac{\partial f}{\partial y} &= \frac{\partial f}{\partial r}\frac{\partial r}{\partial y} + \frac{\partial f}{\partial \theta}\frac{\partial \theta}{\partial y} = \frac{\partial f}{\partial r}\sin\theta + \frac{\partial f}{\partial \theta}\frac{\cos\theta}{r} \end{aligned} \tag{2.15}$$

In matrix form, it becomes

$$\underbrace{\begin{bmatrix} \dfrac{\partial f}{\partial x} \\ \dfrac{\partial f}{\partial y} \end{bmatrix}}_{\nabla_{xy} f} = \underbrace{\begin{bmatrix} \cos(\theta) & -\frac{\sin(\theta)}{r} \\ \sin(\theta) & \frac{\cos(\theta)}{r} \end{bmatrix}}_{\nabla \mathbf{R}} \underbrace{\begin{bmatrix} \dfrac{\partial f}{\partial r} \\ \dfrac{\partial f}{\partial \theta} \end{bmatrix}}_{\nabla_{r\theta} f} \tag{2.16}$$

where $\nabla_{xy} f$ is the gradient of f with respect to x and y, $\nabla_{r\theta} f$ is the gradient of f with respect to r and θ, and $\nabla\mathbf{R}$ is the gradient of the position vector $\mathbf{R}$, which becomes a matrix:

$$\nabla\mathbf{R} = \left.\begin{bmatrix} \dfrac{\partial r(x,y)}{\partial x} & \dfrac{\partial r(x,y)}{\partial y} \\ \dfrac{\partial \theta(x,y)}{\partial x} & \dfrac{\partial \theta(x,y)}{\partial y} \end{bmatrix}\right|_{\substack{x=r\cos\theta \\ y=r\sin\theta}} = \begin{bmatrix} \cos(\theta) & -\frac{\sin(\theta)}{r} \\ \sin(\theta) & \frac{\cos(\theta)}{r} \end{bmatrix} \tag{2.17}$$

As observed, $\nabla\mathbf{R}$ contains the **partial derivatives of the polar coordinates** with respect to the Cartesian coordinates. It serves as the bridge connecting the gradients of the same function expressed in two different **orthogonal coordinate systems**.

2.2.5.2 *The Jacobian matrix*

It is worth noting that the transpose of $\nabla\mathbf{R}$ is, in fact, the **Jacobian matrix** [3]:

$$\underbrace{\begin{bmatrix} \dfrac{\partial f}{\partial x} \\ \dfrac{\partial f}{\partial y} \end{bmatrix}}_{\nabla_{xy} f} = \mathbf{J}^{\top} \underbrace{\begin{bmatrix} \dfrac{\partial f}{\partial r} \\ \dfrac{\partial f}{\partial \theta} \end{bmatrix}}_{\nabla_{r\theta} f} \tag{2.18}$$

In the literature, the Jacobian matrix is more commonly used. Since we have the function `gr.grad_vf()` to compute the gradient of any vector function, it is often employed to compute the Jacobian matrix (by taking its transpose).

2.2.5.3 *Python code for Jacobian matrix*

```
x, y, θ = sp.symbols('x, y, θ', real=True)  # define symbolic variables
X = sp.Matrix([x, y])                        # Position vector in x,y
r = sp.symbols('r', nonnegative=True)

r_xy = sp.sqrt(x**2+y**2)             # r as a function expressed in x,y
θ_xy  = sp.atan(y/x)                  # θ as a function expressed in x,y

R_xy = sp.Matrix([r_xy, θ_xy])        # Position vector R written in x,y
gradR = sp.simplify(gr.grad_vf(R_xy, X).subs({x:r*cos(θ), y:r*sin(θ)}))
display(Math(f" \\text{{Jacobian matrix J (from polar to Cartesian)}}\
                                                   ={latex(gradR.T)}"))
```

$$\text{Jacobian matrix } \mathbf{J} \text{ (from polar to Cartesian)} = \begin{bmatrix} \cos(\theta) & \sin(\theta) \\ -\frac{\sin(\theta)}{r} & \frac{\cos(\theta)}{r} \end{bmatrix}$$

In the reverse direction:

```
x_rθ = r*sp.cos(θ)                    # x as a function expressed in polar
y_rθ = r*sp.sin(θ)

X_rθ = sp.Matrix([x_rθ, y_rθ])      # vector function of x and y in polar
gradX = sp.simplify(gr.grad_vf(X_rθ, R).subs({x:r*cos(θ), y:r*sin(θ)}))
display(Math(f" \\text{{Jacobian matrix J (from Cartesian to polar)}}\
                                         ={latex(gradX.T)}"))
```

$$\text{Jacobian matrix } \mathbf{J} \text{ (from Cartesian to polar)} = \begin{bmatrix} \cos(\theta) & -r\sin(\theta) \\ \sin(\theta) & r\cos(\theta) \end{bmatrix}$$

The Jacobian matrices for the forward and inverse transformations are mutual inverses; therefore, their product is the identity matrix:

```
(gradR@gradX).simplify()
```

$$\begin{bmatrix} 1 & 0 \\ 0 & 1 \end{bmatrix}$$

It is important to note that the Jacobian matrix is, in general, **not unitary**, which distinguishes it from the coordinate transformation matrix **T** discussed earlier. Consider the following clarification:

> The coordinate transformation matrix **T** is used to transform **vectors** between coordinate systems, whereas the Jacobian matrix **J** is used to transform the **gradients of a scalar function** (also a vector) between coordinates.

In **SymPy**, the method `sp.jacobian()` can be used to compute the **Jacobian matrix**:

```
# J for polar to Cartesian, using the position vector R:
Jp_C = R_xy.jacobian([x,y])
display(Math(f" \\text{{Jacobian matrix Jp_C in $x$ and $y$}}=\
                                        {latex(Jp_C)}"))
```

$$\text{Jacobian matrix Jp\_C in } x \text{ and } y = \begin{bmatrix} \frac{x}{\sqrt{x^2+y^2}} & \frac{y}{\sqrt{x^2+y^2}} \\ -\frac{y}{x^2\left(1+\frac{y^2}{x^2}\right)} & \frac{1}{x\left(1+\frac{y^2}{x^2}\right)} \end{bmatrix}$$

This is written in x and y. It can be written in r and θ, which can be done by substitution:

```
Jp_C = simplify(Jp_C.subs({x:r*sp.cos(θ),y:r*sp.sin(θ)}))
display(Math(f" \\text{{Jacobian matrix Jp_C in $r$ and θ}}=\
                                              {latex(Jp_C)}"))
```

$$\text{Jacobian matrix Jp\_C in } r \text{ and } \theta = \begin{bmatrix} \cos(\theta) & \sin(\theta) \\ -\frac{\sin(\theta)}{r} & \frac{\cos(\theta)}{r} \end{bmatrix}$$

2.2.5.4 *Inverse of the Jacobian matrix*

We can obtain the inverse of the Jacobian matrix using `sp.inv()`:

```
display(Math(f" \\text{{ Inverse of Jp_C in $r$ and θ}}=\
                      {latex(sp.simplify(Jp_C.inv()))}"))
```

$$\text{Inverse of Jp\_C in } r \text{ and } \theta = \begin{bmatrix} \cos(\theta) & -r\sin(\theta) \\ \sin(\theta) & r\cos(\theta) \end{bmatrix}$$

The inverse of this Jacobian matrix can also be obtained directly by expressing x and y in terms of r and θ, and then computing the Jacobian matrix. Let us verify this:

```
# J formula for polar to Cartesian, using the position vector X:
JC_p = X_rθ.jacobian([r,θ])
display(Math(f" \\text{{Jacobian matrix JC_p in $r$ and θ}}=\
                                              {latex(JC_p)}"))
```

$$\text{Jacobian matrix JC\_p in } r \text{ and } \theta = \begin{bmatrix} \cos(\theta) & -r\sin(\theta) \\ \sin(\theta) & r\cos(\theta) \end{bmatrix}$$

The inverse Jacobian $\mathbf{J}^{-1}$ is used to compute $\nabla_{r\theta} f$ for given $\nabla_{xy} f$:

$$\underbrace{\begin{bmatrix} \frac{\partial f}{\partial r} \\ \frac{\partial f}{\partial \theta} \end{bmatrix}}_{\nabla_{r\theta} f} = \mathbf{J}^{-\top} \underbrace{\begin{bmatrix} \frac{\partial f}{\partial x} \\ \frac{\partial f}{\partial y} \end{bmatrix}}_{\nabla_{xy} f} \tag{2.19}$$

2.2.6 *General definition of the Jacobian matrix*

The Jacobian matrix contains the partial derivatives of one set of coordinates with respect to another. It arises whenever we deal with a pair of different coordinate systems that are locally orthogonal and, therefore, will be used frequently throughout this book. We introduce it here in a general form.

Consider a general vector of differentiable functions, $\mathbf{R}$, of several variables. The gradient of such a vector is a matrix [5] containing all its first-order partial derivatives. Suppose the vector function $\mathbf{R}$ is defined as a mapping: $\mathbb{R}^n \rightarrow \mathbb{R}^m$, meaning it has m component functions, each depending on n independent variables. It can be expressed as

$$\mathbf{R} = \begin{bmatrix} r_1(x_1, x_2, ..., x_n) \\ r_2(x_1, x_2, ..., x_n) \\ \vdots \\ r_m(x_1, x_2, ..., x_n) \end{bmatrix} \tag{2.20}$$

where $r_i; (i = 1, 2, \ldots, m)$ are the m component functions and $x_i; (i = 1, 2, \ldots, n)$ are the n independent variables.

The Jacobian matrix of $\mathbf{R}$ with respect to x_i is defined as an $m \times n$ matrix, denoted by $\mathbf{J}$:

$$\begin{aligned} \mathbf{J} = [J_{ij}] = \left[\frac{\partial r_i}{\partial x_j}\right] &= \begin{bmatrix} \frac{\partial \mathbf{R}}{\partial x_1} & \frac{\partial \mathbf{R}}{\partial x_2} & \cdots & \frac{\partial \mathbf{R}}{\partial x_n} \end{bmatrix} \\ = \begin{bmatrix} \nabla^\top r_1 \\ \nabla^\top r_2 \\ \vdots \\ \nabla^\top r_m \end{bmatrix} &= \begin{bmatrix} \frac{\partial r_1}{\partial x_1} & \frac{\partial r_1}{\partial x_2} & \cdots & \frac{\partial r_1}{\partial x_n} \\ \frac{\partial r_2}{\partial x_1} & \frac{\partial r_2}{\partial x_2} & \cdots & \frac{\partial r_2}{\partial x_n} \\ \vdots & \vdots & \ddots & \vdots \\ \frac{\partial r_m}{\partial x_1} & \frac{\partial r_m}{\partial x_2} & \cdots & \frac{\partial r_m}{\partial x_n} \end{bmatrix} \end{aligned} \tag{2.21}$$

where ∇ stands for the gradient operator vector (with partial derivative operators): $\nabla^\top = \left[\frac{\partial}{\partial x_1}, \frac{\partial}{\partial x_2}, \ldots, \frac{\partial}{\partial x_n}\right]$.

In the special case of

$$\mathbf{R} = \begin{bmatrix} r(x, y) \\ \theta(x, y) \end{bmatrix}$$

we obtain Eq. (2.17).

2.2.7 *Strains in the polar coordinate system*

Strains are defined in terms of the derivatives of displacements, commonly referred to as the strain–displacement relations (see Chapter 4 of Ref. [5]):

$$\begin{aligned} \varepsilon_{xx} &= \frac{\partial u_x}{\partial x}, \quad \varepsilon_{yy} = \frac{\partial u_y}{\partial y} \\ \varepsilon_{xy} &= \frac{1}{2}\left(\frac{\partial u_x}{\partial y} + \frac{\partial u_y}{\partial x}\right) \end{aligned} \tag{2.22}$$

We can now derive the expressions for the strains in the polar coordinate system, ε_{rr}, $\varepsilon_{\theta\theta}$, and $\varepsilon_{r\theta}$, in terms of the displacement components u_r and u_θ and their partial derivatives. Since this process involves derivatives of displacement functions expressed in two different coordinate systems, we will make use of the **Jacobian matrix**.

The overall strategy is as follows:

1. **Express** the Cartesian displacement components u_x and u_y in terms of the polar components u_r and u_θ. This can be achieved using the coordinate transformation of displacement vectors given by Eq. (2.8), which involves the transformation matrix $\mathbf{T}$.
2. **Differentiate** u_x and u_y with respect to r and θ to express their partial derivatives in terms of u_r, u_θ, and their derivatives. This step employs the Jacobian matrix through Eq. (2.18) for each displacement component.
3. **Relate** the strain components in the two coordinate systems using Eq. (2.13) and/or Eq. (2.14). This step yields the strain expressions in polar coordinates in terms of u_r, u_θ, and their derivatives.

2.2.7.1 *Deriving* ε_{rr}

Since all strains in the polar coordinate system must be expressed in terms of u_r and u_θ, we begin by defining symbolic functions for these displacement components.

To derive the normal strain ε_{rr}, we first consider the displacement component u_x, which can be expressed using the first equation in Eq. (2.8):

$$u_x = u_r \cos\theta - u_\theta \sin\theta \tag{2.23}$$

Next, by taking the partial derivatives of u_x with respect to r and θ, we obtain $\frac{\partial u_x}{\partial r}$ and $\frac{\partial u_x}{\partial \theta}$. The following code snippet illustrates this computation.

2.2.7.2 *SymPy code for deriving* ε_{rr}

```
x, y, θ = sp.symbols('x, y, θ', real=True)  # define symbolic variables
r = sp.symbols('r', nonnegative=True)
ur, uθ = sp.symbols('u_r, u_θ', real=True)      # displacements in polar
ur = sp.Function('u_r')(r, θ)          # symbolic functions for ur and uθ
uθ = sp.Function('u_θ')(r, θ)

ux = ur*sp.cos(θ) - uθ*sp.sin(θ)                        # ux written in polar
ux_r, ux_θ = diff(ux,r), diff(ux,θ)      # diffs of ux wrt polar coords.
display(Math(f" \\frac{{∂u_x}}{{∂r}} = {latex(ux_r)}"))
display(Math(f" \\frac{{∂u_x}}{{∂θ}} = {latex(ux_θ)}"))
```

$$\frac{\partial u_x}{\partial r} = -\sin(\theta)\frac{\partial}{\partial r}u_\theta(r,\theta) + \cos(\theta)\frac{\partial}{\partial r}u_r(r,\theta)$$

$$\frac{\partial u_x}{\partial \theta} = -u_r(r,\theta)\sin(\theta) - u_\theta(r,\theta)\cos(\theta) - \sin(\theta)\frac{\partial}{\partial \theta}u_\theta(r,\theta) + \cos(\theta)\frac{\partial}{\partial \theta}u_r(r,\theta)$$

Using Eq. (2.18) for function u_x, we should have

$$\begin{bmatrix} \dfrac{\partial u_x}{\partial x} \\ \dfrac{\partial u_x}{\partial y} \end{bmatrix} = \mathbf{J}^\top \begin{bmatrix} \dfrac{\partial u_x}{\partial r} \\ \dfrac{\partial u_x}{\partial \theta} \end{bmatrix} \tag{2.24}$$

The first equation in Eq. (2.24) can be used to compute $\dfrac{\partial u_x}{\partial x}$:

```
# Jacobian matrix from polar to Cartesian (expressed in r and θ):
Jp_C = sp.Matrix([[cos(θ), sin(θ)], [-sin(θ)/r, cos(θ)/r]])

ux_rθ = Matrix([ux_r, ux_θ])                          # form vector function
ux_x = (Jp_C.T@ux_rθ)[0].expand().simplify().expand()     # use Jacobian
#print('frac{∂u_x}{∂x}=',latex(ux_x))       # used to produce the next Eq.
```

The foregoing code cell produces a lengthy expression. It is manually simplifies as

$$\begin{aligned}\frac{\partial u_x}{\partial x} = &-\frac{\sin(2\theta)\frac{\partial}{\partial r}u_\theta}{2} + \frac{\cos(2\theta)\frac{\partial}{\partial r}u_r}{2} + \frac{\frac{\partial}{\partial r}u_r}{2} + \frac{u_r\sin^2(\theta)}{r} \\ &+ \frac{u_\theta\sin(2\theta)}{2r} + \frac{\sin^2(\theta)\frac{\partial}{\partial \theta}u_\theta}{r} - \frac{\sin(2\theta)\frac{\partial}{\partial \theta}u_r}{2r}\end{aligned} \tag{2.25}$$

This is the general expression for $\frac{\partial u_x}{\partial x} = \varepsilon_{xx}$, representing the **normal strain** in Cartesian coordinates for an arbitrary angle θ.

To relate this to the strains in the polar coordinate system, we use Eq. (2.13), which shows that $\varepsilon_{rr} = \varepsilon_{xx}$ when $\theta = 0$. This corresponds to substituting $\theta = 0$ into the transformation matrix $\mathbf{T}$. Therefore, the strain component ε_{rr} can be obtained simply by setting $\theta = 0$ in all coefficients of the preceding expression. We then have

$$\varepsilon_{rr} = \varepsilon_{xx}\Big|_{\mathbf{T}(\theta=0)} = \frac{\partial u_x}{\partial x}\Big|_{\theta=0}$$

which can be found using the following snippet:

```
# Set θ=0 for the coefficients to obtain ε_xx which is ε_rr:
dic_0 = {sin(θ):0, sin(2*θ):0, cos(θ):1, cos(2*θ):1}        # set θ = 0
εrr = ux_x.subs(dic_0)
display(Math(f" \\text{{ε$_{{rr}}$}}={latex(εrr)}"))
```

$$\varepsilon_{rr} = \frac{\partial}{\partial r} u_r(r, \theta)$$

Next, setting $\theta = \pi/2$ and using Eq. (2.14), we can conclude that $\varepsilon_{\theta\theta} = \varepsilon_{xx}|_{\mathbf{T}(\theta=\pi/2)} = \frac{\partial u_x}{\partial x}|_{\theta=\pi/2}$.

```
# Set θ=π/2 for the coefficients to obtain ε_xx which gives ε_θθ:
dic_pi2 = {sin(θ):1, sin(2*θ):0, cos(θ):0, cos(2*θ):-1}   # set θ = π/2
ε_θθ = ux_x.subs(dic_pi2)
display(Math(f" \\text{{ε$_{{θθ}}$}} = {latex(ε_θθ)}"))
```

$$\varepsilon_{\theta\theta} = \frac{u_r(r, \theta)}{r} + \frac{\frac{\partial}{\partial \theta} u_\theta(r, \theta)}{r}$$

2.2.7.3 *Alternative means for deriving $\varepsilon_{\theta\theta}$*

Alternatively, using the second equation in Eq. (2.8), we have

$$u_y = u_r \sin\theta + u_\theta \cos\theta \tag{2.26}$$

Performing the differentiation of u_y with respect to r and θ gives $\frac{\partial u_y}{\partial r}$ and $\frac{\partial u_y}{\partial \theta}$. The snippet is as follows:

```
uy = ur*sp.sin(θ) + uθ*sp.cos(θ)                # disp. uy written in polar
uy_r, uy_θ = diff(uy,r), diff(uy,θ)         # diffs of uy wrt polar coords.
display(Math(f" \\frac{{∂u_y}}{{∂r}} = {latex(uy_r)}"))
display(Math(f" \\frac{{∂u_y}}{{∂θ}} = {latex(uy_θ)}"))
```

$$\frac{\partial u_y}{\partial r} = \sin(\theta)\frac{\partial}{\partial r}u_r(r,\theta) + \cos(\theta)\frac{\partial}{\partial r}u_\theta(r,\theta)$$

$$\frac{\partial u_y}{\partial \theta} = u_r(r,\theta)\cos(\theta) - u_\theta(r,\theta)\sin(\theta) + \sin(\theta)\frac{\partial}{\partial \theta}u_r(r,\theta) + \cos(\theta)\frac{\partial}{\partial \theta}u_\theta(r,\theta)$$

Now, using Eq. (2.18) for u_y, we have

$$\begin{bmatrix} \dfrac{\partial u_y}{\partial x} \\ \dfrac{\partial u_y}{\partial y} \end{bmatrix} = \mathbf{J}^\top \begin{bmatrix} \dfrac{\partial u_y}{\partial r} \\ \dfrac{\partial u_y}{\partial \theta} \end{bmatrix} \tag{2.27}$$

The second equation in Eq. (2.27) can be used to compute $\dfrac{\partial u_y}{\partial y}$:

```
uy_rθ= Matrix([uy_r, uy_θ])                      # form vector function
uy_y = (Jp_C.T@uy_rθ)[1].simplify().expand()   # use the second component
```

This gives the general expression for $\frac{\partial u_y}{\partial y}$. Since $\frac{\partial u_y}{\partial y} = \varepsilon_{yy}$, using Eq. (2.13) we find $\varepsilon_{\theta\theta} = \varepsilon_{yy}|_{\mathbf{T}(\theta=0)}$. This is done using the following snippet:

```
εθθ = uy_y.subs(dic_θ).simplify().expand()                    # set θ = 0
display(Math(f" \\text{{ε$_{{θθ}}$}}={latex(εθθ)}"))
```

$$\varepsilon_{\theta\theta} = \frac{u_r(r,\theta)}{r} + \frac{\frac{\partial}{\partial \theta}u_\theta(r,\theta)}{r}$$

This gives $\varepsilon_{\theta\theta} = \frac{u_r}{r} + \frac{1}{r}\frac{\partial u_\theta}{\partial \theta}$.

In comparison, $\varepsilon_{yy} = \frac{\partial u_y}{\partial y}$ depends only on a single displacement component u_y, whereas $\varepsilon_{\theta\theta}$ involves both u_r and u_θ. This difference arises because, when transforming the derivatives of a vector function, both the coordinate system and the components of the vector function change simultaneously.

2.2.7.4 *SymPy code for deriving* $\varepsilon_{r\theta}$

Next, let us derive the expression for $\varepsilon_{r\theta}$. For this purpose, we first determine ε_{xy}, as indicated in Eq. (2.13). The procedure is similar to the previous derivation but involves two steps:

Step 1: Using the second equation in Eq. (2.24), compute $\dfrac{\partial u_x}{\partial y}$. Then, substitute $\theta = 0$ into all coefficients of the resulting expression so that $\frac{\partial u_x}{\partial y}$ is expressed in terms of displacements in the polar coordinate system.

The following snippet finds the expression:

```
ux_y  = (Jp_C.T@ux_rθ)[1].expand().simplify().expand() # use second ux_rθ
ux_y_ = ux_y.subs(dic_0                                # set θ =0
display(Math(f" \\frac{{∂u_x}}{{∂y}}={latex(ux_y_)}"))
```

$$\frac{\partial u_x}{\partial y} = -\frac{u_\theta(r,\theta)}{r} + \frac{\frac{\partial}{\partial \theta}u_r(r,\theta)}{r}$$

Step 2: Using the first equation in Eq. (2.27), compute $\dfrac{\partial u_y}{\partial x}$. Then, substitute $\theta = 0$ into all the coefficients of the resulting expression so that $\frac{\partial u_y}{\partial x}$ is expressed in terms of displacements in the polar coordinate system.

The following snippet produces the expression:

```
uy_x = (Jp_C.T@uy_rθ)[0].simplify().expand()             # use Jacobian
#print('frac{∂u_x}{∂x}=',latex(ux_x))                    # to produce Eq.
uy_x = uy_x.subs(dic_0).simplify().expand()              # set θ = 0
display(Math(f" \\frac{{∂u_y}}{{∂x}} = {latex(uy_x)}"))
```

$$\frac{\partial u_y}{\partial x} = \frac{\partial}{\partial r}u_\theta(r,\theta)$$

Setting $\theta = 0$, using Eq. (2.13), and combining these two steps, we have $\varepsilon_{r\theta} = \varepsilon_{xy}$:

```
uxy_θr = ((Jp_C.T@ux_rθ)[1] + (Jp_C.T@uy_rθ)[0])/2
ε_rθ = uxy_θr.subs(dic_0).expand()    # or use dic_pi2 (revers the sign)
display(Math(f" \\text{{ε$_{{rθ}}$}}={latex(ε_rθ)}"))
```

$$\varepsilon_{r\theta} = \frac{\frac{\partial}{\partial r}u_\theta(r,\theta)}{2} - \frac{u_\theta(r,\theta)}{2r} + \frac{\frac{\partial}{\partial \theta}u_r(r,\theta)}{2r}$$

The strain–displacement relations in the polar coordinate system are summarized as follows for future reference:

$$\begin{aligned} \varepsilon_{rr} &= \frac{\partial u_r}{\partial r} \\ \varepsilon_{\theta\theta} &= \frac{u_r}{r} + \frac{\partial u_\theta}{r\partial \theta} \\ 2\varepsilon_{r\theta} &= \frac{\partial u_r}{r\partial \theta} + \frac{\partial u_\theta}{\partial r} - \frac{u_\theta}{r} \end{aligned} \tag{2.28}$$

Readers may note the **dimensionless nature** of each term, as all of them represent strains. The length measure associated with the θ-direction is properly adjusted by the scale factor r.

For **axisymmetric problems**, all strain components are independent of θ, and $\varepsilon_{r\theta} = 0$. Equation (2.28) therefore simplifies to

$$\begin{aligned} \varepsilon_{rr} &= \frac{\partial u_r}{\partial r} \\ \varepsilon_{\theta\theta} &= \frac{u_r}{r} \end{aligned} \tag{2.29}$$

This expression clearly shows that the radial displacement u_r induces a circumferential (hoop) strain $\varepsilon_{\theta\theta}$, which leads to the development of **hoop stress** in axisymmetric problems.

Although the derivation of the strain–displacement relations is somewhat detailed, it serves as an excellent example of how **coordinate transformations for vectors and tensors**, together with the **Jacobian matrix**, can be effectively applied.

A more general and systematic approach for deriving strain–displacement relations uses **orthogonal curvilinear coordinates**. A comprehensive discussion can be found in Appendix 2A of Ref. [2]. This approach yields a unified form of the strain–displacement relations, applicable to **polar, cylindrical, and spherical** coordinate systems, and it will be used directly in later sections.

2.3 Cylindrical coordinate system

2.3.1 *Coordinates and Jacobian matrix*

In many applications, it is necessary to express various orders of derivatives of a scalar function originally defined in Cartesian coordinates in terms of **cylindrical (or polar)** coordinates. This is particularly useful, for instance, when deriving **stress components** from a **stress function** formulated in the polar coordinate system.

This section presents the techniques for performing such transformations by introducing the **cylindrical coordinate system**, as illustrated in Fig. 2.3.

The cylindrical coordinate system can be regarded as an **extension of the polar coordinate system** with an additional z-axis that is perpendicular to the polar plane. Consequently, all equations derived in this section

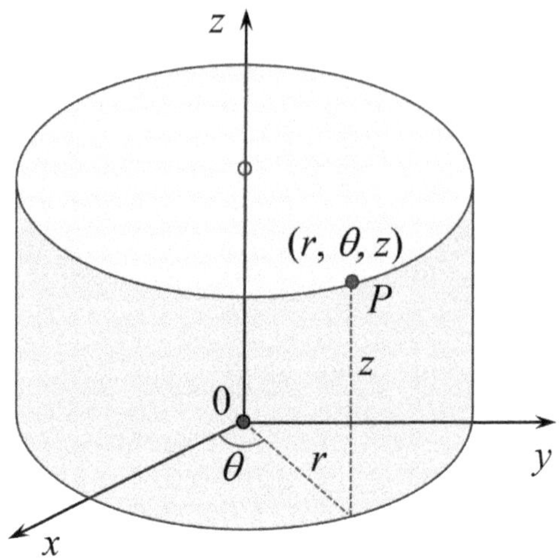

Figure 2.3. Cylindrical coordinates with the radial axis r, polar angle θ, and vertical axis z.

also apply directly to the polar coordinate system if the terms involving the z-axis are omitted.

We now define the cylindrical coordinates in **SymPy** as follows:

```
x, y, z = sp.symbols("x, y, z")        # symbolic variables for Cartesian
X = sp.Matrix([x, y, z])                  # Position vector in Cartesian

r, θ, z = sp.symbols("r, θ, z ")     # symbolic variables for cylindrical
R = sp.Matrix([r, θ, z])                       # Position vector in polar

# Cylindrical to Cartesian
X_ = vector.transform_coords("cylindrical", (r, θ, z))
X_ = sp.Matrix(X_)                      # X generated by the `vector` module
display(Math(f"\\text{{Cylindrical to Cartesian:}}\;\; \
                                             {latex(X)} = {latex(X_)}"))
```

$$\text{Cylindrical to Cartesian:}\;\; \begin{bmatrix} x \\ y \\ z \end{bmatrix} = \begin{bmatrix} r\cos(\theta) \\ r\sin(\theta) \\ z \end{bmatrix}$$

The cylindrical coordinate system is a typical example of a curvilinear coordinate system. It possesses scale factors relative to the Cartesian coordinate system, which represent the physical lengths corresponding to infinitesimal displacements along the coordinate directions r, θ, and z. Let us now print these scale factors:

```
cyl_scale = vector.scale_coeff_coords("cylindrical", (r, θ, z))
cyl_scale                                  # scale factors: (α, β, γ)
```

$(1,\ r,\ 1)$

It is clear that only θ needs a scale factor r to produce a length in the θ-direction.

Let us compute the gradient of the vector of coordinates defined earlier:

```
gradf = vector.grad(X_, coords=(r, θ, z), h_vec=cyl_scale)
gradf.T
```

$$\left[\begin{bmatrix}\cos(\theta)\\ \sin(\theta)\\ 0\end{bmatrix}\quad \begin{bmatrix}-r\sin(\theta)\\ \cos(\theta)\\ 0\end{bmatrix}\quad \begin{bmatrix}0\\ 0\\ 1\end{bmatrix}\right]$$

```
# Alternative: using `gr.grad_vf()` with a transpose:
gr_grad_XR = gr.grad_vf(X_, R).T
gr_grad_XR
```

$$\begin{bmatrix}\cos(\theta) & -r\sin(\theta) & 0\\ \sin(\theta) & r\cos(\theta) & 0\\ 0 & 0 & 1\end{bmatrix}$$

Note that when using the `vector` module, we need to rearrange `gradf` to form a proper shape as a Jacobian matrix:

```
J_XR = Matrix([gradf[0].T,gradf[1].T,gradf[2].T]).T
J_XR
```

$$\begin{bmatrix}\cos(\theta) & -r\sin(\theta) & 0\\ \sin(\theta) & r\cos(\theta) & 0\\ 0 & 0 & 1\end{bmatrix}$$

```
# The inverse Jacobian:
J_RX = simplify(J_XR.inv())
J_RX
```

$$\begin{bmatrix}\cos(\theta) & \sin(\theta) & 0\\ -\frac{\sin(\theta)}{r} & \frac{\cos(\theta)}{r} & 0\\ 0 & 0 & 1\end{bmatrix}$$

This Jacobian matrix can be employed to transform the gradient of an arbitrary scalar function between the cylindrical and Cartesian coordinate systems.

2.3.2 *Gradient of a scalar function*

We are now prepared to compute the gradient of an arbitrary differentiable scalar function in the cylindrical coordinate system, utilizing the previously derived Jacobian matrix.

```
ϕ = sp.symbols("ϕ", cls=sp.Function)      # define an arbitrary function

# compute the gradient of the function in cylindrical coordinates:
gradϕ_R = vector.grad(ϕ(r, θ, z), coords=(r, θ, z),h_vec=cyl_scale)

gradϕ_X = J_RX.T@gradϕ_R                                # transfer the gradient
display(Math(f" \\text{{First derivatives $[∂ϕ/dx, ∂ϕ/dy, ∂ϕ/dz]^T$}}=\
                                           \\small{latex(gradϕ_X)}"))
```

First derivatives $[\partial\phi/dx, \partial\phi/dy, \partial\phi/dz]^T$

$$= \begin{bmatrix} \cos(\theta)\frac{\partial}{\partial r}\phi(r,\theta,z) - \frac{\sin(\theta)\frac{\partial}{\partial \theta}\phi(r,\theta,z)}{r} \\ \sin(\theta)\frac{\partial}{\partial r}\phi(r,\theta,z) + \frac{\cos(\theta)\frac{\partial}{\partial \theta}\phi(r,\theta,z)}{r} \\ \frac{\partial}{\partial z}\phi(r,\theta,z) \end{bmatrix}$$

We obtain the gradient (first derivatives) of an arbitrary scalar function ϕ with respect to the cylindrical coordinates r and θ:

$$\begin{bmatrix} \frac{\partial\phi}{\partial x} \\ \frac{\partial\phi}{\partial y} \\ \frac{\partial\phi}{\partial z} \end{bmatrix} = \underbrace{\begin{bmatrix} \cos\theta & -\frac{\sin\theta}{r} & 0 \\ \sin\theta & \frac{\cos\theta}{r} & 0 \\ 0 & 0 & 1 \end{bmatrix}}_{\mathbf{J}^\top} \begin{bmatrix} \frac{\partial\phi}{\partial r} \\ \frac{\partial\phi}{\partial \theta} \\ \frac{\partial\phi}{dz} \end{bmatrix} = \begin{bmatrix} \cos\theta\frac{\partial\phi}{\partial r} - \frac{\sin\theta}{r}\frac{\partial\phi}{\partial\theta} \\ \sin\theta\frac{\partial\phi}{\partial r} + \frac{\cos\theta}{r}\frac{\partial\phi}{\partial\theta} \\ \frac{\partial\phi}{\partial z} \end{bmatrix} \tag{2.30}$$

Each component in the vector on the right-hand side of Eq. (2.30) can be regarded as a new scalar function of r, θ, and z.

When $\theta = 0$, meaning that the radial axis of the cylindrical coordinate system is aligned with the x-axis of the Cartesian coordinate system, we have

$$\begin{bmatrix} \frac{\partial \phi}{\partial x} \\ \frac{\partial \phi}{\partial y} \\ \frac{\partial \phi}{\partial z} \end{bmatrix} = \begin{bmatrix} \frac{\partial \phi}{\partial r} \\ \frac{1}{r}\frac{\partial \phi}{\partial \theta} \\ \frac{\partial \phi}{\partial z} \end{bmatrix} \tag{2.31}$$

This is the relationship of the first derivatives between these two coordinate systems.

2.3.3 *Strains in cylindrical coordinates*

To obtain the strain–displacement relations in the cylindrical coordinate system, we employ the fact that the strain tensor is the symmetric part of the displacement gradient. The general formula for forming the symmetric tensor is given in Section 4.6 of Ref. [5]:

$$\varepsilon = \frac{1}{2}\left[\nabla \mathbf{u}_C + (\nabla \mathbf{u}_C)^\top\right] \tag{2.32}$$

where $\mathbf{u}_C$ denotes the displacement vector expressed in Cartesian coordinates. The strains in the cylindrical coordinate system can now be derived by evaluating the displacement gradient according to the following steps:

1. Express the displacement vector in Cartesian coordinates, $\mathbf{u}_C$, in terms of its components in the cylindrical coordinates, $\mathbf{u}_R$.
2. Compute the symmetric gradient of $\mathbf{u}_C$ using Eq. (2.32) with respect to (r, θ, z).
3. Apply Eq. (2.13) and set $\theta = 0$ to obtain the strain tensor in the cylindrical coordinate system.

The corresponding SymPy code snippet is given as follows:

```
# Define the displacements in Cylindrical coordinates:
ur = sp.Function('u_r')(r, θ, z)
uθ = sp.Function('u_θ')(r, θ, z)
uz = sp.Function('u_z')(r, θ, z)
uR = Matrix([ur, uθ, uz])                        # form vector function

T = Matrix([[ sp.cos(θ), sp.sin(θ), 0],      # T: Transformation matrix
            [-sp.sin(θ), sp.cos(θ), 0],           # Cartesian to Polar
            [     0    ,     0    , 1]])
```

```
# Step-1: Express uC in terms of uP in r,θ,z:
#ux = ur*sp.cos(θ) - uθ*sp.sin(θ)                    # ux written in polar
#uy = ur*sp.sin(θ) + uθ*sp.cos(θ)                    # uy written in polar
uC=T.T @ uR # vector transformation gives Matrix([ux, uy, uz]) in r,θ,z

# Step-2: Compute the gradient of vector uC:
grad_uC = vector.grad(uC, coords=(r, θ, z), h_vec=cyl_scale)
grad_uC = Matrix([grad_uC[0].T,grad_uC[1].T,grad_uC[2].T]).T  # reshape
sym_grad_uC = ((J_RX@grad_uC.T).T+J_RX@grad_uC.T)/2
#sym_grad_uC          # lengthy, grad of uC in Cartesian written in r,θ

# Step-3: Coordinate transformation
dic_0 = {sin(θ):0, cos(θ):1}                                  # set θ = 0
ε = sym_grad_uC.subs(dic_0)
ε                                                        # symmetric tensor
```

$$\begin{bmatrix} \frac{\partial}{\partial r}u_r(r,\theta,z) & \frac{\frac{\partial}{\partial r}u_\theta(r,\theta,z)}{2} + \frac{-u_\theta(r,\theta,z)+\frac{\partial}{\partial \theta}u_r(r,\theta,z)}{2r} & \frac{\frac{\partial}{\partial z}u_r(r,\theta,z)}{2} + \frac{\frac{\partial}{\partial r}u_z(r,\theta,z)}{2} \\ \frac{\frac{\partial}{\partial r}u_\theta(r,\theta,z)}{2} + \frac{-u_\theta(r,\theta,z)+\frac{\partial}{\partial \theta}u_r(r,\theta,z)}{2r} & \frac{u_r(r,\theta,z)+\frac{\partial}{\partial \theta}u_\theta(r,\theta,z)}{r} & \frac{\frac{\partial}{\partial z}u_\theta(r,\theta,z)}{2} + \frac{\frac{\partial}{\partial \theta}u_z(r,\theta,z)}{2r} \\ \frac{\frac{\partial}{\partial z}u_r(r,\theta,z)}{2} + \frac{\frac{\partial}{\partial r}u_z(r,\theta,z)}{2} & \frac{\frac{\partial}{\partial z}u_\theta(r,\theta,z)}{2} + \frac{\frac{\partial}{\partial \theta}u_z(r,\theta,z)}{2r} & \frac{\partial}{\partial z}u_z(r,\theta,z) \end{bmatrix}$$

The above can also be done simply using `vector.sym_grad()`:

```
vector.sym_grad(uR, coords=(r, θ, z), h_vec = cyl_scale)
```

$$\begin{bmatrix} \frac{\partial}{\partial r}u_r(r,\theta,z) & \frac{\frac{\partial}{\partial r}u_\theta(r,\theta,z)}{2} - \frac{u_\theta(r,\theta,z)}{2r} + \frac{\frac{\partial}{\partial \theta}u_r(r,\theta,z)}{2r} & \frac{\frac{\partial}{\partial z}u_r(r,\theta,z)}{2} + \frac{\frac{\partial}{\partial r}u_z(r,\theta,z)}{2} \\ \frac{\frac{\partial}{\partial r}u_\theta(r,\theta,z)}{2} - \frac{u_\theta(r,\theta,z)}{2r} + \frac{\frac{\partial}{\partial \theta}u_r(r,\theta,z)}{2r} & \frac{u_r(r,\theta,z)}{r} + \frac{\frac{\partial}{\partial \theta}u_\theta(r,\theta,z)}{r} & \frac{\frac{\partial}{\partial z}u_\theta(r,\theta,z)}{2} + \frac{\frac{\partial}{\partial \theta}u_z(r,\theta,z)}{2r} \\ \frac{\frac{\partial}{\partial z}u_r(r,\theta,z)}{2} + \frac{\frac{\partial}{\partial r}u_z(r,\theta,z)}{2} & \frac{\frac{\partial}{\partial z}u_\theta(r,\theta,z)}{2} + \frac{\frac{\partial}{\partial \theta}u_z(r,\theta,z)}{2r} & \frac{\partial}{\partial z}u_z(r,\theta,z) \end{bmatrix}$$

We have obtained all the strain components in the cylindrical coordinates, which can be rewritten as follows:

$$\begin{aligned}
\varepsilon_{rr} &= \frac{\partial u_r}{\partial r};\ \varepsilon_{\theta\theta} = \frac{u_r}{r} + \frac{\partial u_\theta}{r\partial\theta};\ \varepsilon_{zz} = \frac{\partial u_z}{\partial z} \\
2\varepsilon_{rz} &= \frac{\partial u_r}{\partial z} + \frac{\partial u_z}{\partial r} \\
2\varepsilon_{\theta z} &= \frac{\partial u_\theta}{\partial z} + \frac{\partial u_z}{r\partial\theta} \\
2\varepsilon_{r\theta} &= \frac{\partial u_r}{r\partial\theta} + \frac{\partial u_\theta}{\partial r} - \frac{u_\theta}{r}
\end{aligned} \tag{2.33}$$

It is seen that the strains in the cylindrical coordinate system are a direct extension of those in the polar coordinate system Eq. (2.28), since the z coordinate remains identical in both the 3D Cartesian and cylindrical systems.

For **axisymmetric problems**, none of the strain components vary with θ, and $\varepsilon_{r\theta} = \varepsilon_{z\theta} = 0$. Equation (2.33) then reduces to

$$\begin{aligned}
\varepsilon_{rr} &= \frac{\partial u_r}{\partial r};\ \varepsilon_{\theta\theta} = \frac{u_r}{r};\ \varepsilon_{zz} = \frac{\partial u_z}{\partial z} \\
2\varepsilon_{rz} &= \frac{\partial u_r}{\partial z} + \frac{\partial u_z}{\partial r}
\end{aligned} \tag{2.34}$$

The following snippet derives the formulas for the transformation of the strain tensor from polar to Cartesian coordinates:

```
εxx, εyy, εzz, εyz, εxz, εxy=symbols("ε_xx, ε_yy, ε_zz, ε_yz, ε_xz, ε_xy")
εrr, εθθ, εzz, εθz, εrz, εrθ=symbols("ε_rr, ε_θθ, ε_zz, ε_θz, ε_rz, ε_rθ")

ε_C = Matrix([[εxx, εxy, εxz],              # ε: strain tensor in Cartesian
              [εxy, εyy, εyz],
              [εxz, εyz, εzz]])

ε_R_ = Matrix([[εrr, εrθ, εrz],                # ε: strain tensor in polar
               [εrθ, εθθ, εθz],
               [εrz, εθz, εzz]])

ε_R = (T@ε_C@T.T).applyfunc(expand).applyfunc(simplify)
display(Math(f"{latex(εrr)}={latex(ε_R[0,0])}"))   # Cartesian to Polar
display(Math(f"{latex(εθθ)}={latex(ε_R[1,1])}"))
display(Math(f"{latex(εzz)}={latex(ε_R[2,2])}"))
display(Math(f"{latex(εθz)}={latex(ε_R[1,2])}"))
display(Math(f"{latex(εrz)}={latex(ε_R[0,2])}"))
display(Math(f"{latex(εrθ)}={latex(ε_R[0,1])}"))

dic_0 = {sin(θ):0, sin(2*θ):0, cos(θ):1, cos(2*θ):1}          # set θ = 0
display(Math(f"\\text{{Identity at $θ = 0$: }}\\small {latex(ε_R_)} = \
                                      \\small {latex(ε_R.subs(dic_0))}"))
```

$$\varepsilon_{rr} = \varepsilon_{xx}\cos^2(\theta) + \varepsilon_{xy}\sin(2\theta) + \varepsilon_{yy}\sin^2(\theta)$$
$$\varepsilon_{\theta\theta} = \varepsilon_{xx}\sin^2(\theta) - \varepsilon_{xy}\sin(2\theta) + \varepsilon_{yy}\cos^2(\theta)$$
$$\varepsilon_{zz} = \varepsilon_{zz}$$
$$\varepsilon_{\theta z} = -\varepsilon_{xz}\sin(\theta) + \varepsilon_{yz}\cos(\theta)$$
$$\varepsilon_{rz} = \varepsilon_{xz}\cos(\theta) + \varepsilon_{yz}\sin(\theta)$$
$$\varepsilon_{r\theta} = -\frac{\varepsilon_{xx}\sin(2\theta)}{2} + \varepsilon_{xy}\cos(2\theta) + \frac{\varepsilon_{yy}\sin(2\theta)}{2}$$

Identity at $\theta = 0$: $\begin{bmatrix} \varepsilon_{rr} & \varepsilon_{r\theta} & \varepsilon_{rz} \\ \varepsilon_{r\theta} & \varepsilon_{\theta\theta} & \varepsilon_{\theta z} \\ \varepsilon_{rz} & \varepsilon_{\theta z} & \varepsilon_{zz} \end{bmatrix} = \begin{bmatrix} \varepsilon_{xx} & \varepsilon_{xy} & \varepsilon_{xz} \\ \varepsilon_{xy} & \varepsilon_{yy} & \varepsilon_{yz} \\ \varepsilon_{xz} & \varepsilon_{yz} & \varepsilon_{zz} \end{bmatrix}$

2.3.4 *Hessian matrix in cylindrical and polar coordinates*

As discussed in Chapters 2 and 4 of Ref. [1], it is often necessary to compute the second derivatives of a differentiable scalar function (e.g., a stress function) in Cartesian coordinates. The same requirement also arises when dealing with mechanics problems formulated in non-Cartesian coordinates. This section derives expressions for the second derivatives of a scalar function in different coordinate systems and examines how these derivatives are related.

The second derivatives of a scalar function can be obtained by taking the gradient of its gradient (vector), which leads to the **Hessian matrix** of the scalar function ϕ, for example, in (r, θ, z) coordinates. This Hessian matrix [3] can then be transformed into Cartesian coordinates using the (transpose) Jacobian matrix. The implementation in SymPy is provided in the following code snippet. However, since each term in the Hessian matrix is quite lengthy, only the first term, i.e., the $[0, 0]$ component, is displayed as an example:

```
d2ϕ_X = vector.grad(gradϕ_X, coords=(r,θ,z)) # gradient in rθz, Lengthy
d2ϕ_X = simplify(d2ϕ_X)

H = (J_RX.T@d2ϕ_X)                              # transfer to Cartesian xyz
H = Matrix([H[0].T,H[1].T,H[2].T]).T        # re-arrange to Hessian matrix
#gr.printM(H[0,0],'The [0,0] term in the Hessian matrix:')      # Lengthy
#gr.printM(H[1,1],'The [1,1] term in the Hessian matrix:')      # Lengthy
#gr.printM(H[0,1],'The [0,1] term in the Hessian matrix:')      # Lengthy
#gr.printM(H[2,2],'The [2,2] term in the Hessian matrix:')
#gr.printM(H[0,2],'The [0,2] term in the Hessian matrix:')
gr.printM(H[1,2],'The [1,2] term in the Hessian matrix:')
```

The [1,2] term in the Hessian matrix:

$$\sin(\theta)\frac{\partial^2}{\partial z\partial r}\phi(r,\theta,z) + \frac{\cos(\theta)\frac{\partial^2}{\partial\theta\partial z}\phi(r,\theta,z)}{r}$$

Readers may uncomment the last line in the foregoing code cell. The [0,0] term in the Hessian matrix is still too long to display properly in SymPy format. We therefore rewrite it manually as follows:

$$\begin{aligned}
\frac{\partial^2\phi}{\partial x^2} &= \left(\cos\theta\frac{\partial^2}{\partial r^2}\phi - \frac{\sin\theta\frac{\partial^2}{\partial\theta\partial r}\phi}{r} + \frac{\sin\theta\frac{\partial}{\partial\theta}\phi}{r^2}\right)\cos\theta \\
&\quad - \frac{\left(-r\sin\theta\frac{\partial}{\partial r}\phi + r\cos\theta\frac{\partial^2}{\partial\theta\partial r}\phi - \sin\theta\frac{\partial^2}{\partial\theta^2}\phi - \cos\theta\frac{\partial}{\partial\theta}\phi\right)\sin\theta}{r^2} \\
\frac{\partial^2\phi}{\partial y^2} &= \left(\sin\theta\frac{\partial^2}{\partial r^2}\phi + \frac{\cos\theta\frac{\partial^2}{\partial\theta\partial r}\phi}{r} - \frac{\cos\theta\frac{\partial}{\partial\theta}\phi}{r^2}\right)\sin\theta \\
&\quad + \frac{\left(r\left(\sin\theta\frac{\partial^2}{\partial\theta\partial r}\phi + \cos\theta\frac{\partial}{\partial r}\phi\right) - \sin\theta\frac{\partial}{\partial\theta}\phi + \cos\theta\frac{\partial^2}{\partial\theta^2}\phi\right)\cos\theta}{r^2} \\
\frac{\partial^2\phi}{\partial x\partial y} &= \left(\cos\theta\frac{\partial^2}{\partial r^2}\phi - \frac{\sin\theta\frac{\partial^2}{\partial\theta\partial r}\phi}{r} + \frac{\sin\theta\frac{\partial}{\partial\theta}\phi}{r^2}\right)\sin\theta \\
&\quad + \frac{\left(r\left(-\sin\theta\frac{\partial}{\partial r}\phi + \cos\theta\frac{\partial^2}{\partial\theta\partial r}\phi\right) - \sin\theta\frac{\partial^2}{\partial\theta^2}\phi - \cos\theta\frac{\partial}{\partial\theta}\phi\right)\cos\theta}{r^2} \\
\frac{\partial^2\phi}{\partial x\partial z} &= \cos\theta\frac{\partial^2}{\partial z\partial r}\phi - \frac{\sin\theta\frac{\partial^2}{\partial\theta\partial z}\phi}{r} \\
\frac{\partial^2\phi}{\partial y\partial z} &= \sin\theta\frac{\partial^2}{\partial z\partial r}\phi + \frac{\cos\theta\frac{\partial^2}{\partial\theta\partial z}\phi}{r}
\end{aligned} \tag{2.35}$$

When $\theta = 0$, meaning that the problem is axisymmetric and hence θ-independent, the matrix can be displayed in full, as shown in the following snippet:

```
d2ϕ_X = vector.grad(gradϕ_X, coords=(r,θ,z)) # gradient in rθz, Lengthy
dic_0 = {sin(θ):0, sin(2*θ):0, cos(θ):1, cos(2*θ):1}        # set θ = 0
d2ϕ_X0 = simplify(d2ϕ_X).subs(dic_0)

H = (J_RX.T@d2ϕ_X0).subs(dic_0)                 # transfer to Cartesian xyz
H = Matrix([H[0].T,H[1].T,H[2].T]).T       # re-arrange in Hessian matrix
gr.printM(H,'The Hessian matrix with all second derivatives of ϕ \
when setting (θ = 0):')
```

The Hessian matrix with all second derivatives of ϕ when setting (θ = 0):

$$\begin{bmatrix} \frac{\partial^2}{\partial r^2}\phi(r,\theta,z) & \frac{r\frac{\partial^2}{\partial\theta\partial r}\phi(r,\theta,z)-\frac{\partial}{\partial\theta}\phi(r,\theta,z)}{r^2} & \frac{\partial^2}{\partial z\partial r}\phi(r,\theta,z) \\ \frac{\frac{\partial^2}{\partial\theta\partial r}\phi(r,\theta,z)}{r} - \frac{\frac{\partial}{\partial\theta}\phi(r,\theta,z)}{r^2} & \frac{r\frac{\partial}{\partial r}\phi(r,\theta,z)+\frac{\partial^2}{\partial\theta^2}\phi(r,\theta,z)}{r^2} & \frac{\frac{\partial^2}{\partial\theta\partial z}\phi(r,\theta,z)}{r} \\ \frac{\partial^2}{\partial z\partial r}\phi(r,\theta,z) & \frac{\frac{\partial^2}{\partial\theta\partial z}\phi(r,\theta,z)}{r} & \frac{\partial^2}{\partial z^2}\phi(r,\theta,z) \end{bmatrix}$$

The Hessian matrix of a scalar function ϕ contains all possible second derivatives. It is summarized as follows for both Cartesian and polar coordinate systems, assuming $\theta = 0$:

$$\begin{bmatrix} \frac{\partial^2\phi}{\partial x^2} & \frac{\partial^2\phi}{\partial y\partial x} & \frac{\partial^2\phi}{\partial z\partial x} \\ \frac{\partial^2\phi}{\partial x\partial y} & \frac{\partial^2\phi}{\partial y^2} & \frac{\partial^2\phi}{\partial z\partial y} \\ \frac{\partial^2\phi}{\partial x\partial z} & \frac{\partial^2\phi}{\partial y\partial z} & \frac{\partial^2\phi}{\partial z^2} \end{bmatrix} = \begin{bmatrix} \frac{\partial^2\phi}{\partial r^2} & \frac{1}{r}\frac{\partial^2\phi}{\partial\theta\partial r} - \frac{1}{r^2}\frac{\partial\phi}{\partial\theta} & \frac{\partial^2\phi}{\partial z\partial r} \\ \frac{1}{r}\frac{\partial^2\phi}{\partial\theta\partial r} - \frac{1}{r^2}\frac{\partial\phi}{\partial\theta} & \frac{1}{r}\frac{\partial\phi}{\partial r} + \frac{1}{r^2}\frac{\partial^2\phi}{\partial\theta^2} & \frac{1}{r}\frac{\partial^2\phi}{\partial\theta\partial z} \\ \frac{\partial^2\phi}{\partial z\partial r} & \frac{1}{r}\frac{\partial^2\phi}{\partial\theta\partial z} & \frac{\partial^2\phi}{\partial z^2} \end{bmatrix} \tag{2.36}$$

This establishes a correspondence between the second derivatives in the Cartesian and cylindrical (or polar) coordinate systems. We note that the Hessian matrix is symmetric.

The Hessian matrix in the Cartesian and 2D polar systems (when $\theta = 0$) can be obtained by simply dropping all terms related to the z-axis:

$$\begin{bmatrix} \frac{\partial^2\phi}{\partial x^2} & \frac{\partial^2\phi}{\partial y\partial x} \\ \frac{\partial^2\phi}{\partial x\partial y} & \frac{\partial^2\phi}{\partial y^2} \end{bmatrix} = \begin{bmatrix} \frac{\partial^2\phi}{\partial r^2} & \frac{1}{r}\frac{\partial^2\phi}{\partial\theta\partial r} - \frac{1}{r^2}\frac{\partial\phi}{\partial\theta} \\ \frac{1}{r}\frac{\partial^2\phi}{\partial\theta\partial r} - \frac{1}{r^2}\frac{\partial\phi}{\partial\theta} & \frac{1}{r}\frac{\partial\phi}{\partial r} + \frac{1}{r^2}\frac{\partial^2\phi}{\partial\theta^2} \end{bmatrix} \tag{2.37}$$

2.3.5 *Laplace operator in cylindrical and polar coordinates*

The Laplacian of the function, $\nabla^2\phi$, is the trace of the Hessian matrix, which interestingly has a much simpler expression. It can also be obtained using `vector.lap()`:

```
#lap = sp.simplify(H[0,0]+H[1,1]+H[2,2])           # trace of Hessian
# Alternatively:
lap = vector.lap(ϕ(r, θ, z), coords=(r, θ, z),h_vec=cyl_scale)
display(Math(f"\\text{{ }}$$∇^2$$ϕ = {latex(simplify(lap).expand())}"))
```

$$\nabla^2\phi = \frac{\partial^2}{\partial r^2}\phi(r,\theta,z) + \frac{\partial^2}{\partial z^2}\phi(r,\theta,z) + \frac{\frac{\partial}{\partial r}\phi(r,\theta,z)}{r} + \frac{\frac{\partial^2}{\partial\theta^2}\phi(r,\theta,z)}{r^2}$$

It is interesting to observe that each of the second derivatives of ϕ in the Hessian matrix depends on r, θ, and z and hence has a lengthy expression. The trace of the Hessian matrix (the Laplacian), however, is independent of r, θ, and z. The root reason is that the trace of a matrix is an invariant, meaning that it does not change when the coordinate system rotates (see Section 3.9 of Ref. [5]).

In summary, the Laplacian of a function, $\nabla^2\phi$, has the following relations in Cartesian and polar coordinates:

$$\nabla^2\phi = \frac{\partial^2\phi}{\partial x^2} + \frac{\partial^2\phi}{\partial y^2} + \frac{\partial^2\phi}{\partial z^2} = \frac{\partial^2\phi}{\partial r^2} + \frac{1}{r}\frac{\partial\phi}{\partial r} + \frac{1}{r^2}\frac{\partial^2\phi}{\partial\theta^2} + \frac{\partial^2\phi}{\partial z^2} \tag{2.38}$$

The Laplace operator ∇^2 can be written as

$$\nabla^2 = \frac{\partial^2}{\partial x^2} + \frac{\partial^2}{\partial y^2} + \frac{\partial^2}{\partial z^2} = \frac{\partial^2}{\partial r^2} + \frac{1}{r}\frac{\partial}{\partial r} + \frac{1}{r^2}\frac{\partial^2}{\partial\theta^2} + \frac{\partial^2}{\partial z^2} \tag{2.39}$$

In the polar coordinate system, we have

$$\nabla^2 = \frac{\partial^2}{\partial x^2} + \frac{\partial^2}{\partial y^2} = \frac{\partial^2}{\partial r^2} + \frac{1}{r}\frac{\partial}{\partial r} + \frac{1}{r^2}\frac{\partial^2}{\partial\theta^2} \tag{2.40}$$

These formulas will be frequently used throughout this book.

2.3.6 *Biharmonic operator in cylindrical and polar coordinates*

The biharmonic (or double Laplace) operator appears in the governing equations for the stress function [1] in the Cartesian coordinate system. A similar requirement arises when formulating solid mechanics problems in non-Cartesian coordinate systems.

When the biharmonic (or double Laplace) operator in cylindrical or polar coordinates is needed, one can either apply the Laplace operator twice or directly use `vector.biharmonic()` with cylindrical coordinates:

```
ϕ = sp.symbols("ϕ", cls=sp.Function)(r, θ, z)   # an arbitrary function

bi_h = vector.biharmonic(ϕ, coords=(r, θ, z),h_vec=cyl_scale)
#sp.simplify(bi_h)
```

The foregoing code cell produces the biharmonic operator in cylindrical coordinates, which is rewritten as

$$\begin{aligned}
\nabla^4\phi = \nabla^2\nabla^2\phi &= \left(\frac{\partial^2}{\partial r^2} + \frac{1}{r}\frac{\partial}{\partial r} + \frac{1}{r^2}\frac{\partial^2}{\partial \theta^2} + \frac{\partial^2}{\partial z^2}\right) \\
&\quad \times \left(\frac{\partial^2\phi}{\partial r^2} + \frac{1}{r}\frac{\partial\phi}{\partial r} + \frac{1}{r^2}\frac{\partial^2\phi}{\partial \theta^2} + \frac{\partial^2\phi}{\partial z^2}\right) \\
&= \frac{\partial^4\phi}{\partial x^4} + \frac{\partial^4\phi}{\partial y^4} + \frac{\partial^4\phi}{\partial z^4} + 2\frac{\partial^4\phi}{\partial y^2\partial z^2} + 2\frac{\partial^4\phi}{\partial x^2\partial z^2} + 2\frac{\partial^4\phi}{\partial x^2\partial y^2} \\
&= \frac{\partial^4\phi}{\partial r^4} + \frac{\partial^4\phi}{\partial z^4} + 2\frac{\partial^4\phi}{\partial z^2\partial r^2} + \frac{2}{r}\frac{\partial^3\phi}{\partial r^3} + \frac{2}{r}\frac{\partial^3\phi}{\partial z^2\partial r} - \frac{1}{r^2}\frac{\partial^2\phi}{\partial r^2} \\
&\quad + \frac{2}{r^2}\frac{\partial^4\phi}{\partial \theta^2\partial r^2} + \frac{2}{r^2}\frac{\partial^4\phi}{\partial \theta^2\partial z^2} + \frac{1}{r^3}\frac{\partial\phi}{\partial r} - \frac{2}{r^3}\frac{\partial^3\phi}{\partial \theta^2\partial r} + \frac{4}{r^4}\frac{\partial^2\phi}{\partial \theta^2} + \frac{1}{r^4}\frac{\partial^4\phi}{\partial \theta^4}
\end{aligned} \tag{2.41}$$

When the problem is axisymmetric and hence independent of θ, the biharmonic operator in cylindrical coordinates becomes

$$\begin{aligned}
\nabla^4\phi = \nabla^2\nabla^2\phi &= \left(\frac{\partial^2}{\partial r^2} + \frac{1}{r}\frac{\partial}{\partial r} + \frac{\partial^2}{\partial z^2}\right)\left(\frac{\partial^2\phi}{\partial r^2} + \frac{1}{r}\frac{\partial\phi}{\partial r} + \frac{\partial^2\phi}{\partial z^2}\right) \\
&= \frac{\partial^4\phi}{\partial r^4} + \frac{\partial^4\phi}{\partial z^4} + 2\frac{\partial^4\phi}{\partial z^2\partial r^2} + \frac{2}{r}\frac{\partial^3\phi}{\partial r^3} + \frac{2}{r}\frac{\partial^3\phi}{\partial z^2\partial r} - \frac{1}{r^2}\frac{\partial^2\phi}{\partial r^2} + \frac{1}{r^3}\frac{\partial\phi}{\partial r}
\end{aligned} \tag{2.42}$$

By simply removing all the terms related to z in Eq. (2.41), we obtain the biharmonic operator in the 2D polar coordinates as

$$\begin{aligned}
\nabla^4\phi = \nabla^2\nabla^2\phi &= \left(\frac{\partial^2}{\partial r^2} + \frac{1}{r}\frac{\partial}{\partial r} + \frac{1}{r^2}\frac{\partial^2}{\partial \theta^2}\right)\left(\frac{\partial^2\phi}{\partial r^2} + \frac{1}{r}\frac{\partial\phi}{\partial r} + \frac{1}{r^2}\frac{\partial^2\phi}{\partial \theta^2}\right) \\
&= \frac{\partial^4\phi}{\partial r^4} + \frac{2}{r}\frac{\partial^3\phi}{\partial r^3} - \frac{1}{r^2}\frac{\partial^2\phi}{\partial r^2} + \frac{2}{r^2}\frac{\partial^4\phi}{\partial \theta^2\partial r^2} \\
&\quad + \frac{1}{r^3}\frac{\partial\phi}{\partial r} - \frac{2}{r^3}\frac{\partial^3\phi}{\partial \theta^2\partial r} + \frac{4}{r^4}\frac{\partial^2\phi}{\partial \theta^2} + \frac{1}{r^4}\frac{\partial^4\phi}{\partial \theta^4}
\end{aligned} \tag{2.43}$$

Further, if the problem is axisymmetric and hence independent of θ, the biharmonic operator in polar coordinates becomes

$$\begin{aligned}
\nabla^4\phi = \nabla^2\nabla^2\phi &= \left(\frac{\partial^2}{\partial r^2} + \frac{1}{r}\frac{\partial}{\partial r}\right)\left(\frac{\partial^2\phi}{\partial r^2} + \frac{1}{r}\frac{\partial\phi}{\partial r}\right) \\
&= \frac{\partial^4\phi}{\partial r^4} + \frac{2}{r}\frac{\partial^3\phi}{\partial r^3} - \frac{1}{r^2}\frac{\partial^2\phi}{\partial r^2} + \frac{1}{r^3}\frac{\partial\phi}{\partial r}
\end{aligned} \tag{2.44}$$

These are also useful in our later operations in this book.

2.3.7 *Stresses in the cylindric and polar coordinates*

We can now look at the stresses in the cylindrical coordinate system. We define first the stress tensor in SymPy:

```
Srr, Srθ, Srz = symbols("\u03C3_rr \u03C3_rθ \u03C3_rz", cls=Function)
Sθr, Sθθ, Sθz = symbols("\u03C3_θr \u03C3_θθ \u03C3_θz", cls=Function)
Srz, Sθz, Szz = symbols("\u03C3_rz \u03C3_θz \u03C3_zz", cls=Function)

S_cyl = Matrix([                              # stress in cylindrical cords.
    [Srr(r, θ, z), Srθ(r, θ, z), Srz(r, θ, z)],
    [Sθr(r, θ, z), Sθθ(r, θ, z), Sθz(r, θ, z)],
    [Srz(r, θ, z), Sθz(r, θ, z), Szz(r, θ, z)]])
S_cyl          # a tensor of 9 stresses each is a function of (r, θ, z)
```

$$\begin{bmatrix} \sigma_{rr}(r,\theta,z) & \sigma_{r\theta}(r,\theta,z) & \sigma_{rz}(r,\theta,z) \\ \sigma_{\theta r}(r,\theta,z) & \sigma_{\theta\theta}(r,\theta,z) & \sigma_{\theta z}(r,\theta,z) \\ \sigma_{rz}(r,\theta,z) & \sigma_{\theta z}(r,\theta,z) & \sigma_{zz}(r,\theta,z) \end{bmatrix}$$

As seen, each stress component is a function of (r, θ, z). In the 2D polar coordinate system, the z coordinate is omitted, and the stress tensor reduces to

$$\sigma(r,\theta) = \begin{bmatrix} \sigma_{rr}(r,\theta) & \sigma_{r\theta}(r,\theta) \\ \sigma_{r\theta}(r,\theta) & \sigma_{\theta\theta}(r,\theta) \end{bmatrix} \tag{2.45}$$

The transformation of the stress tensor between Cartesian and cylindrical coordinate systems follows the same procedure [5].

2.3.8 *Equilibrium equations in cylindrical and polar coordinates*

As discussed in Chapter 6 of Ref. [5], the divergence of the stress vector (formed by stress components in any of the x, y, or z-directions) at a point in the solid, combined with the applied body force at that point, leads to the equilibrium equations. The same approach can be applied to establish the equilibrium equations in the cylindrical coordinate system. This process can be efficiently implemented using the following code snippets:

```
# Compute the divergence of the stress vector:
div_Scyl = vector.div_tensor(S_cyl.T, coords=(r, θ, z),h_vec=cyl_scale)

div_Scyl = div_Scyl.expand()
div_Scyl
```

$$\begin{bmatrix} \frac{\partial}{\partial r}\sigma_{rr}(r,\theta,z) + \frac{\partial}{\partial z}\sigma_{rz}(r,\theta,z) + \frac{\sigma_{rr}(r,\theta,z)}{r} - \frac{\sigma_{\theta\theta}(r,\theta,z)}{r} + \frac{\frac{\partial}{\partial\theta}\sigma_{\theta r}(r,\theta,z)}{r} \\ \frac{\partial}{\partial r}\sigma_{r\theta}(r,\theta,z) + \frac{\partial}{\partial z}\sigma_{\theta z}(r,\theta,z) + \frac{2\sigma_{r\theta}(r,\theta,z)}{r} + \frac{\frac{\partial}{\partial\theta}\sigma_{\theta\theta}(r,\theta,z)}{r} \\ \frac{\partial}{\partial r}\sigma_{rz}(r,\theta,z) + \frac{\partial}{\partial z}\sigma_{zz}(r,\theta,z) + \frac{\sigma_{rz}(r,\theta,z)}{r} + \frac{\frac{\partial}{\partial\theta}\sigma_{\theta z}(r,\theta,z)}{r} \end{bmatrix}$$

The equilibrium equations in the cylindrical coordinate system become

$$\begin{bmatrix} \frac{\partial\sigma_{rr}}{\partial r} + \frac{1}{r}\frac{\partial\sigma_{\theta r}}{\partial\theta} + \frac{\sigma_{rr}-\sigma_{\theta\theta}}{r} + \frac{\partial\sigma_{zr}}{\partial z} \\ \frac{\partial\sigma_{r\theta}}{\partial r} + \frac{1}{r}\frac{\partial\sigma_{\theta\theta}}{\partial\theta} + \frac{2\sigma_{r\theta}}{r} + \frac{\partial\sigma_{z\theta}}{\partial z} \\ \frac{\partial\sigma_{rz}}{\partial r} + \frac{1}{r}\frac{\partial\sigma_{\theta z}}{\partial\theta} + \frac{\sigma_{rz}}{r} + \frac{\partial\sigma_{zz}}{\partial z} \end{bmatrix} + \begin{bmatrix} b_r \\ b_\theta \\ b_z \end{bmatrix} = \mathbf{0} \tag{2.46}$$

In polar coordinates, all terms related to z are omitted, and hence Eq. (2.46) reduces to

$$\begin{bmatrix} \frac{\partial\sigma_{rr}}{\partial r} + \frac{1}{r}\frac{\partial\sigma_{\theta r}}{\partial\theta} + \frac{\sigma_{rr}-\sigma_{\theta\theta}}{r} \\ \frac{\partial\sigma_{r\theta}}{\partial r} + \frac{1}{r}\frac{\partial\sigma_{\theta\theta}}{\partial\theta} + \frac{2\sigma_{r\theta}}{r} \end{bmatrix} + \begin{bmatrix} b_r \\ b_\theta \end{bmatrix} = \mathbf{0} \tag{2.47}$$

We will solve Eq. (2.47) for several practically important problems in later chapters.

If the problem is **axisymmetric**, all variables are independent of θ, and the non-axisymmetric stress components vanish: $\sigma_{r\theta} = \sigma_{z\theta} = 0$. In this case, the equilibrium equations in cylindrical coordinates reduce to

$$\begin{aligned} \frac{\partial\sigma_{rr}}{\partial r} + \frac{\partial\sigma_{zr}}{\partial z} + \frac{\sigma_{rr}-\sigma_{\theta\theta}}{r} + b_r &= 0 \\ \frac{\partial\sigma_{rz}}{\partial r} + \frac{\sigma_{rz}}{r} + \frac{\partial\sigma_{zz}}{\partial z} + b_z &= 0 \end{aligned} \tag{2.48}$$

2.3.9 *Stress–displacement relations in cylindrical and polar coordinates*

The constitutive equation relates the strains ε_{ij} $(i = x, y, z)$ to the stresses σ_{ij}. It can be written as (see Chapter 5 of Ref. [5])

$$\sigma_{ij} = \lambda\delta_{ij}e + 2G\varepsilon_{ij} \tag{2.49}$$

where G is the shear modulus, λ is the Lamé constant of the material, δ_{ij} is the Kronecker delta, and

$$e = \frac{\partial u_x}{\partial x} + \frac{\partial u_y}{\partial y} + \frac{\partial u_z}{\partial z} = \varepsilon_{xx} + \varepsilon_{yy} + \varepsilon_{zz} \tag{2.50}$$

is the trace of the strain tensor, also known as the **volume strain**.

In this study, we assume all material constants are known. Since the constitutive equation (2.49) describes the intrinsic physical properties of the material, it is **independent of the coordinate system** used to describe the solid mechanics problem. Therefore, Eq. (2.49) applies to any orthogonal coordinate system, including Cartesian, polar, cylindrical, and spherical (to be discussed).

For isotropic materials, substituting the strain–displacement relations from Eq. (2.33) into Eq. (2.49) yields the **stress–displacement relations** in cylindrical coordinates:

$$
\begin{aligned}
\sigma_{rr} &= \lambda e + 2G\frac{\partial u_r}{\partial r} \\
\sigma_{\theta\theta} &= \lambda e + 2G\left(\frac{u_r}{r} + \frac{\partial u_\theta}{r\partial\theta}\right) \\
\sigma_{zz} &= \lambda e + 2G\frac{\partial u_z}{\partial z} \\
\sigma_{rz} &= G\left(\frac{\partial u_r}{\partial z} + \frac{\partial u_z}{\partial r}\right) \\
\sigma_{r\theta} &= G\left(\frac{\partial u_r}{r\partial\theta} + \frac{\partial u_\theta}{\partial r} - \frac{u_\theta}{r}\right)
\end{aligned}
\tag{2.51}
$$

where

$$
e = \frac{\partial u_r}{\partial r} + \frac{u_r}{r} + \frac{\partial u_\theta}{r\partial\theta} + \frac{\partial u_z}{\partial z} \tag{2.52}
$$

If the problem is **axisymmetric**, all terms involving the partial derivative with respect to θ should be removed.

Substituting Eq. (2.51) into Eq. (2.48), we obtain the equilibrium equations in terms of displacements, which become

$$
\begin{aligned}
(G+\lambda)\frac{\partial e}{\partial r} + G\left(\nabla^2 u_r - \frac{u_r}{r^2}\right) + b_r = 0 \\
(G+\lambda)\frac{\partial e}{\partial z} + G\nabla^2 u_z + b_z = 0
\end{aligned}
\tag{2.53}
$$

where ∇^2 is the Laplace operator in cylindrical coordinates for axisymmetric problems:

$$
\nabla^2 = \frac{\partial^2}{\partial r^2} + \frac{\partial}{r\partial r} + \frac{\partial^2}{\partial z^2} \tag{2.54}
$$

Again, for **axially symmetric problems** in the polar coordinate system, all terms in the foregoing equations related to z should be omitted.

2.4 Spherical coordinate system

2.4.1 *Definition*

Consider the spherical coordinates (r, θ, φ), representing the **radial distance**, **polar angle**, and **azimuthal angle**, respectively, as illustrated in Fig. 2.4.

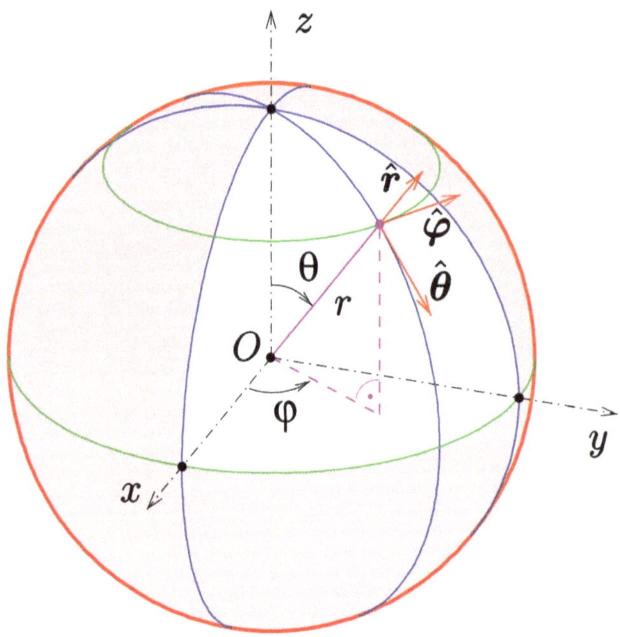

Figure 2.4. Spherical coordinates with radial axis r, polar angle θ, and azimuthal angle φ. Image from en.wikipedia by Ag2gaeh under the CC BY-SA 4.0 license.

Let us now define these coordinates in Python:

```
x, y, z = sp.symbols("x, y, z", real=True)
r = sp.symbols("r", ϕpositive = True)
θ, ϕ = sp.symbols("θ, ϕ", real=True)   # use ϕ for φ to avoid conflict

X_rθϕ = vector.transform_coords("spherical", (r, θ, ϕ))
X_rθϕ = Matrix(X_rθϕ)

X = sp.Matrix([x, y, z])                # position vector in Cartesian
R = sp.Matrix([r, θ, ϕ])                # position vector in spherical
```

```
display(Math(f" \\text{{Spherical to Cartesian: }} \
                                        {latex(X)}={latex(X_rθϕ)}"))
fr = sp.sqrt(x**2 + y**2 + z**2)
fθ = sp.acos(z/sp.sqrt(x**2 + y**2 + z**2))
fϕ = sp.acos(x/sp.sqrt(x**2 + y**2) )
R_xyz = sp.Matrix([fr, fθ, fϕ])
display(Math(f" \\text{{Cartesian to Spherical: }} \
                                        {latex(R)}={latex(R_xyz)}"))
```

$$\text{Spherical to Cartesian: } \begin{bmatrix} x \\ y \\ z \end{bmatrix} = \begin{bmatrix} r\sin(\theta)\cos(\phi) \\ r\sin(\theta)\sin(\phi) \\ r\cos(\theta) \end{bmatrix}$$

$$\text{Cartesian to spherical: } \begin{bmatrix} r \\ \theta \\ \phi \end{bmatrix} = \begin{bmatrix} \sqrt{x^2+y^2+z^2} \\ \operatorname{acos}\left(\frac{z}{\sqrt{x^2+y^2+z^2}}\right) \\ \operatorname{acos}\left(\frac{x}{\sqrt{x^2+y^2}}\right) \end{bmatrix}$$

The scale factor of the spherical coordinates with respect to the Cartesian coordinates can be found as follows:

```
# Compute the scale factor:
sph_scale = vector.scale_coeff_coords("spherical", (r, θ, ϕ))
display(Math(f" \\text{{Scalars (α, β, γ) = }} {latex(sph_scale)}"))
```

$$\text{Scalars } (\alpha, \beta, \gamma) = (1,\ r,\ r\sin(\theta))$$

The Jacobian matrices between the spherical coordinates and the Cartesian coordinates are computed as follows:

```
J_XR = X_rθϕ.jacobian([r, θ, ϕ])
display(Math(f"\\text{{Jacobian, }}\\frac{{∂X}}{{∂R}}={latex(J_XR)}"))
#gradf = vector.grad(X_rθϕ, coords=(r, θ, ϕ),h_vec=sph_scale)
#J_XR = Matrix([gradf[0].T,gradf[1].T,gradf[2].T]).T

J_RX = sp.simplify(J_XR.inv())
display(Math(f"\\text{{Inverse, }}\\frac{{∂R}}{{∂X}}={latex(J_RX)}"))
```

$$\text{Jacobian, } \frac{\partial \mathbf{X}}{\partial \mathbf{R}} = \begin{bmatrix} \sin(\theta)\cos(\phi) & r\cos(\theta)\cos(\phi) & -r\sin(\theta)\sin(\phi) \\ \sin(\theta)\sin(\phi) & r\sin(\phi)\cos(\theta) & r\sin(\theta)\cos(\phi) \\ \cos(\theta) & -r\sin(\theta) & 0 \end{bmatrix}$$

$$\text{Inverse, } \frac{\partial \mathbf{R}}{\partial \mathbf{X}} = \begin{bmatrix} \sin(\theta)\cos(\phi) & \sin(\theta)\sin(\phi) & \cos(\theta) \\ \frac{\cos(\theta)\cos(\phi)}{r} & \frac{\sin(\phi)\cos(\theta)}{r} & -\frac{\sin(\theta)}{r} \\ -\frac{\sin(\phi)}{r\sin(\theta)} & \frac{\cos(\phi)}{r\sin(\theta)} & 0 \end{bmatrix}$$

2.4.2 *Strains in the spherical coordinate system*

To derive the strain–displacement relations in the spherical coordinate system, we use the fact that the strain tensor is the symmetric part of the displacement gradient.

In this case, we use the method `vector.sym_grad()` to compute the strain components in terms of the derivatives of the displacements, carefully accounting for the geometric properties of spherical coordinates.

For validation and comparison, we also compute the strain tensor in Cartesian coordinates, which allows us to directly check the consistency between the two coordinate systems.

The following snippet illustrates this procedure in Python:

```
# Define the displacements in Spherical coordinates:
ur = sp.Function('u_r')(r, θ, ϕ)
uθ = sp.Function('u_θ')(r, θ, ϕ)
uϕ = sp.Function('u_ϕ')(r, θ, ϕ)
uS = Matrix([ur, uθ, uϕ])                          # form vector function

ux = sp.Function('u_x')(x, y, z)
uy = sp.Function('u_y')(x, y, z)
uz = sp.Function('u_z')(x, y, z)
uC = Matrix([ux, uy, uz])

ε_Cart = vector.sym_grad(uC, coords=(x, y, z), h_vec = (1, 1, 1))
display(Math(f"\\text{{Strain tensor in Cartesian coordinates: }} "))
display(Math(f"\\text{{ }} \\small {latex(ε_Cart)} "))

ε_sph = vector.sym_grad(uS, coords=(r, θ, ϕ), h_vec = sph_scale)
display(Math(f"\\text{{Strain tensor in spherical coordinates: }} "))
display(Math(f"\\text{{ }} \\scriptsize {latex(ε_sph)} "))
```

Strain tensor in Cartesian coordinates:

$$\begin{bmatrix} \frac{\partial}{\partial x}u_x(x,y,z) & \frac{\frac{\partial}{\partial y}u_x(x,y,z)}{2} + \frac{\frac{\partial}{\partial x}u_y(x,y,z)}{2} & \frac{\frac{\partial}{\partial z}u_x(x,y,z)}{2} + \frac{\frac{\partial}{\partial x}u_z(x,y,z)}{2} \\ \frac{\frac{\partial}{\partial y}u_x(x,y,z)}{2} + \frac{\frac{\partial}{\partial x}u_y(x,y,z)}{2} & \frac{\partial}{\partial y}u_y(x,y,z) & \frac{\frac{\partial}{\partial z}u_y(x,y,z)}{2} + \frac{\frac{\partial}{\partial y}u_z(x,y,z)}{2} \\ \frac{\frac{\partial}{\partial z}u_x(x,y,z)}{2} + \frac{\frac{\partial}{\partial x}u_z(x,y,z)}{2} & \frac{\frac{\partial}{\partial z}u_y(x,y,z)}{2} + \frac{\frac{\partial}{\partial y}u_z(x,y,z)}{2} & \frac{\partial}{\partial z}u_z(x,y,z) \end{bmatrix}$$

Strain tensor in spherical coordinates:

$$\left[\begin{matrix}\frac{\partial}{\partial r} u_{r}(r, \theta, \phi) & \frac{\frac{\partial}{\partial r} u_{\theta}(r,\theta,\phi)}{2} - \frac{u_{\theta}(r,\theta,\phi)}{2r} + \frac{\frac{\partial}{\partial \theta} u_{r}(r,\theta,\phi)}{2r} \\ \frac{\frac{\partial}{\partial r} u_{\theta}(r,\theta,\phi)}{2} - \frac{u_{\theta}(r,\theta,\phi)}{2r} + \frac{\frac{\partial}{\partial \theta} u_{r}(r,\theta,\phi)}{2r} & \frac{u_{r}(r,\theta,\phi)}{r} + \frac{\frac{\partial}{\partial \theta} u_{\theta}(r,\theta,\phi)}{r} \\ \frac{\frac{\partial}{\partial r} u_{\phi}(r,\theta,\phi)}{2} - \frac{u_{\phi}(r,\theta,\phi)}{2r} + \frac{\frac{\partial}{\partial \phi} u_{r}(r,\theta,\phi)}{2r \sin(\theta)} & -\frac{u_{\phi}(r,\theta,\phi)\cos(\theta)}{2r \sin(\theta)} + \frac{\frac{\partial}{\partial \theta} u_{\phi}(r,\theta,\phi)}{2r} + \frac{\frac{\partial}{\partial \phi} u_{\theta}(r,\theta,\phi)}{2r \sin(\theta)}\end{matrix}\right.$$

$$\left.\begin{matrix}\frac{\frac{\partial}{\partial r} u_{\phi}(r,\theta,\phi)}{2} - \frac{u_{\phi}(r,\theta,\phi)}{2r} + \frac{\frac{\partial}{\partial \phi} u_{r}(r,\theta,\phi)}{2r \sin(\theta)} \\ -\frac{u_{\phi}(r,\theta,\phi)\cos(\theta)}{2r \sin(\theta)} + \frac{\frac{\partial}{\partial \theta} u_{\phi}(r,\theta,\phi)}{2r} + \frac{\frac{\partial}{\partial \phi} u_{\theta}(r,\theta,\phi)}{2r \sin(\theta)} \\ \frac{u_{r}(r,\theta,\phi)}{r} + \frac{u_{\theta}(r,\theta,\phi)\cos(\theta)}{r \sin(\theta)} + \frac{\frac{\partial}{\partial \phi} u_{\phi}(r,\theta,\phi)}{r \sin(\theta)}\end{matrix}\right]$$

We rewrite the formulas of the strain tensor in Cartesian coordinates in a more concise form for future reference:

$$\begin{aligned}
\varepsilon_{rr} &= \frac{\partial u_r}{\partial r} \\
\varepsilon_{\theta\theta} &= \frac{u_r}{r} + \frac{\partial u_\theta}{r \partial \theta} \\
\varepsilon_{\varphi\varphi} &= \frac{1}{r \sin\theta} \frac{\partial u_\varphi}{\partial \varphi} + \frac{u_\theta}{r} \cot\theta + \frac{u_r}{r} \\
2\varepsilon_{\theta\varphi} &= \frac{1}{r}\left(\frac{\partial u_\varphi}{\partial \theta} - u_\varphi \cot\theta\right) + \frac{1}{r \sin\theta} \frac{\partial u_\theta}{\partial \varphi} \\
2\varepsilon_{r\varphi} &= \frac{1}{r \sin\theta} \frac{\partial u_r}{\partial \varphi} + \frac{\partial u_\varphi}{\partial r} - \frac{u_\varphi}{r} \\
2\varepsilon_{r\theta} &= \frac{\partial u_r}{r \partial \theta} + \frac{\partial u_\theta}{\partial r} - \frac{u_\theta}{r}
\end{aligned} \tag{2.55}$$

Here, we replaced ϕ with the more commonly used φ for the azimuthal angle.

2.4.3 *Laplace operator in spherical coordinates*

If only the **Laplace operator** of a function, $\nabla^2 F$, is required, it can be computed directly using `vector.lap()`:

```
F = sp.symbols("F", cls=sp.Function)(r, θ, ϕ)  # an arbitrary function
lap = vector.lap(F, coords=(r, θ, ϕ), h_vec = sph_scale)
display(Math(f"\\text{{}}$$ ^2$$F = {latex(simplify(lap).expand())} "))
```

$$\nabla^2 F = \frac{\partial^2}{\partial r^2}F(r,\theta,\phi) + \frac{2\frac{\partial}{\partial r}F(r,\theta,\phi)}{r} + \frac{\frac{\partial^2}{\partial \theta^2}F(r,\theta,\phi)}{r^2} + \frac{\frac{\partial}{\partial \theta}F(r,\theta,\phi)}{r^2\tan(\theta)} + \frac{\frac{\partial^2}{\partial \phi^2}F(r,\theta,\phi)}{r^2\sin^2(\theta)}$$

2.4.4 *Stresses in the spherical coordinates*

We can now examine the stresses in the spherical coordinate system. First, we define the **stress tensor** symbolically in SymPy:

```
Srr, Srθ, Srϕ = symbols("\u03C3_rr \u03C3_rθ \u03C3_rϕ", cls=Function)
Sθr, Sθθ, Sθϕ = symbols("\u03C3_θr \u03C3_θθ \u03C3_θϕ", cls=Function)
Sϕr, Sϕθ, Sϕϕ= symbols("\u03C3_ϕr \u03C3_ϕθ \u03C3_ϕϕ", cls=Function)

Ss = Matrix([
    [Srr(r, θ, ϕ), Srθ(r, θ, ϕ), Srϕ(r, θ, ϕ)],
    [Sθr(r, θ, ϕ), Sθθ(r, θ, ϕ), Sθϕ(r, θ, ϕ)],
    [Sϕr(r, θ, ϕ), Sϕθ(r, θ, ϕ), Sϕϕ(r, θ, ϕ)]])
    # a tensor of 9 stress components, each is a function of (r, θ, ϕ)
display(Math(f" \\text{{Stress tensor}}={latex(Ss)}"))
```

$$\text{Stress tensor} = \begin{bmatrix} \sigma_{rr}(r,\theta,\phi) & \sigma_{r\theta}(r,\theta,\phi) & \sigma_{r\phi}(r,\theta,\phi) \\ \sigma_{\theta r}(r,\theta,\phi) & \sigma_{\theta\theta}(r,\theta,\phi) & \sigma_{\theta\phi}(r,\theta,\phi) \\ \sigma_{\phi r}(r,\theta,\phi) & \sigma_{\phi\theta}(r,\theta,\phi) & \sigma_{\phi\phi}(r,\theta,\phi) \end{bmatrix}$$

2.4.5 *Equilibrium equations in spherical coordinates*

To establish the **equilibrium equations** in the spherical coordinate system, we compute the divergence of the stress vector, following the same procedure as for the cylindrical coordinate system:

```
# Compute the divergence of the stress vector:
div_Ss = vector.div_tensor(Ss.T, coords=(r, θ, ϕ), h_vec=sph_scale)

div_Ss = div_Ss.expand()
div_Ss                                  # divergence of the stress vector
```

$$
\begin{bmatrix}
\frac{\partial}{\partial r}\sigma_{rr}(r,\theta,\phi) + \frac{2\sigma_{rr}(r,\theta,\phi)}{r} + \frac{\sigma_{\theta r}(r,\theta,\phi)\cos(\theta)}{r\sin(\theta)} - \frac{\sigma_{\theta\theta}(r,\theta,\phi)}{r} \\
-\frac{\sigma_{\phi\phi}(r,\theta,\phi)}{r} + \frac{\frac{\partial}{\partial\theta}\sigma_{\theta r}(r,\theta,\phi)}{r} + \frac{\frac{\partial}{\partial\phi}\sigma_{\phi r}(r,\theta,\phi)}{r\sin(\theta)} \\
\frac{\partial}{\partial r}\sigma_{r\theta}(r,\theta,\phi) + \frac{3\sigma_{r\theta}(r,\theta,\phi)}{r} + \frac{\sigma_{\theta\theta}(r,\theta,\phi)\cos(\theta)}{r\sin(\theta)} - \frac{\sigma_{\phi\phi}(r,\theta,\phi)\cos(\theta)}{r\sin(\theta)} \\
+\frac{\frac{\partial}{\partial\theta}\sigma_{\theta\theta}(r,\theta,\phi)}{r} + \frac{\frac{\partial}{\partial\phi}\sigma_{\phi\theta}(r,\theta,\phi)}{r\sin(\theta)} \\
\frac{\partial}{\partial r}\sigma_{r\phi}(r,\theta,\phi) + \frac{2\sigma_{r\phi}(r,\theta,\phi)}{r} + \frac{\sigma_{\theta\phi}(r,\theta,\phi)\cos(\theta)}{r\sin(\theta)} \\
+\frac{\frac{\partial}{\partial\theta}\sigma_{\theta\phi}(r,\theta,\phi)}{r} + \frac{\frac{\partial}{\partial\phi}\sigma_{\phi\phi}(r,\theta,\phi)}{r\sin(\theta)}
\end{bmatrix}
$$

As discussed in Ref. [5], a divergence is always associated with a source, which represents the governing physical law for a given problem. In solid mechanics, the divergence of the stress tensor, together with a source term (body forces), yields the equilibrium equations. In the spherical coordinate system, these equations are expressed as follows:

$$
\begin{bmatrix}
\frac{\partial\sigma_{rr}}{\partial r} + \frac{1}{r}\frac{\partial\sigma_{\theta r}}{\partial\theta} + \frac{1}{r\sin\theta}\frac{\partial\sigma_{\varphi r}}{\partial\varphi} + \frac{1}{r}\left(2\sigma_{rr} - \sigma_{\theta\theta} - \sigma_{\varphi\varphi} + \sigma_{\theta r}\cot\theta\right) \\
\frac{\partial\sigma_{r\theta}}{\partial r} + \frac{1}{r}\frac{\partial\sigma_{\theta\theta}}{\partial\theta} + \frac{1}{r\sin\theta}\frac{\partial\sigma_{\varphi\theta}}{\partial\varphi} + \frac{1}{r}\left((\sigma_{\theta\theta} - \sigma_{\varphi\varphi})\cot\theta + 3\sigma_{r\theta}\right) \\
\frac{\partial\sigma_{r\varphi}}{\partial r} + \frac{1}{r}\frac{\partial\sigma_{\theta\varphi}}{\partial\theta} + \frac{1}{r\sin\theta}\frac{\partial\sigma_{\varphi\varphi}}{\partial\varphi} + \frac{1}{r}\left(2\sigma_{r\varphi} + \sigma_{\theta\varphi}\cot\theta\right)
\end{bmatrix}
+ \begin{bmatrix} b_r \\ b_\theta \\ b_\varphi \end{bmatrix} = 0 \tag{2.56}
$$

Note that the last term in the parentheses, $\frac{1}{r}\left(2\sigma_{rp} + \sigma_{\theta p}\cot\theta\right)$, differs from that in Ref. [2] and the online article (https://en.wikipedia.org/wiki/Linear_elasticity), which gives $\frac{1}{r}\left(3\sigma_{rp} + 2\sigma_{\theta p}\cot\theta\right)$. The reason for this discrepancy is unclear, and the correct form is uncertain. Caution is advised when using this term. The third equation will not be used in this book.

2.5 Other useful equations in polar coordinates

To solve problems in the polar coordinate system, a set of additional equations expressed in polar coordinates is needed. These equations can be derived from the corresponding equations in the Cartesian coordinate system [1] using the relations developed in the preceding sections.

2.5.1 *Stress and stress functions in the polar coordinates*

As discussed in Chapter 4 of Ref. [1], a stress function $\phi(x, y)$ can be introduced, allowing the stress solution to be expressed in the following form:

$$\sigma_{xx} - \varphi = \frac{\partial^2 \phi}{\partial y^2}, \quad \sigma_{yy} - \varphi = \frac{\partial^2 \phi}{\partial x^2}, \quad \sigma_{xy} = -\frac{\partial^2 \phi}{\partial x \partial y} \tag{2.57}$$

where φ is the potential function for the body force, $[b_x \; b_y]^\top$, applied over the solid:

$$b_x = -\frac{\partial \varphi}{\partial x}, \quad b_y = -\frac{\partial \varphi}{\partial y} \tag{2.58}$$

The stress components in the polar coordinates in terms of the stress function can be written as

$$\begin{aligned}
\sigma_{rr} &= \frac{1}{r}\frac{\partial \phi}{\partial r} + \frac{1}{r^2}\frac{\partial^2 \phi}{\partial \theta^2} + \varphi \\
\sigma_{\theta\theta} &= \frac{\partial^2 \phi}{\partial r^2} + \varphi \\
\sigma_{r\theta} &= \frac{1}{r^2}\frac{\partial \phi}{\partial \theta} - \frac{1}{r}\frac{\partial^2 \phi}{\partial \theta \partial r} = -\frac{\partial}{\partial r}\left(\frac{1}{r}\frac{\partial \phi}{\partial \theta}\right)
\end{aligned} \tag{2.59}$$

where the stress function ϕ is defined in the polar system (r, θ) as $\phi(r, \theta)$. Equation (2.59) is derived via the following procedure:

- The expression for σ_{rr} is obtained using the fact that $\sigma_{rr} = \sigma_{xx}|_{\theta=0} = \frac{\partial^2 \phi}{\partial y^2}|_{\theta=0} + \varphi$, in which $\frac{\partial^2 \phi}{\partial y^2}|_{\theta=0}$ is given in Eq. (2.37).
- The expression for $\sigma_{\theta\theta}$ is obtained using $\sigma_{\theta\theta} = \sigma_{yy}|_{\theta=0} = \frac{\partial^2 \phi}{\partial x^2}|_{\theta=0} + \varphi$, where $\frac{\partial^2 \phi}{\partial x^2}|_{\theta=0}$ is also given in Eq. (2.37).
- The expression for $\sigma_{r\theta}$ is obtained using $\sigma_{r\theta} = \sigma_{xy}|_{\theta=0} = \frac{\partial^2 \phi}{\partial x \partial x}|_{\theta=0}$, where $\frac{\partial^2 \phi}{\partial x \partial y}|_{\theta=0}$ is given in Eq. (2.37).

Similarly, the body force, $[b_r \; b_\theta]^\top$, in the polar system can be obtained as

$$b_r = -\frac{\partial \varphi}{\partial r}, \quad b_\theta = -\frac{\partial \varphi}{r \partial \theta} \tag{2.60}$$

in which we used the similar fact that $b_r = b_x|_{\theta=0} = -\frac{\partial\varphi}{\partial x^2}|_{\theta=0}$ and $b_\theta = b_y|_{\theta=0} = -\frac{\partial\varphi}{\partial y}|_{\theta=0}$. These partial derivatives are given in Eq. (2.31).

The biharmonic operator in the polar system is given in Eq. (2.43).

2.5.2 *Compatibility equations in stresses in polar coordinates*

The compatibility equation expressed in terms of stresses in Cartesian coordinates is

$$\nabla^2(\sigma_{xx} + \sigma_{yy}) = -\kappa'_\nu \left(\frac{\partial b_x}{\partial x} + \frac{\partial b_y}{\partial y}\right) \tag{2.61}$$

where the right-hand side is the divergence of the body force vector field. It is scaled by the constant κ_ν, which depends only on Poisson's ratio of the material, and is given by [1]

$$\begin{aligned} \kappa_\nu &= 1 + \nu \quad \text{plane stress} \\ \kappa_\nu &= \frac{1}{1-\nu} \quad \text{plane strain} \end{aligned} \tag{2.62}$$

Since the divergence of a vector field is a physical property of the field, its value should not be affected by the choice of coordinates. The divergence for a given force vector field in the polar coordinate system can be obtained by first computing it in the more general cylindrical coordinates and then reducing the results to the polar coordinate system. The snippet is as follows:

```
# define a force vector in cylindrical coordinates:
br = Function("b_r ")(r, θ, z)
bθ = Function("b_θ ")(r, θ, z)
bz = Function("b_z ")(r, θ, z)
bv = Matrix([br, bθ, bz])                          # body force vector

# Compute the divergence of the force vector in cylindrical coordinates:
div_bv = vector.div(bv, coords=(r, θ, z), h_vec=cyl_scale).expand()
display(Math(f"\\text{{div}}(\\scriptsize {latex(bv)})={latex(div_bv)}"))
```

$$\text{div}\left(\begin{bmatrix} b_r(r,\theta,z) \\ b_\theta(r,\theta,z) \\ b_z(r,\theta,z) \end{bmatrix}\right) = \frac{\partial}{\partial r}b_r(r,\theta,z) + \frac{\partial}{\partial z}b_z(r,\theta,z) + \frac{b_r(r,\theta,z)}{r} + \frac{\frac{\partial}{\partial\theta}b_\theta(r,\theta,z)}{r}$$

Removing the term related to z from the foregoing result gives

$$\left(\frac{\partial b_x}{\partial x} + \frac{\partial b_y}{\partial y}\right) = \frac{\partial b_r}{\partial r} + \frac{b_r}{r} + \frac{1}{r}\frac{\partial b_\theta}{\partial \theta} \tag{2.63}$$

Using Eq. (2.61), the compatibility equations in terms of stresses in polar coordinates become

$$\left(\frac{\partial^2}{\partial r^2} + \frac{1}{r}\frac{\partial}{\partial r} + \frac{1}{r^2}\frac{\partial^2}{\partial \theta^2}\right)(\sigma_{rr} + \sigma_{\theta\theta}) = -\kappa_\nu'\left(\frac{\partial b_r}{\partial r} + \frac{b_r}{r} + \frac{1}{r}\frac{\partial b_\theta}{\partial \theta}\right) \tag{2.64}$$

2.5.3 *Compatibility equations in stress function in polar coordinates*

The compatibility equation written in terms of the stress function becomes

$$\nabla^4\phi = -\kappa_\nu'\left(\frac{\partial b_r}{\partial r} + \frac{b_r}{r} + \frac{1}{r}\frac{\partial b_\theta}{\partial \theta}\right) \tag{2.65}$$

where the biharmonic $\nabla^4\phi$ is defined in Eq. (2.43). If $b_r = 0$ and b_θ is constant, we have

$$\nabla^4\phi = 0 \tag{2.66}$$

2.6 Solution strategy for solid mechanics problems

The solution strategy for solving solid mechanics problems in non-Cartesian coordinates is largely the same as the one discussed in Ref. [1] and is outlined as follows.

2.6.1 *Stress methods*

To solve problems in polar coordinates using the stress method, the procedure is as follows:

1. **Find a stress function** that satisfies the compatibility Eqs. (2.65) or (2.66), which generally contains integral constants.
2. **Determine the stresses**: Use Eq. (2.59) to compute the stresses.
3. **Relate strains and stresses**: Use the constitutive equation (the same as in Cartesian coordinates) to relate the strains to the stresses.
4. **Express displacements in terms of strains**: Write the displacements (and their differentials) in terms of the strains using Eq. (2.28), and then express them in terms of stresses via the constitutive equations. This set of equations consists of first-order partial differential equations.
5. **Solve for displacements** by integration and using the separation of variables technique. This leads to a general displacement solution with constants that need to be determined.

6. **Apply boundary conditions**: Use the stress boundary conditions (SBCs) and/or the displacement boundary conditions (DBCs) to determine these constants.
7. **Apply integral equilibrium equations** on the boundary of a free body diagram, which relate the applied force to the stresses on the boundary, also known as the force boundary conditions (FBCs). This further determines the remaining unknown constants.

The key step in the stress method is assuming the stress function. Since the stress function is governed by the biharmonic equation, we can either solve the biharmonic equation directly or use a linear combination of basis functions that satisfy it. The main challenge in the stress method lies in solving for displacements via integration, which can introduce integral functions and/or integral constants. A separation of variables technique may be needed, sometimes requiring manual operations. Additionally, the procedure can fail if the antiderivative of a function cannot be found.

The sequence of imposing these conditions to determine the unknown constants can vary depending on implementation convenience. In the following chapters, this strategy will be used to solve a number of problems.

2.6.2 *Displacement methods*

In contrast to the stress method, the displacement method generally follows the procedure below, as discussed in Ref. [4]:

1. **Assume the displacements** using appropriate basis functions that contain unknown constants.
2. **Compute the strains** using the displacement expressions via the strain–displacement relations in Eq. (2.28).
3. **Compute the stresses** using the strain expressions and the constitutive relations.
4. **Apply differential equilibrium equations**, such as Eq. (2.46) in the domain, which relate the body force to the stresses. This step helps determine some of the unknown constants.
5. **Apply integral equilibrium equations** on the boundary of a free body diagram, which relate the applied force to the stresses on the boundary (FBCs). This further determines unknown constants.
6. **Use SBCs** to determine additional unknown constants.
7. **Use DBCs** to determine the remaining unknown constants.

The key step is assuming the displacement expressions, which is also the most challenging part. The partial differential equations governing the displacements are often difficult to solve in general. Therefore, displacements are frequently assumed as a linear combination of a set of basis functions [6]. The method will fail if such a set of basis functions cannot be found (i.e., the displacement functions do not span the space of the basis functions), just as not all functions have a closed-form antiderivative.

If a suitable set of basis functions can be found, the displacement method is usually straightforward: no indefinite integration is required, there is no need to handle integral constants or functions, and it can often be automated. This is one reason why displacement methods are more commonly used in numerical methods based on domain discretization for complex problems. Proper domain discretization can make the displacement approximation straightforward [7–9].

The sequence of imposing these conditions to determine the unknown constants can vary depending on implementation convenience. In the following chapters, the displacement method will also be used to solve a number of problems.

The next chapter studies stress methods for solid mechanics problems that possess certain symmetry features.

References

[1] G. R. Liu, *Solid Mechanics with Python: Cartesian Coordinates*, World Scientific, New Jersey, 2025.

[2] A. P. Boresi, K. Chong and J. D. Lee, *Elasticity in Engineering Mechanics*, John Wiley & Sons, New York, 2010.

[3] G. R. Liu, *Calculus: A Practical Course with Python*, World Scientific, New Jersey, 2025.

[4] G. R. Liu, *Vector Calculus: Formulations, Applications, and Python Codes*, World Scientific, New Jersey, 2025.

[5] G. R. Liu, *Mechanics of Materials: Formulations and Solutions with Python*, World Scientific, New Jersey, 2024.

[6] G. R. Liu, *Numbers and Functions: Theory, Formulation, and Python Codes*, World Scientific, New Jersey, 2024.

[7] G. R. Liu and S. S. Quek, *The Finite Element Method: A Practical Course*, Butterworth-Heinemann, London, 2013.
[8] G. R. Liu and T. T. Nguyen, *Smoothed Finite Element Methods*, CRC Press, New York, 2010.
[9] G. R. Liu, *Mesh Free Methods: Moving Beyond the Finite Element Method*, CRC Press, Taylor and Francis Group, New York, 2010.

Chapter 3

Stress Methods for Problems with Symmetry

Let us import the necessary external modules or dependencies for later use in this chapter:

```
# Place cursor in this cell, and press Ctrl+Enter to import dependencies.
import sys                              # for accessing the computer system
sys.path.append('../grbin/')              # Add in the path to your system

from commonImports import *        # Import dependencies from '../grbin/'
import grcodes as gr                     # Import the module of the author
importlib.reload(gr)                 # When grcodes is modified, reload it

init_printing(use_unicode=True)          # For latex-like quality printing
np.set_printoptions(precision=4,suppress=True)    # Digits in print-outs
```

In engineering applications, many structural components are designed with some degree of symmetry. A typical example is an axisymmetric structural component, such as a disk, cylinder, shaft, or any component with a partially circular shape. This geometric symmetry plays a crucial role in reducing centrifugal forces when a disk is spinning, thereby improving both performance and safety.

When computing stresses and displacements in solids, this symmetry can often be exploited. In the stress method, for instance, the stress function can be assumed to take a specific form to simplify the solution process. This chapter discusses stress methods for solving problems with symmetry, beginning with axisymmetric problems.

In stress methods, stress functions are typically employed, and the stresses are expressed in terms of the second derivatives of the stress function.

The key task in such methods is to determine the appropriate stress function. This concept is discussed in detail in Ref. [1] for problems in the Cartesian coordinate system. Here, we apply the same concept in the polar coordinate system.

3.1 Axisymmetric problems

For solids with axisymmetric geometry, we use the polar coordinate system with its origin at the center of the component. This allows the geometric axisymmetry to be fully utilized and ensures that the radial coordinate is independent of the tangential coordinate, θ.

If the loading is also axisymmetric, we obtain an axisymmetric mechanics problem, where the variables simplify as follows:

1. **Independence from** θ: All variables are independent of the tangential polar coordinate θ.
2. **Body forces**: The body force in the tangential direction, b_θ, is zero, and any nonzero body force can only act in the radial direction, b_r.
3. **Displacement components**: The displacement component in the tangential direction, u_θ, vanishes because it is not axisymmetric. The only nonzero displacement occurs in the radial direction, u_r.
4. **Shear stresses**: The shear stress $\sigma_{r\theta}$ vanishes, as it is not axisymmetric.

3.1.1 *Basic equations for axisymmetric problems*

Under these conditions, the equilibrium equation (2.47) (with $\sigma_{r\theta} = \sigma_{z\theta} = 0$ and $\frac{\partial \sigma_{\theta\theta}}{\partial \theta} = 0$) reduces to a single equation:

$$\frac{\partial \sigma_{rr}}{\partial r} + \frac{\sigma_{rr} - \sigma_{\theta\theta}}{r} + b_r = 0 \tag{3.1}$$

The unknowns are two stresses, σ_{rr} and $\sigma_{\theta\theta}$. The relationship between these stresses and the stress function is simplified from Eq. (2.59) to a set of two equations:

$$\begin{aligned} \sigma_{rr} &= \frac{1}{r}\frac{\partial \phi}{\partial r} + \varphi \\ \sigma_{\theta\theta} &= \frac{\partial^2 \phi}{\partial r^2} + \varphi \end{aligned} \tag{3.2}$$

where φ is the potential function for a given body force b_r:

$$b_r = -\frac{\partial \varphi}{\partial r} \tag{3.3}$$

Therefore, φ can always be found, provided that b_r is integrable with respect to r.

3.1.2 *Axisymmetric problems without body force*

When the body force is zero, we have $b_r = 0$, and φ can only be a constant, which has no effect on the body force. As a result, the system of equations can be further simplified, and the solution will then be governed solely by the boundary conditions.

First, the strain–displacement relationship becomes (see Eq. (2.34))

$$\varepsilon_{rr} = \frac{\partial u_r}{\partial r}; \quad \varepsilon_{\theta\theta} = \frac{u_r}{r}; \quad \varepsilon_{r\theta} = 0 \tag{3.4}$$

The stress function is governed by the biharmonic equation (also known as the compatibility equation) and is therefore often referred to as the biharmonic function [1]. For an axisymmetric problem in polar coordinates, the stress function φ depends only on r, and the biharmonic compatibility equation for zero body-force problems becomes

$$\begin{aligned} \nabla^4\phi = \nabla^2\nabla^2\phi &= \left(\frac{\partial^2}{\partial r^2} + \frac{1}{r}\frac{\partial}{\partial r}\right)\left(\frac{\partial^2\phi}{\partial r^2} + \frac{1}{r}\frac{\partial\phi}{\partial r}\right) \\ &= \frac{\partial^4\phi}{\partial r^4} + \frac{2}{r}\frac{\partial^3\phi}{\partial r^3} - \frac{1}{r^2}\frac{\partial^2\phi}{\partial r^2} + \frac{1}{r^3}\frac{\partial\phi}{\partial r} = 0 \end{aligned} \tag{3.5}$$

which is Eq. (2.44). This is a 1D fourth-order differential equation (DE) with only one independent variable r. It has an equivalent chained differential form:

$$\nabla^4\phi = \frac{d}{dr}\left\{r\frac{d}{dr}\left[\frac{1}{r}\frac{d}{dr}\left(r\frac{d\phi}{dr}\right)\right]\right\} = 0 \tag{3.6}$$

This chained structure allows us to integrate the equation in stages to obtain the solution for the stress function, which can be carried out by hand. It is even easier, however, when using SymPy.

3.1.3 *General solution for the 1D biharmonic equation*

We use SymPy to directly obtain the general solution for the one-dimensional fourth-order biharmonic differential equation, Eq. (3.5):

```
# Solve ∇^4ϕ=0 (zero body force) for general solution:
r = sp.symbols('r', positive=True)              # define symbolic variable
ϕ = sp.Function('ϕ')(r)                        # define the stress function

lap = diff(diff(ϕ,r),r)+(1/r)*diff(ϕ, r)    # define Laplacian operator
bih = diff(diff(lap,r),r)+ (1/r)*diff(lap,r)          #biharmonic operator

diff_eq = sp.Eq(bih, 0)                      # define differential equation
g_sln = sp.dsolve(diff_eq, ϕ)             # solve the differential equation

display(Math(f" \\text{{The general solution to $∇^4ϕ=0$ is: }} \
                                                            {latex(g_sln)}"))
```

The general solution to $\nabla^4\phi = 0$ is

$$\phi(r) = C_1 + C_2r^2 + C_3r^2 \log(r) + C_4 \log(r)$$

This is the general expression for the stress function for problems without body force. It contains four arbitrary constants. We can rename these constants for consistency with the author's notation convention and to facilitate tracking (readers may choose to handle this differently or leave it as is). Since these constants will be determined later, their specific notation is immaterial. Thus, we write

$$\phi(r) = c_0 \log(r) + c_1r^2 \log(r) + c_2r^2 + c_3 \tag{3.7}$$

The Python snippet to redefine the stress function is follows:

```
c0, c1, c2, c3=symbols('c0, c1, c2, c3') # arbitrary integral constants
ϕ = c0*sp.log(r)+c1*r**2*sp.log(r)+c2*r**2+c3              #reset constants
ϕ
```

$c_0 \log(r) + c_1r^2 \log(r) + c_2r^2 + c_3$

These four arbitrary constants can be determined based on the specific setup of the actual axisymmetric problem, including the strain–displacement relations, stress boundary conditions (SBCs), and displacement boundary conditions (DBCs).

3.1.4 *General formulas for stresses*

We can now compute the stresses using Eq. (3.2) without considering body forces. This process involves only differentiation and can be carried out straightforwardly:

```
σ_rr = (diff(ϕ,r)/r).expand()                          # differentiate once
display(Math(f" \\text{{ Stress $σ_{{rr}}$}}={latex(σ_rr)}"))

σ_θθ = (diff(diff(ϕ,r),r)).expand()                    # differentiate twice
display(Math(f" \\text{{ Stress $σ_{{θθ}}$}}={latex(σ_θθ)}"))
```

Stress $\sigma_{rr} = \frac{c_0}{r^2} + 2c_1 \log(r) + c_1 + 2c_2$

Stress $\sigma_{\theta\theta} = -\frac{c_0}{r^2} + 2c_1 \log(r) + 3c_1 + 2c_2$

We obtain the general formula for stresses for any axisymmetric problem without body force:

$$\begin{aligned}\sigma_{rr} &= \frac{c_0}{r^2} + 2c_1 \log(r) + c_1 + 2c_2 \\ \sigma_{\theta\theta} &= -\frac{c_0}{r^2} + 2c_1 \log(r) + 3c_1 + 2c_2 \\ \sigma_{r\theta} &= 0\end{aligned} \tag{3.8}$$

The general stress solution contains three constants, c_0, c_1, and c_2, which must be determined using the boundary conditions for a specific problem.

3.1.5 *General formulas for strains*

Consider a plane stress problem, such as a thin disk. We can now obtain the general formulas for the strains using the constitutive equation (assuming zero shear stress):

$$\begin{aligned}\varepsilon_{rr} &= \frac{1}{E}(\sigma_{rr} - \nu\sigma_{\theta\theta}) \\ \varepsilon_{\theta\theta} &= \frac{1}{E}(\sigma_{\theta\theta} - \nu\sigma_{rr}) \\ \varepsilon_{r\theta} &= 0\end{aligned} \tag{3.9}$$

where E is Young's modulus and ν is Poisson's ratio. Note that the constitutive equation in polar coordinates is the same as in Cartesian coordinates, as it represents only a relationship between material mechanical properties. The following code snippets derive these expressions for the strains, also in terms of the unknown constants:

```
E, ν = symbols('E, ν', positive=True)                  # material constants
ε_rr = (σ_rr - ν*σ_θθ)/E
display(Math(f" \\text{{ Strain $ε_{{rr}}$}}={latex(ε_rr)}"))

ε_θθ = (σ_θθ - ν*σ_rr)/E
display(Math(f" \\text{{ Strain $ε_{{θθ}}$}}={latex(ε_θθ)}"))
```

$$\text{Strain } \varepsilon_{rr} = \frac{\frac{c_0}{r^2} + 2c_1 \log(r) + c_1 + 2c_2 - \left(-\frac{c_0}{r^2} + 2c_1 \log(r) + 3c_1 + 2c_2\right)}{E}$$

$$\text{Strain } \varepsilon_{\theta\theta} = \frac{-\frac{c_0}{r^2} + 2c_1 \log(r) + 3c_1 + 2c_2 - \nu\left(\frac{c_0}{r^2} + 2c_1 \log(r) + c_1 + 2c_2\right)}{E}$$

3.1.6 *General formulas for displacements*

To obtain the displacement expression, we use the strain–displacement relationship. From the first equation in Eq. (3.4), we have $\frac{\partial u_r}{\partial r} = \varepsilon_{rr}$, and hence u_r can be obtained by a single integration:

```
C = symbols('C')
u_r = sp.integrate(ε_rr,(r)) + C                    # C is an integral constant
display(Math(f"\\text{{Displacement $u_{{r}}$}}=\small {latex(u_r)}"))
```

Displacement
$$u_r = C + \frac{-r\left(c_1\nu + c_1 + 2c_2\nu - 2c_2\right) - \frac{c_0\nu + c_0}{r}}{E} + \frac{\left(-2c_1 r\nu + 2c_1 r\right)\log(r)}{E}$$

Next, using the second equation in Eq. (3.4), we need $r\varepsilon_{\theta\theta} - u_r = 0$. The term $r\varepsilon_{\theta\theta} - u_r$ can be found as follows:

```
(r*ε_θθ-u_r).simplify()
```

$$-C + \frac{4c_1 r}{E}$$

This expression must always be zero for any r:

$$-C + \frac{4c_1 r}{E} = 0 \tag{3.10}$$

which is possible only if both C and c_1 are zero:

$$C = c_1 = 0 \tag{3.11}$$

Therefore, the general solution for the stresses is updated as

$$\begin{aligned} \sigma_{rr} &= \frac{c_0}{r^2} + 2c_2 \\ \sigma_{\theta\theta} &= -\frac{c_0}{r^2} + 2c_2 \\ \sigma_{r\theta} &= 0 \end{aligned} \tag{3.12}$$

Using Eq. (3.11), the general solution for the displacement can now be found using the following snippet:

```
u_rg = u_r.subs({C:0, c1:0}).expand()
u_rg
```

$$-\frac{c_0\nu}{Er}-\frac{c_0}{Er}-\frac{2c_2r\nu}{E}+\frac{2c_2r}{E}$$

Finally, we obtain the general expression for displacement, which is rewritten as

$$u_r=-c_0\frac{(1+\nu)}{E}\frac{1}{r}+2c_2\frac{(1-\nu)}{E}r \tag{3.13}$$

There are a total of two unknown constants, c_0 and c_2. These must be determined using the boundary conditions (BCs).

Note that Eq. (3.13) applies to 2D plane stress problems. If the problem involves plane strain, such as in a long cylindrical pipe, a similar set of solutions can be obtained by modifying the constitutive equation. Alternatively, we can use Eq. (3.13) with the following substitutions [2]:

$$\begin{aligned}&\text{first: substitute } \nu \text{ by } \frac{\nu}{(1-\nu)}\\&\text{then: substitute } E \text{ by } \frac{E}{(1-\nu^2)}\\&\text{no change in } G\end{aligned} \tag{3.14}$$

With these substitutions, all expressions derived for the plane stress problem can be applied to the corresponding plane strain problem.

We are now ready to find solutions for specific problems.

3.1.7 *Example: Pressurized disks*

Consider a disk subjected to internal pressure p_i and external pressure p_o on the outer surface, as shown in Fig. 3.1. Assuming zero body force, determine the stresses and displacement within the disk.

This problem is clearly axisymmetric, and the formulas presented in the previous section can be applied. We can now consider several specific cases.

3.1.8 *Case I: The simplest case*

If the axisymmetric domain does not have a hole at the center, c_0 in Eq. (3.12) must be zero to ensure a meaningful **finite** solution. Otherwise, the stress becomes infinite at $r=0$, which is not physical.

Setting $c_0=0$, the solution for the stresses becomes

$$\sigma_{rr}=\sigma_{\theta\theta}=2c_2 \tag{3.15}$$

Both stresses are constant. If the externally applied uniform stress is σ, we have $\sigma_{rr}=\sigma_{\theta\theta}=\sigma$. The constant c_2 is then determined as $c_2=\sigma/2$.

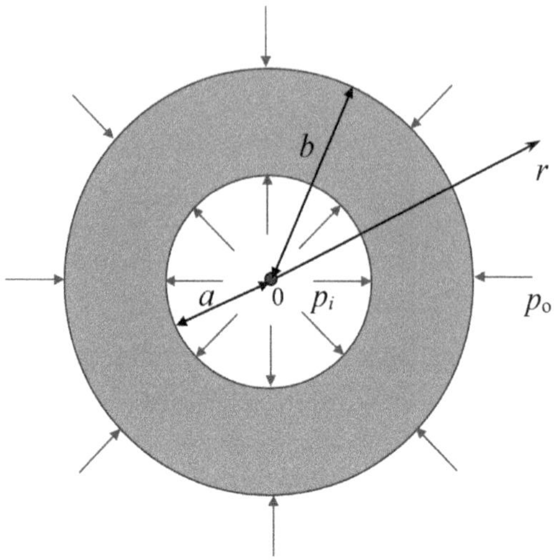

Figure 3.1. A disk subjected to pressure on the inner and outer surfaces.

The displacement is given by

$$u_r = \frac{(1-\nu)\sigma}{E} r \tag{3.16}$$

It grows linearly with r. Equation (3.16) is very similar to that of the displacement of a 1D bar fixed at one end, $u_x = \frac{\sigma}{E}x$. The detailed derivation can be found in Ref. [2]. The only difference is due to Poisson's effect: because the disk is constrained within its plane, the displacement is reduced by a factor of $(1 - \nu)$.

3.1.9 *Case II: Pressurized disks*

3.1.9.1 *Stress solution*

If there is a hole at the origin ($r = 0$), other solutions are possible because c_0 in Eq. (3.12) cannot be simply set to zero. We write the following code to determine these two constants, c_0 and c_2, using the stress boundary conditions. Using Eq. (3.12), we have:

```
r = sp.symbols("r", positive=True)
c0, c2 = sp.symbols("c0, c2")
σ_rr = c0/r**2 + 2*c2                        # the general solution for stress
display(Math(f"\\text{{general solution of $σ_{{rr}}$}}={latex(σ_rr)}"))

σ_θθ = -c0/r**2 + 2*c2                       # the general solution for stress
display(Math(f"\\text{{general solution of $σ_{{θθ}}$}}={latex(σ_θθ)}"))
```

general solution of $\sigma_{rr} = \frac{c_0}{r^2} + 2c_2$

general solution of $\sigma_{\theta\theta} = -\frac{c_0}{r^2} + 2c_2$

Since the disk is pressurized on both the inner ($r = a$) and outer ($r = b$) surfaces, we have two stress boundary conditions to determine c_0 and c_2:

```
a, b = sp.symbols("a, b", positive=True)
pi, po = sp.symbols("p_i, p_o")
SBC1 = σ_rr.subs(r,a)+pi                    # stress boundary conditions
SBC2 = σ_rr.subs(r,b)+po
cs = sp.solve([SBC1, SBC2],[c0, c2])               # solve for c0 and c2
display(Math(f"\\text{{Solution of the constants}}={latex(cs)}"))
```

$$\text{Solution of the constants} = \left\{c_0 : \frac{a^2b^2p_i - a^2b^2p_o}{a^2 - b^2},\ c_2 : \frac{-a^2p_i + b^2p_o}{2a^2 - 2b^2}\right\}$$

We have now determined the two constants that satisfy the conditions on both surfaces. By substituting these constants back into the stress expressions, we obtain:

```
σ_rr = σ_rr.subs(cs).expand().collect(pi).collect(po).simplify()
σ_θθ = σ_θθ.subs(cs).expand().collect(pi).collect(po).simplify()
#display(σ_rr, σ_θθ); print(σ_rr, σ_θθ)     # to get the formulas below
```

The stresses in a disk pressurized from both the inner ($r = a$) and outer ($r = b$) surfaces are found as follows:

$$\begin{aligned} \sigma_{rr} &= \left(\frac{a^2p_i\left(b^2 - r^2\right) - b^2p_o\left(a^2 - r^2\right)}{r^2\left(a^2 - b^2\right)}\right) \\ \sigma_{\theta\theta} &= \left(\frac{-a^2p_i\left(b^2 + r^2\right) + b^2p_o\left(a^2 + r^2\right)}{r^2\left(a^2 - b^2\right)}\right) \end{aligned} \tag{3.17}$$

with $\sigma_{r\theta} = 0$. In this equation, p_i is the pressure on the inner surface, and p_o is the pressure on the outer surface.

3.1.9.2 *Some findings from the stress solution*

We now write down the stresses in the disk for further examination. When $p_i = 0$, meaning that the disk is subjected to pressure only on the outer surface, we have

$$\begin{aligned}
\sigma_{rr} &= -\frac{b^2(r^2-a^2)}{r^2(b^2-a^2)}p_o \quad \text{always negative (in compression)} \\
\sigma_{\theta\theta} &= -\frac{b^2(r^2+a^2)}{r^2(b^2-a^2)}p_o \quad \text{always negative (in compression)}
\end{aligned} \tag{3.18}$$

This indicates that the entire disk is under compression, as expected. This solution will be used later when studying stress concentration around a hole in plates.

When $p_o = 0$, meaning that the disk is subjected to pressure only on the inner surface, we have

$$\begin{aligned}
\sigma_{rr} &= -\frac{a^2(b^2-r^2)}{r^2(b^2-a^2)}p_i \quad \text{always negative (in compression)} \\
\sigma_{\theta\theta} &= \frac{a^2(b^2+r^2)}{r^2(b^2-a^2)}p_i \quad \text{always positive (in tension)}
\end{aligned} \tag{3.19}$$

3.1.9.3 *Plots of stress distribution*

Let us plot the distribution of the stresses in a disk with the given specific parameters:

```
# Define these useful formulas:
σ_rrpi = a**2*pi*(b**2 - r**2)/(r**2*(a**2 - b**2))
σ_θθpi =-a**2*pi*(b**2 + r**2)/(r**2*(a**2 - b**2))
σ_rrpo =-b**2*po*(a**2 - r**2)/(r**2*(a**2 - b**2))
σ_θθpo = b**2*po*(a**2 + r**2)/(r**2*(a**2 - b**2))
```

```
# SymPy expressions to Python functions:
dic_ab = {a:1, b:5}
func1 = sp.lambdify(r, σ_rrpi.subs(dic_ab).subs({pi:1}), 'numpy')
func2 = sp.lambdify(r, σ_θθpi.subs(dic_ab).subs({pi:1}), 'numpy')
func3 = sp.lambdify(r, σ_rrpo.subs(dic_ab).subs({po:1}), 'numpy')
func4 = sp.lambdify(r, σ_θθpo.subs(dic_ab).subs({po:1}), 'numpy')

labels=[r'$σ_{rr},a=1,b=5,pi=1,po=0$',r'$σ_{θθ},a=1,b=5,pi=1,po=0$',
        r'$σ_{rr},a=1,b=5,pi=0,po=1$',r'$σ_{θθ},a=1,b=5,pi=0,po=1$']

r_data = np.linspace(1, 5, 400)                  #  r values for the plot
x_data = [r_data, r_data, r_data, r_data]
y_data = [func1(r_data), func2(r_data), func3(r_data), func4(r_data)]
                                                          # function values
#gr.plot_fs(x_data, y_data,labels,'r','stresses','imagesSMp/Disk_by_pi')
```

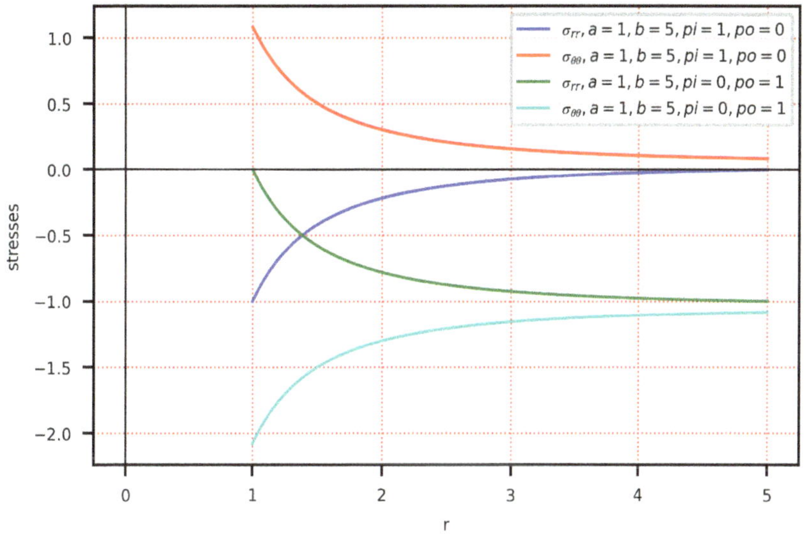

Figure 3.2. Stress distributions in a thin disk with an inner radius of 1 and an outer radius of 5, subject to a unit pressure.

The profiles of the stress distributions in a thin disk with an inner radius of 1 and an outer radius of 5, subject to a unit pressure, are shown in Fig. 3.2. It is observed that when the disk is pressurized from the outer surface by p_o, both stress components are negative (in compression). The hoop stress $\sigma_{\theta\theta}$ is always larger in magnitude than the applied pressure. Its maximum occurs on the inner surface, exceeding twice the applied pressure. The radial stress σ_{rr} is always smaller than the applied pressure, with its maximum value on the outer surface, equal to the applied pressure, consistent with the stress boundary condition used to obtain the solution. It decays quadratically to zero at the inner surface.

When the disk is pressurized from the inner surface by p_i, the hoop stress $\sigma_{\theta\theta}$ is always positive (tensile). Its maximum occurs on the inner surface, exceeding the applied pressure. The radial stress σ_{rr} is always negative and smaller in magnitude than the applied pressure. Its maximum occurs on the inner surface, equal to the applied pressure, consistent with the stress boundary condition. It decays quadratically to zero toward the outer surface.

From Eq. (3.19) and Fig. 3.2, we summarize the following findings for an internally pressurized disk:

1. σ_{rr} is always negative, indicating compression in the radial direction.
2. $\sigma_{\theta\theta}$, known as the **hoop stress**, is always positive, indicating tension in the hoop direction.

3. $\sigma_{\theta\theta}$ reaches a maximum at the inner surface:

$$\sigma_{\theta\theta}^{\max} = \frac{(a^2+b^2)p_i}{b^2-a^2} \tag{3.20}$$

4. $\sigma_{\theta\theta}^{\max}$ is always larger than p_i. It decreases as b increases, but will never fall below p_i, regardless of how much material is added outside.
5. $\sigma_{\theta\theta}^{\max}$ approaches ∞ as $b \to a$, meaning the hoop stress grows rapidly with internal pressure when the disk is very thin. This explains why a balloon can be effectively inflated with internal pressure.

3.1.9.4 *Displacement solution*

By substituting c_0 and c_2 into Eq. (3.13), we obtain the displacement formula:

```
u_rpipo = u_rg.subs({c0:cs[c0], c2:cs[c2]}).expand()
u_rpipo = u_rpipo.factor(pi, po)
u_rpipo_ = (pi*(-a**2*b**2*ν - a**2*b**2 + a**2*r**2*ν - a**2*r**2) \
         + po*( a**2*b**2*ν + a**2*b**2 - b**2*r**2*ν + b**2*r**2))\
                                         /(E*r*(a**2 - b**2))
u_rpipo # print(u_rpipo)
```

$$\frac{p_i\left(-a^2b^2\nu - a^2b^2 + a^2r^2\nu - a^2r^2\right) + p_o\left(a^2b^2\nu + a^2b^2 - b^2r^2\nu + b^2r^2\right)}{Er\left(a^2-b^2\right)}$$

Rearranging the terms in the resulting expression, we have a more readable formula:

$$u_r = \frac{(1-\nu)r}{E(b^2-a^2)}\left[p_i a^2\left(1+\frac{(1+\nu)}{(1-\nu)}\frac{b^2}{r^2}\right) - p_o b^2\left(1+\frac{(1+\nu)}{(1-\nu)}\frac{a^2}{r^2}\right)\right] \tag{3.21}$$

This formula allows us to conveniently set either p_i or p_o to zero. The outer pressure contributes negatively, while the inner pressure contributes positively to the radial displacement in the disk.

3.2 Anti-axisymmetric problems

3.2.1 *Spinning disk with angular acceleration*

This section discusses axisymmetric problems with body forces. A typical example is a spinning disk with an angular velocity that may include angular acceleration. The major difference from the previous section is that we now need to determine the potential function for the body force, and the biharmonic equation for the stress function becomes nonhomogeneous.

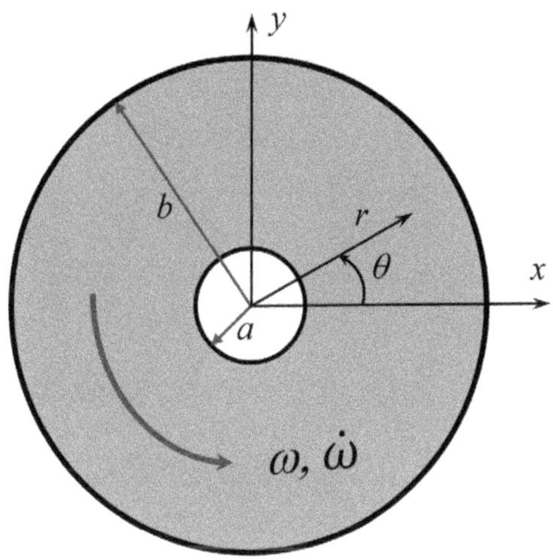

Figure 3.3. A thin disk of uniform thickness. It rotates with an angular velocity and constant angular acceleration, resulting in body forces throughout the disk.

Setting of the problem

Consider a thin disk of uniform thickness with an inner radius a and an outer radius b. It is fixed on a stiff shaft of radius a and rotates together with the shaft at an angular velocity ω with a given angular acceleration $\dot{\omega}$, as shown in Fig. 3.3. The outer edge of the disk is free of forces. Polar coordinates (r, θ) are defined with the origin at the centroid of the disk.

The disk rotates with a constant angular acceleration and hence experiences body forces over its domain, given by

$$b_r = \rho r \omega^2, \quad b_\theta = -\rho r \dot{\omega} \tag{3.22}$$

Note that both force components are independent of θ but depend on r.

We aim to derive a complete set of solutions, including:

1. the expression for the stress function that satisfies the compatibility equation,
2. expressions for stresses and strains,
3. expressions for displacements.

3.2.2 *Formulation of the problem*

Since the disk is rotating, it is not strictly axisymmetric. It represents a mixed axisymmetric problem with respect to θ, as illustrated in Fig. 3.4.

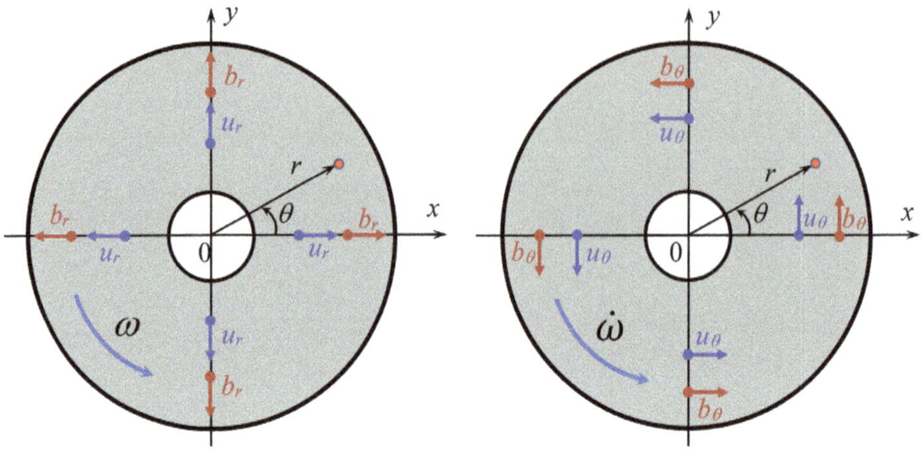

Figure 3.4. A thin disk of uniform thickness. It rotates with an angular velocity and constant angular acceleration, resulting in body forces throughout the disk.

This implies that the displacement component u_r is axisymmetric with respect to θ, induced by the axisymmetric body force b_r, while the displacement component u_θ is anti-axisymmetric with respect to θ, induced by the anti-axisymmetric body force b_θ given in Eq. (3.22). In either case, the stresses and displacements are independent of the θ-direction; however, the anti-axisymmetric displacement component u_θ and the stress component $\sigma_{r\theta}$ may be nonzero due to the presence of anti-axisymmetric body forces. Under these conditions, the equilibrium equation (Eq. 2.47) reduces to

$$\begin{aligned}\frac{\partial \sigma_{rr}}{\partial r} + \frac{\sigma_{rr} - \sigma_{\theta\theta}}{r} + b_r = 0\\ \frac{\partial \sigma_{r\theta}}{\partial r} + \frac{2\sigma_{r\theta}}{r} + b_\theta = 0\end{aligned} \tag{3.23}$$

In this case, we still have two equations, but they are now independent of each other and can therefore be solved separately.

Note that if b_r arises from the centrifugal force, which is linear in r, the displacement component u_r must be antisymmetric with respect to r (increasing away from or decreasing toward the origin). Similarly, if b_θ results from a constant angular acceleration of rotation, it is also linear in r, so the displacement component u_θ must likewise be antisymmetric with respect to r. It increases as it moves away from the origin and decreases as it approaches the origin. This understanding is important when devising a displacement method for this problem in the following chapter.

The strain–displacement relation Eq. (2.28) is reduced to

$$\begin{aligned} \varepsilon_{rr} &= \frac{\partial u_r}{\partial r} \\ \varepsilon_{\theta\theta} &= \frac{u_r}{r} \\ 2\varepsilon_{r\theta} &= \frac{\partial u_\theta}{\partial r} - \frac{u_\theta}{r} \end{aligned} \tag{3.24}$$

Consider a plane stress (thin disk) problem. The constitutive equation is given as follows:

$$\begin{aligned} \varepsilon_{rr} &= \frac{1}{E}(\sigma_{rr} - \nu\sigma_{\theta\theta}) \\ \varepsilon_{\theta\theta} &= \frac{1}{E}(\sigma_{\theta\theta} - \nu\sigma_{rr}) \\ 2\varepsilon_{r\theta} &= \frac{1}{G}\sigma_{r\theta} \end{aligned} \tag{3.25}$$

3.2.3 *Strategy to solve the problem*

Since the two equations in Eq. (3.23) are decoupled, they can be solved independently. Our strategy is as follows:

1. Solve the first equation using the stress-function approach, as in the previous section.
2. Solve the second differential (first-order) equation for $\sigma_{r\theta}$ independently.
3. Apply the constitutive equations to obtain strains.
4. Use the strain–displacement relations to solve for the displacements.

We begin with the first equation in Eq. (3.23). The relationship between the stresses and the stress function is given in Eq. (3.2). Using the potential function φ, the body force b_r can be expressed in polar coordinates as

$$b_r = -\frac{\partial \varphi}{\partial r} \tag{3.26}$$

The body force component b_θ is still given by Eq. (3.22), since in our problem, b_θ does not vary with θ. The compatibility equation for the stress function ϕ, given in Eq. (2.65), is reduced to

$$\nabla^4\phi = -\kappa'_\nu\left(\frac{\partial b_r}{\partial r} + \frac{b_r}{r} + \frac{1}{r}\frac{\partial b_\theta}{\partial \theta}\right) = -\kappa'_\nu\left(\frac{\partial b_r}{\partial r} + \frac{b_r}{r}\right) = -\underbrace{2\kappa'_\nu\rho\omega^2}_{k} \tag{3.27}$$

where $\kappa'_\nu = \nu - 1$ (for the plane stress problem), and we have used Eq. (3.22). Note that k is a constant. The biharmonic equation for the stress function is no longer homogeneous but has a constant right-hand side.

3.2.4 *Python codes to find the solutions*

We implement the following code to derive the required formulas, following the strategy outlined above:

```
# Solve ∇^4ϕ = - f for general solution:
ω, ωt = sp.symbols('ω, ω_t')                        # define symbolic variables
ρ, E, ν = symbols('ρ, E, ν', positive=True)             # material constants
r, a, b = sp.symbols('r, a, b', positive=True)
θ = sp.symbols('θ', real=True)

σrr, σθθ, σrθ = sp.symbols('σ_rr, σ_θθ, σ_rθ')

ϕ = sp.Function('ϕ')(r)                              # define the stress function
lap = diff(diff(ϕ,r),r)+(1/r)*diff(ϕ, r)     # define Laplacian operator
bih = diff(diff(lap,r),r)+ (1/r)*diff(lap,r)          #bi harmonic operator

G = E/(2*(1+ν))
kνp = ν - 1                                         # kνp for plane stress problems
k = 2*kνp*ρ*ω**2
```

```
diff_eq = sp.Eq(bih + k, 0)                        # define differential equation

g_sln = sp.dsolve(diff_eq, ϕ)                     # solve the differential equation

g_sln                                           # the general solution with constants
```

$$\phi(r) = C_1 + C_2 r^2 + C_3 r^2 \log(r) + C_4 \log(r) - \frac{r^4 \nu \rho \omega^2}{32} + \frac{r^4 \rho \omega^2}{32}$$

This is the general expression for the stress function for problems with the given body force. It contains four arbitrary constants which can be redefined. Therefore, we write

$$\phi(r) = c_0 + c_1 r^2 + c_2 r^2 \log(r) + c_3 \log(r) + (1 - \nu)\frac{r^4 \rho \omega^2}{32} \tag{3.28}$$

The following Python snippet redefines the stress function in SymPy:

```
c0, c1, c2, c3, c4, c5, c6 = symbols('c0, c1, c2, c3, c4, c5, c6')
ϕ = c0+c1*r**2+c2*r**2*sp.log(r)+c3*sp.log(r)+(1-ν)*ρ*ω**2*r**4/32
```

These four arbitrary constants can be determined through the following process, based on the setup of our problem. Due to the presence of body forces, we first find the potential function for the body force, φ, by integrating Eq. (3.26):

```
br = ρ*ω**2*r
p = integrate(-br, r) + c4              # use p for φ, to avoid conflict
display(p)
```

$$c_4 - \frac{r^2\rho\omega^2}{2}$$

Since the force potential can be zero at $r = 0$, we have $c_4 = 0$. The force potential is rewritten as follows:

```
# The force potential φ found:
p = - r**2*ρ*ω**2/2
```

3.2.5 *General formulas for stresses*

We can now compute the stresses using Eq. (3.2), which involves only differentiations:

```
σrr = (diff(ϕ,r)/r).expand() + p                         # differentiate once
display(Math(f" \\text{{ General stress $σ_{{rr}}$}}={latex(σrr)}"))

σθθ = diff(ϕ,(r,2)).expand() + p                         # differentiate twice
σθθ =σθθ.subs(c4, 0)                                     # c4 = 0, found earlier
display(Math(f" \\text{{ General stress $σ_{{θθ}}$}}={latex(σθθ)}"))
```

$$\text{General stress } \sigma_{rr} = 2c_1 + 2c_2\log(r) + c_2 + \frac{c_3}{r^2} + \frac{r^2\nu\rho\omega^2}{8} - \frac{3r^2\rho\omega^2}{8}$$

$$\text{General stress } \sigma_{\theta\theta} = 2c_1 + 2c_2\log(r) + 3c_2 - \frac{c_3}{r^2} - \frac{3r^2\nu\rho\omega^2}{8} - \frac{r^2\rho\omega^2}{8}$$

We obtain the general formulas for the two normal stresses. It is easy to confirm that these stresses satisfy the first equation in Eq. (3.23):

```
eq1 = (σrr.diff(r) + (σrr-σθθ)/r + br).simplify()
print(f" Is the first Equation satisfied? {eq1==0}")
```

```
Is the first Equation satisfied? True
```

Next, let us solve the second equation in Eq. (3.23), using `sp.dsolve()`:

```
ωt = sp.symbols('ω_t')
σrθ = sp.Function("σ_rθ")(r)
bθ = -ρ*ωt*r
sp.dsolve(σrθ.diff(r,1)+2*σrθ/r+bθ, σrθ).expand()
```

$$\sigma_{r\theta}(r) = \frac{C_1}{r^2} + \frac{r^2\rho\omega_t}{4}$$

We have obtained the general expression for $\sigma_{r\theta}$, which contains a new integration constant. For convenience in subsequent operations, we redefine this expression in Python:

```
# redefine σrθ:
σrθg = r**2*ρ*ωt/4 + c5/r**2
σrθg
```

$$\frac{c_5}{r^2} + \frac{r^2 \rho \omega_t}{4}$$

3.2.6 *General formulas for strains*

After all the stresses are obtained, we can now compute the strains using Eq. (3.25):

```
# Use the first equation:
εrr = ((σrr - ν*σθθ)/E).expand()
display(Math(f" \\text{{ General formula for strain $ε_{{rr}}$}}:"))
display(Math(f" \\text{{}}\\scriptsize {latex(εrr)}"))

# Use the second equation:
εθθ = ((σθθ - ν*σrr)/E).expand()
display(Math(f" \\text{{ General formula for strain $ε_{{θθ}}$}}:"))
display(Math(f" \\text{{}}\\scriptsize {latex(εθθ)}"))

# Use the third equation:
εrθ =(σrθg/G/2).expand()
display(Math(f" \\text{{ General formula for strain $ε_{{rθ}}$}}: \
                                        \\scriptsize {latex(εrθ)}"))
```

General formula for strain ε_{rr}:

$$-\frac{2c_1\nu}{E} + \frac{2c_1}{E} - \frac{2c_2\nu\log(r)}{E} - \frac{3c_2\nu}{E} + \frac{2c_2\log(r)}{E} + \frac{c_2}{E} + \frac{c_3\nu}{Er^2} + \frac{c_3}{Er^2} + \frac{3r^2\nu^2\rho\omega^2}{8E} - \frac{3r^2\rho\omega^2}{8E}$$

General formula for strain $\varepsilon_{\theta\theta}$:

$$-\frac{2c_1\nu}{E} + \frac{2c_1}{E} - \frac{2c_2\nu\log(r)}{E} - \frac{c_2\nu}{E} + \frac{2c_2\log(r)}{E} + \frac{3c_2}{E} - \frac{c_3\nu}{Er^2} - \frac{c_3}{Er^2} + \frac{r^2\nu^2\rho\omega^2}{8E} - \frac{r^2\rho\omega^2}{8E}$$

General formula for strain $\varepsilon_{r\theta}$: $\frac{c_5\nu}{Er^2} + \frac{c_5}{Er^2} + \frac{r^2\nu\rho\omega_t}{4E} + \frac{r^2\rho\omega_t}{4E}$

3.2.7 General formulas for displacements

To obtain the displacements, we use Eq. (3.24) along with the previously obtained strains. From the first equation in Eq. (3.24), we have $\frac{\partial u_r}{\partial r} = \varepsilon_{rr}$, and hence u_r can be found by integrating once:

```
C = symbols('C')                              # integral constants for disp.
ur, uθ = symbols('u_r, u_θ')
ur = sp.integrate(εrr,(r)) + C                 # A is an integral constant
display(Math(f" \\text{{General formula for displacement $u_{{r}}$}}:"))
display(Math(f" \\text{{}}\\scriptsize {latex(ur)}"))
```

General formula for displacement u_r:

$$C + \frac{(-2c_2r\nu + 2c_2r)\log(r)}{E}$$
$$+ \frac{r^3(\nu^2\rho\omega^2 - \rho\omega^2) + r(-16c_1\nu + 16c_1 - 8c_2\nu - 8c_2) + \frac{-8c_3\nu - 8c_3}{r}}{8E}$$

Next, using the second equation in Eq. (3.24), we have $r\varepsilon_{\theta\theta} - u_r = 0$, where $r\varepsilon_{\theta\theta} - u_r$ is found as follows:

```
(r*εθθ-ur).simplify()
```

$$-C + \frac{4c_2r}{E}$$

Since this expression must always be zero for any r, we must have

$$-C + \frac{4c_2r}{E} = 0 \tag{3.29}$$

which is possible only if both C and c_2 are zero:

$$C = c_2 = 0 \tag{3.30}$$

We can now substitute these two zero constants back into the formulas derived so far to simplify them:

```
urg = ur.subs({C:0, c2:0}).expand()
display(Math(f" \\text{{Current Displacement $u_{{r}}$}}={latex(urg)}"))

σrrg = σrr.subs({C:0, c2:0}).expand()
display(Math(f" \\text{{Current stress $σ_{{rr}}$}}={latex(σrrg)}"))

σθθg = σθθ.subs({C:0, c2:0}).expand()
display(Math(f" \\text{{ Current stress $σ_{{θθ}}$}}={latex(σθθg)}"))
```

Current displacement $u_r = -\frac{2c_1r\nu}{E} + \frac{2c_1r}{E} - \frac{c_3\nu}{Er} - \frac{c_3}{Er} + \frac{r^3\nu^2\rho\omega^2}{8E} - \frac{r^3\rho\omega^2}{8E}$

Current stress $\sigma_{rr} = 2c_1 + \frac{c_3}{r^2} - \frac{r^2\nu\rho\omega^2}{8} - \frac{3r^2\rho\omega^2}{8}$

Current stress $\sigma_{\theta\theta} = 2c_1 - \frac{c_3}{r^2} - \frac{3r^2\nu\rho\omega^2}{8} - \frac{r^2\rho\omega^2}{8}$

There are still two unknown constants in these expressions.

We are now ready to solve the third equation in Eq. (3.24) for u_θ:

```
uθ = sp.Function("u_θ")(r)                           # define the unknown function
uθg = sp.dsolve(uθ.diff(r,1) - uθ/r - 2*εrθ, uθ).expand()
uθg
```

$$u_\theta(r) = C_1 r - \frac{c_5\nu}{Er} - \frac{c_5}{Er} + \frac{r^3\nu\rho\omega_t}{4E} + \frac{r^3\rho\omega_t}{4E}$$

```
# reset the formula:
c6 = symbols('c6')
uθg = c6*r - c5*ν/(E*r) - c5/(E*r) + (1+ν)*r**3*ρ*ωt/(4*E)
display(Math(f" \\text{{General Displacement $u_{{θ}}$}}={latex(uθg)}"))
```

General displacement $u_\theta = c_6 r - \frac{c_5\nu}{Er} - \frac{c_5}{Er} + \frac{r^3\rho\omega_t(\nu+1)}{4E}$

We obtain the general formula for the displacement u_θ, which contains two integration constants. In total, there are four unknown constants: c_1, c_3, c_5, c_6. We have exactly four boundary conditions to determine them:

$$\begin{aligned} u_r = 0;\ \ u_\theta = 0 \quad &\text{at} \quad r = a \\ \sigma_{rr} = 0;\ \ \sigma_{r\theta} = 0 \quad &\text{at} \quad r = b \end{aligned} \tag{3.31}$$

The following snippet imposes these conditions and finds the final solution for these constants:

```
cs = [c1, c3, c5, c6]          # put all the unknown constants in a list

# enforce other conditions simultaneously:
cond1 = urg.subs({r:a})
cond2 = uθg.subs({r:a})
cond3 = σrrg.subs({r:b})
cond4 = σrθg.subs({r:b})
sln_cs = sp.solve([cond1, cond2, cond3, cond4], cs)

# simplify the expressions in the solution:
cs_final = {k: v.simplify().collect(a).collect(b) for k, v in \
                                                        sln_cs.items()}
sln_cs4 = gr.split_dict_n(cs_final, 4)  # split the solution dic into 4
display(sln_cs4[0])
display(sln_cs4[1])
display((sln_cs4[2],sln_cs4[3]))
```

$$\left\{c_1: \frac{\rho\omega^2\left(a^4\left(1-\nu^2\right)+b^4\left(\nu^2+4\nu+3\right)\right)}{16\left(a^2\left(1-\nu\right)+b^2\left(\nu+1\right)\right)}\right\}$$
$$\left\{c_3: \frac{a^2b^2\rho\omega^2\left(a^2\left(\nu^2-1\right)+b^2\left(-\nu^2-2\nu+3\right)\right)}{8\left(a^2\left(1-\nu\right)+b^2\left(\nu+1\right)\right)}\right\}$$
$$\left(\left\{c_5: -\frac{b^4\rho\omega_t}{4}\right\}, \left\{c_6: \frac{\rho\omega_t\left(a^4\left(-\nu-1\right)+b^4\left(-\nu-1\right)\right)}{4Ea^2}\right\}\right)$$

Finally, we substitute these constants back into the general expressions, yielding the final solutions for the displacements and stresses, followed by carefully applied simplifications. The code snippets are as follows:

```
# Compute the final solutions:

ur_final = urg.subs(cs_final).factor()
ur_final = ur_final.collect(a**2).collect(b**2).simplify().collect(r**2)
display(Math(f" \\text{{ Final solution for displacement $u_{{r}}$}}:"))
display(Math(f" \\text{{}}\\scriptsize {latex(ur_final)}"))

uθ_final = uθg.subs(cs_final).factor()
display(Math(f" \\text{{ Final solution for displacement $u_{{θ}}$}}=\
                                        \\scriptsize {latex(uθ_final)}"))
tmp = -r**2*ρ*ω**2/8
σrr_final = σrrg.subs(cs_final).collect(a**4).collect(b**4)
σrr_final = σrr_final.collect(tmp).collect(ρ*ω**2/8)
display(Math(f" \\text{{ Final solution for stress $σ_{{rr}}$}}:"))
display(Math(f" \\text{{}}\\scriptsize {latex(σrr_final)}"))

σθθ_final = σθθg.subs(cs_final).collect(a**4).collect(b**4)

σθθ_final = σθθ_final.collect(tmp).collect(ρ*ω**2/8)
display(Math(f" \\text{{ Final solution for stress $σ_{{θθ}}$}}:"))
display(Math(f" \\text{{}}\\scriptsize {latex(σθθ_final)}"))

σrθ_final = σrθg.subs(cs_final).factor(E)
display(Math(f" \\text{{ Final solution for stress $σ_{{rθ}}$}}=\
                                       \\scriptsize {latex(σrθ_final)}"))
```

Final solution for displacement u_r:

$$\frac{\rho\omega^2\left(a-r\right)\left(a+r\right)\left(\nu-1\right)\left(\nu+1\right)\left(a^2\left(b^2\left(\nu+1\right)+r^2\left(1-\nu\right)\right)-b^4\left(\nu+3\right)+b^2r^2\left(\nu+1\right)\right)}{8Er\left(a^2\left(\nu-1\right)-b^2\left(\nu+1\right)\right)}$$

Final solution for displacement

$$u_\theta = \frac{\rho\omega_t\left(-a+r\right)\left(a+r\right)\left(\nu+1\right)\left(ar-b^2\right)\left(ar+b^2\right)}{4Ea^2r}$$

Final solution for stress σ_{rr}:

$$\frac{\rho\omega^2\left(\frac{a^2b^2\left(a^2\left(\nu^2-1\right)+b^2\left(-\nu^2-2\nu+3\right)\right)}{r^2\left(a^2(1-\nu)+b^2(\nu+1)\right)}-r^2\left(\nu+3\right)\right)}{8}$$
$$+\frac{\rho\omega^2\left(a^4\left(1-\nu^2\right)+b^4\left(\nu^2+4\nu+3\right)\right)}{8a^2\left(1-\nu\right)+8b^2\left(\nu+1\right)}$$

Final solution for stress $\sigma_{\theta\theta}$:

$$\frac{\rho\omega^2\left(-\frac{a^2b^2\left(a^2\left(\nu^2-1\right)+b^2\left(-\nu^2-2\nu+3\right)\right)}{r^2\left(a^2(1-\nu)+b^2(\nu+1)\right)}-r^2\left(3\nu+1\right)\right)}{8}$$
$$+\frac{\rho\omega^2\left(a^4\left(1-\nu^2\right)+b^4\left(\nu^2+4\nu+3\right)\right)}{8a^2\left(1-\nu\right)+8b^2\left(\nu+1\right)}$$

Final solution for stress $\sigma_{r\theta}=\frac{\rho\omega_t\left(-b^4+r^4\right)}{4r^2}$

3.2.8 *Verification of solutions*

We have obtained all the solutions for the displacements and stresses. The strains can be readily calculated by applying the same substitutions. This set of solutions is consistent with the results reported in Ref. [3]. The following cell compares the displacement u_r and the two stress components with the corresponding expressions from Ref. [3]:

```
ur_spinDisk=ρ*ω**2*(a**2-r**2)*(ν**2-1)*(a**2*(b**2*(ν+1)+r**2*(1-ν))\
            -b**4*(ν+3)+b**2*r**2*(ν+1))/(8*E*r*(a**2*(ν-1)-b**2*(ν+1)))

uθ_spinDisk = ρ*ωt*(r**2-a**2)*(ν+1)*(a*r-b**2)*(a*r+b**2)/(4*E*a**2*r)

K = (b**2-a**2)*((3+ν)*b**2+(1-ν)*a**2)/(a**2*(1-ν) + b**2*(ν + 1))
ur_Ref = (1-ν**2)*ρ*ω**2/8/E/r*(r**2-a**2)*(K-(r**2-a**2))
#display(Math(f" \\text{{$u_r$ from Ref=}} {latex(ur_ref)}"))
display(Math(f" \\text{{Agrees with $u_r$ from Ref.? }}\
                                 {latex((ur_final - ur_Ref).simplify()==0)}"))

σrr_Ref = ρ*ω**2/8/r**2*( (r**2+a**2)*K-(r**2-a**2)\
                                         *((3+ν)*r**2+(1-ν)*a**2-ν*K))
#display(Math(f" \\text{{ $σ_{{rr}}$ from Ref.=}}{latex(σrr_Ref)}"))
display(Math(f" \\text{{Agrees with $σ_{{rr}}$ from Ref.? }}\
                                 {latex((σrr_final-σrr_Ref).simplify()==0)}"))

σθθ_Ref = ρ*ω**2/8/r**2*( (r**2+a**2)*ν*K-(r**2-a**2)\
                                         *((1+3*ν)*r**2-(1-ν)*a**2-K))
#display(Math(f" \\text{{$σ_{{θθ}}$ from Ref.=}}{latex(σθθ_Ref)}"))
display(Math(f" \\text{{Agrees with $σ_{{θθ}}$ from Ref.? }}\
                                 {latex((σθθ_final-σθθ_Ref).simplify()==0)}"))
```

Agrees with u_r from Ref.? True
Agrees with σ_{rr} from Ref.? True
Agrees with $\sigma_{\theta\theta}$ from Ref.? True

3.2.9 *Compute strains and stresses from displacements*

Since we have obtained the displacements, we can compute the strains and stresses using these displacements, essentially performing the reverse procedure. This approach is useful for verifying solutions when no reliable reference solutions are available. If the stresses computed from the displacements match those used in the displacement calculation, it provides a strong indication that the solutions are correct.

Since stresses are derived from the derivatives of the displacements, the computation is straightforward and can be carried out using the algebraic constitutive equations. The following Python functions facilitate these computations:

```
def displacement2StrainPolar(U,R):
    '''Compute strains for given displacement U=[ur, uθ]^T in the
    Polar coordinates (r, θ). In sympy.'''
    r, θ = R
    ur, uθ = U
    εrr = ur.diff(r)
    εθθ = ur/r + uθ.diff(θ)/r
    εrθ = (ur.diff(θ)/r + uθ.diff(r) - uθ/r)/2
    return (εrr, εθθ, εrθ)

def strain2stress2D(E, ν, ε, Ptype= 'stress'):
    '''2D constitutive equation, from strains to stresses for both
       plane-stress and plane-strain problems.'''
    εxx, εyy, εxy = ε
    if Ptype== "stress":
        E_ = E/(1-ν**2)
        σxx = E_*(εxx + ν*εyy).simplify()
        σyy = E_*(εyy + ν*εxx).simplify()
    else:
        E_ = E/((1+ν)*(1-2*ν))
        σxx = E_*((1-ν)*εxx + ν*εyy).simplify()
        σyy = E_*((1-ν)*εyy + ν*εxx).simplify()

    σxy = (E*εxy/(1+ν)).simplify()
    return (σxx,σyy,σxy)
```

Let us compute the stresses and compare them with the stresses obtained previously:

```
R = sp.Matrix([r, θ])
ε = displacement2StrainPolar([ur_final, uθ_final], R)
σrr, σθθ, σrθ = strain2stress2D(E, ν, ε, Ptype= 'stress')

display(Math(f" \\text{{Agrees with $σ_{{rr}}$ from Ref.? }}\
                              {latex((σrr_final-σrr).simplify()==0)}"))
display(Math(f" \\text{{Agrees with $σ_{{θθ}}$ from Ref.? }}\
                              {latex((σθθ_final-σθθ).simplify()==0)}"))
display(Math(f" \\text{{Agrees with $σ_{{rθ}}$ from Ref.? }}\
                              {latex((σrθ_final-σrθ).simplify()==0)}"))
```

Agrees with σ_{rr} from Ref.? True
Agrees with $\sigma_{\theta\theta}$ from Ref.? True
Agrees with $\sigma_{r\theta}$ from Ref.? True

3.3 Stress axisymmetric problems

3.3.1 *Curved beams under pure bending*

Setting and analysis of the problem

Consider a thin, partial circular curved beam of uniform thickness with an inner radius a and an outer radius b. It is subjected to a uniform pure moment M on its cross-section, as shown in Fig. 3.5. The beam is made of a linear elastic material with Young's modulus E and Poisson's ratio ν. Assume that the deformation of the beam is very small. Polar coordinates (r, θ), shown in Fig. 3.5, are defined with the origin at the center of the circles. The applied moment M on the beam cross-section is independent of θ.

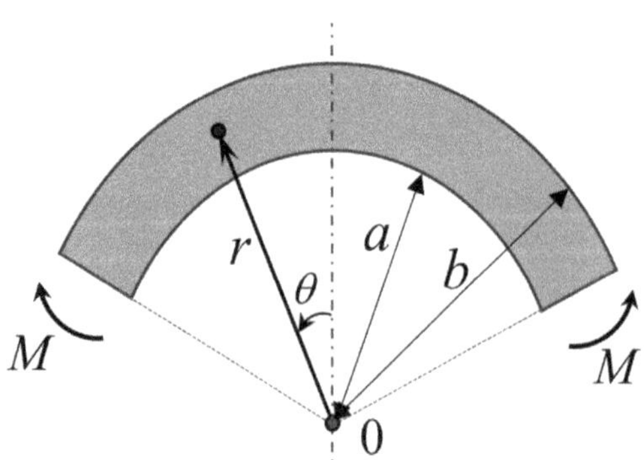

Figure 3.5. A thin partial circular curved beam of uniform thickness loaded by a uniform pure moment M on its cross-section.

The problem has the following features:

1. Since the beam is circular and experiences pure bending, its shape remains circular. The shear stress is zero throughout the beam, which corresponds to the ideal assumption in classic beam theory.
2. Since the applied moment is independent of θ, all stresses and strains are also independent of θ, making the stress solution axisymmetric.
3. The length of the beam does not affect the stress analysis; the solution is valid for curved beams of any length.
4. Most importantly, the displacements of the curved beam are not axisymmetric because the beam has straight edges at the left and right boundaries.

Thus, this is a special mechanics problem: the stress solution can be found assuming axisymmetry, but when determining the displacement solution, the axisymmetry assumption cannot be applied. Special care is required.

3.3.2 *Stresses in a curved beam under pure moment*

Using a stress method, we can exploit the axisymmetric nature of the stresses and strains. The stress function given in Eq. (3.7) can be applied directly and contains four unknown constants.

The Python snippet to redefine the stress function for this curved beam problem is as follows:

```
# Define variables:
E, ν = symbols('E, ν', positive=True)                # material constants
r, a, b = sp.symbols('r, a, b', positive=True)        # radial dimensions
θ = sp.symbols('θ', real=True)
M = sp.symbols('M', real=True)                             # applied moment

σrr, σθθ, σrθ = sp.symbols('σ_rr, σ_θθ, σ_rθ')

c0, c1, c2, c3 = symbols('c0, c1, c2, c3')       # integration constants
cs = sp.Matrix([c0, c1, c2, c3])

# The general form of stress function for axisymmetric problem:
ϕ = c0*sp.log(r)+c1*r**2*sp.log(r)+c2*r**2+c3
display(Math(f" \\text{{ General stress function $ϕ(r)$}}={latex(ϕ)}"))
```

General stress function $\phi(r) = c_0 \log(r) + c_1 r^2 \log(r) + c_2 r^2 + c_3$

These four unknown constants can be determined from the stress boundary conditions and the integral equilibrium equations, following the standard procedure for axisymmetric problems.

3.3.3 *General formulas for stresses*

We can now compute the stresses using Eqs. (3.7) and (3.2) (without body forces). This involves only differentiation and can be carried out straightforwardly:

```
σrrg = (diff(ϕ,r)/r).expand()                         # differentiate once
display(Math(f" \\text{{ General Stress $σ_{{rr}}$}}={latex(σrrg)}"))

σθθg = (diff(diff(ϕ,r),r)).expand()                   # differentiate twice
display(Math(f" \\text{{  General Stress $σ_{{θθ}}$}}={latex(σθθg)}"))
σrθg = sp.S.Zero                                              # axisymmetric
σrrg_  = c0/r**2 + 2*c1*log(r) + c1 + 2*c2
σθθg_ =-c0/r**2 + 2*c1*log(r) + 3*c1 + 2*c2;  σrθg_ = 0
```

$$\text{General Stress } \sigma_{rr} = \frac{c_0}{r^2} + 2c_1 \log{(r)} + c_1 + 2c_2$$

$$\text{General Stress } \sigma_{\theta\theta} = -\frac{c_0}{r^2} + 2c_1 \log{(r)} + 3c_1 + 2c_2$$

```
# Check if the equilibrium equations are satisfied:

br = 0; bθ = 0
σvg = sp.Matrix([σrrg_, σθθg_, σrθg_])
equi1=(σvg[0].diff(r)+σvg[2].diff(θ)/r+(σvg[0]-σvg[1])/r+br)\
                                                  .simplify().factor()
equi2=(σvg[2].diff(r)+σvg[1].diff(θ)/r+2*σvg[2]/r+bθ)\
                                                  .simplify().factor()
#equi1 = equi1.subs(sln_cs).expand().simplify()
#equi2 = equi2.subs(sln_cs).expand().simplify()
equis= [equi1, equi2]            # uncomment this to view these equations
display(equis)
```

[0, 0]

It is confirmed that the equilibrium equations are satisfied.

3.3.4 *Enforcing stress boundary conditions*

Since the shear stresses are zero everywhere, the stress boundary conditions (SBCs) related to shear stress are automatically satisfied. On the inner and outer curved surfaces, the stress boundary conditions for the normal stress σ_{rr} are as follows:

$$\begin{aligned} \sigma_{rr} &= 0 \quad \text{at} \quad r = a \\ \sigma_{rr} &= 0 \quad \text{at} \quad r = b \end{aligned} \tag{3.32}$$

The following snippet enforces these SBCs:

```
# Solve SBCs for some of the unknown constants:
sln_SBC = sp.solve([σrrg.subs(r,a), σrrg.subs(r,b)], cs)

# Simplify these constants:
sln_SBC = {key: value.subs(sln_SBC).simplify() for key, \
                                        value in sln_SBC.items()}
sln_cs = sln_SBC     # keep the constants in a unified dictionary sln_cs
print(f" Current len of sln_cs={len(sln_cs)}; Total cs={len(cs)}")
display(sln_cs)
```

```
Current len of sln_cs=2; Total cs=4
```

$$\left\{c_0: \frac{c_2 \log\left(\left(\frac{b}{a}\right)^{4a^2b^2}\right)}{a^2 - b^2 + \log\left(a^{2a^2}b^{-2b^2}\right)},\ c_1: \frac{2c_2\left(-a^2 + b^2\right)}{a^2 - b^2 + \log\left(a^{2a^2}b^{-2b^2}\right)}\right\}$$

One constant, c_2, remains to be determined. We now apply the integral equilibrium conditions, also known as the force boundary conditions.

3.3.5 *Enforcing integral equilibrium conditions*

At any cross-section of the beam (see Fig. 3.5), the following integral equilibrium conditions must be satisfied:

$$\begin{aligned} \sum F_\theta = 0&: \int_a^b \sigma_{\theta\theta} dr = 0 \\ \sum M_z = 0&: \int_a^b \sigma_{\theta\theta}\left(r - \frac{a+b}{2}\right) dr - M = 0 \end{aligned} \tag{3.33}$$

Note that the moment is measured relative to the centroidal axis of the beam cross-section. The following code snippet enforces these conditions:

```
# Compute the stresses with currently known constants:
display(Math(f" \\text{{ Current stresses: }}"))
display(Math(f" \\scriptsize {latex(σθθg.subs(sln_cs).simplify())}"))

# Compute the sum of the force F_θ on a cross-section via integral:
Fθ = sp.integrate(σθθg.subs(sln_cs), (r, a, b)).simplify()
display(Math(f" \\text{{$∑F_θ$}}=\\small {latex(Fθ)}"))
```

Current stresses:

$$\frac{c_2\left(-2r^2\left(a^2 - b^2\right)\left(\log\left(r^2\right) + 3\right) + 2r^2\left(a^2 - b^2 + \log\left(a^{2a^2}b^{-2b^2}\right)\right) - \log\left(a^{-4a^2b^2}b^{4a^2b^2}\right)\right)}{r^2\left(a^2 - b^2 + \log\left(a^{2a^2}b^{-2b^2}\right)\right)}$$

$$\sum F_\theta = 0$$

The first equation in Eq. (3.33) is satisfied regardless of the value of c_2. We now proceed to enforce the second equation in Eq. (3.33):

```
# Enforce FBCs for the moment:
equI=sp.integrate(σθθg.subs(sln_cs)*(r-(a+b)/2), (r,a,b)).simplify() -M
#display(equI)
sln_equI = sp.solve([equI], cs)                    # solve for the constant

# simplify the expression for the constant:
sln_equI = {key: value.subs(sln_equI).simplify() for key, \
                                        value in sln_equI.items()}
display(Math(f"\\text{{Constant found:}}=\\scriptsize {latex(sln_equI)}"))

Cab = sp.symbols('C_ab')   # defined for further simplifying expressions
c2_numer = (sp.numer(sln_equI[c2])/M).expand().simplify()
c2_denom = sp.denom(sln_equI[c2]).expand().simplify()   #sp.expand_log()

Cab_ = (b**2/a**2 - 1)**2 - 4*b**2/a**2*(log(b/a))**2
display(Math(f" \\text{{ Simplified denominator}}= {latex(Cab_)}"))

dffr = (Cab_*a**4-c2_denom).simplify()        # check the simplification
display(Math(f"\\text{{ Is simplification correct:}}={dffr==0}"))
#executed in 15.9s, finished 15:35:24 2025-02-23
```

$$\text{Constant found:} = \left\{ c_2: \frac{M\left(a^2 - b^2 + \log\left(a^{2a^2} b^{-2b^2}\right)\right)}{a^4 - 4a^2b^2\log(a)^2 - 4a^2b^2\log(b)^2 - 2a^2b^2 + b^4 + \log\left(b^{\log\left(a^{8a^2b^2}\right)}\right)} \right\}$$

$$\text{Simplified denominator} = \left(-1 + \frac{b^2}{a^2}\right)^2 - \frac{4b^2\log\left(\frac{b}{a}\right)^2}{a^2}$$

$$\text{Is simplification correct:} = \text{True}$$

In the cell above, the expressions were simplified due to their length. The constants can now be expressed in the following concise forms:

```
sln_equI[c2] = M*c2_numer/(Cab*a**4)
sln_cs.update(sln_equI)

print(f" Current len of sln_cs={len(sln_cs)}; Total cs={len(cs)}")
sln_cs = {key: sp.expand_log(value.subs(sln_cs).simplify()) for key, \
                                        value in sln_cs.items()}
display(Math(f"\\text{{ Constants found:}}"))
display(Math(f"\\text{{}}\\scriptsize {latex(sln_cs)}"))
c0_ = 4*M*b**2*(log(a) - log(b))/(Cab_*a**2)
c1_ = 2*M*(a**2 - b**2)/(Cab_*a**4)
c2_ = M*(-2*a**2*log(a) - a**2 + 2*b**2*log(b) + b**2)/(Cab_*a**4)
```

```
Current len of sln_cs=3; Total cs=4
```

Constants found:

$$\left\{c_0\colon \frac{4Mb^2\left(-\log\left(a\right)+\log\left(b\right)\right)}{C_{ab}a^2},\ c_1\colon \frac{2M\left(-a^2+b^2\right)}{C_{ab}a^4},\right.$$

$$\left. c_2\colon \frac{M\left(2a^2\log\left(a\right)+a^2-2b^2\log\left(b\right)-b^2\right)}{C_{ab}a^4}\right\}$$

The formulas for the constants are rewritten as follows for future reference:

$$\begin{aligned} c_0 &= -\frac{4Mb^2}{C_{ab}a^2}\log\frac{b}{a}; \quad c_1 = \frac{2M\left(a^2-b^2\right)}{C_{ab}a^4} \\ c_2 &= \frac{M}{C_{ab}a^4}\left(b^2-a^2+2b^2\log b-2a^2\log a\right) \end{aligned} \tag{3.34}$$

3.3.6 *Final formulas for stresses*

Using these constants found, we obtain the final formulas for the stresses:

```
σrr = σrrg.subs(sln_cs).factor()
σθθ = σθθg.subs(sln_cs).simplify()
σrθ = sp.S.Zero
display(Math(f" \\text{{ Final solution for stresses:}}"))
display(Math(f" \\text{{ $σ_{{rr}}$}}=\\scriptsize {latex(σrr)}"))
display(Math(f" \\text{{ $σ_{{θθ}}$}}=\\scriptsize {latex(σθθ)}"))
display(Math(f" \\text{{ $σ_{{rθ}}$}}= {latex(σrθ)}"))
#print('σrr=',σrr); print(); print('σθθ=',σθθ);   # to produce formulas
```

Final solution for stresses:

$$\sigma_{rr} = \frac{\begin{array}{c}4M\left(-a^2b^2\log\left(a\right)+a^2b^2\log\left(b\right)+a^2r^2\log\left(a\right)\right.\\ \left.-a^2r^2\log\left(r\right)-b^2r^2\log\left(b\right)+b^2r^2\log\left(r\right)\right)\end{array}}{C_{ab}a^4r^2}$$

$$\sigma_{\theta\theta} = \frac{4M\left(r^2\left(-a^2+b^2+\log\left(a^{a^2}b^{-b^2}r^{-a^2+b^2}\right)\right)+\log\left(\left(\frac{a}{b}\right)^{a^2b^2}\right)\right)}{C_{ab}a^4r^2}$$

$$\sigma_{r\theta} = 0$$

```
# Define explicitly the formulas for stresses for future use:

C_ab_= (b**2/a**2-1)**2-4*b**2/a**2*(log(b/a))**2

σrr_= 4*M*(-a**2*b**2*log(a)+a**2*b**2*log(b)+a**2*r**2*log(a)-\
   a**2*r**2*log(r)-b**2*r**2*log(b)+b**2*r**2*log(r))/(Cab*a**4*r**2)

σθθ_= 4*M*(r**2*(-a**2+b**2+log(a**(a**2)*r**(-a**2+b**2)/b**(b**2)))+\
                                  log((a/b)**(a**2*b**2)))/(Cab*a**4*r**2)
```

3.3.7 *Final formulas for strains*

After obtaining all the stresses, we compute the strains using Eq. (3.25), assuming that the beam is very thin and the problem corresponds to plane stress:

```
εrr = ((σrr - ν*σθθ)/E).simplify()
εθθ = ((σθθ - ν*σrr)/E).simplify()
εrθ = σrθ/E*(2*(1+ν))
display(Math(f" \\text{{ Final solution for strains}}:"))
display(Math(f" \\text{{$ε_{{rr}}$=}}\\scriptsize {latex(εrr)}"))
display(Math(f" \\text{{$ε_{{θθ}}$=}}\\scriptsize {latex(εθθ)}"))
display(Math(f" \\text{{$ε_{{rθ}}$=}}\\scriptsize {latex(εrθ)}"))
```

Final solution for strains:

$$\varepsilon_{rr} = \frac{\begin{array}{c}4M\left(-\nu\left(r^2\left(-a^2+b^2+\log\left(a^{a^2}b^{-b^2}r^{-a^2+b^2}\right)\right)+\log\left(a^{a^2b^2}b^{-a^2b^2}\right)\right)\\ +\log\left(a^{a^2\left(-b^2+r^2\right)}b^{b^2\left(a^2-r^2\right)}r^{r^2\left(-a^2+b^2\right)}\right)\Big)\end{array}}{C_{ab}Ea^4r^2}$$

$$\varepsilon_{\theta\theta} = \frac{\begin{array}{c}4M\left(r^2\left(-a^2+b^2+\log\left(a^{a^2}b^{-b^2}r^{-a^2+b^2}\right)\right)\right.\\ \left.+\log\left(a^{a^2b^2}b^{-a^2b^2}\left(a^{a^2\left(b^2-r^2\right)}b^{b^2\left(-a^2+r^2\right)}r^{r^2\left(a^2-b^2\right)}\right)^{\nu}\right)\right)\end{array}}{C_{ab}Ea^4r^2}$$

$$\varepsilon_{r\theta} = 0$$

The following snippet confirms that all stresses and strains satisfy the necessary conditions and are thus correct:

```
# Verification of equilibrium constants:

εrrg = ((σrrg - ν*σθθg)/E).simplify()     # compute strain using stresses
εθθg = ((σθθg - ν*σrrg)/E).simplify()
εrθg = σrθg/E*(2*(1+ν))
display(Math(f" \\text{{ General solution for strains}}:"))
display(Math(f" \\text{{$ε_{{rr}}$ = }}\\scriptsize {latex(εrrg)}"))
display(Math(f" \\text{{$ε_{{θθ}}$ = }}\\scriptsize {latex(εθθg)}"))
display(Math(f" \\text{{$ε_{{rθ}}$ = }}\\scriptsize {latex(εrθg)}"))

# Equilibrium Equations (zero body force):
br = 0; bθ = 0
equi1=(σrrg.diff(r)+σrθg.diff(θ)/r+(σrrg-σθθg)/r+br)\
                                                    .simplify().factor()
equi2=(σrθg.diff(r)+σθθg.diff(θ)/r+2*σrθg/r+bθ)\
                                                    .simplify().factor()
equis= [equi1, equi2]
display(Math(f" \\text{{Are the equilibrium equation satisfied? }}\
                {latex(abs(equis[0]+abs(equis[0])).simplify()==0)}"))

# Verification on compatibility condition:
cmptbl = (σrrg+σθθg).diff(r,2)+(σrrg+σθθg).diff(r)/r
display(Math(f" \\text{{Is the compatibility equation satisfied? }}\
                                    {latex(cmptbl.simplify()==0)}"))
```

General solution for strains:

$$\varepsilon_{rr} = \frac{c_0 + r^2\left(2c_1\log\left(r\right) + c_1 + 2c_2\right) - \nu\left(-c_0 + r^2\left(2c_1\log\left(r\right) + 3c_1 + 2c_2\right)\right)}{Er^2}$$

$$\varepsilon_{\theta\theta} = \frac{-c_0 + r^2\left(2c_1\log\left(r\right) + 3c_1 + 2c_2\right) - \nu\left(c_0 + r^2\left(2c_1\log\left(r\right) + c_1 + 2c_2\right)\right)}{Er^2}$$

$$\varepsilon_{r\theta} = 0$$

Are the equilibrium equation satisfied? True

Is the compatibility equation satisfied? True

It is observed that all necessary stress conditions are satisfied regardless of the values of the constants c_0, c_1, and c_2. These strains can now be used to compute the displacements. We use the general constants instead of the specific ones found earlier because the expressions are simpler for subsequent derivations. Additionally, the resulting formulas are more general.

3.3.8 *Displacements in a curved beam under pure moment*

As discussed earlier, the displacement is not axisymmetric. Therefore, the strain–displacement relation in Eq. (3.24) cannot be applied. We must use the more general formulas (Eq. 2.28), in which $\frac{\partial u_r}{\partial r} = \varepsilon_{rr}$. Consequently, u_r can be obtained by integrating once, using the previously determined strain ε_{rr}. The procedure is similar to that in the previous section, but somewhat more involved. This integration introduces an unknown function:

```
g = sp.Function('g')(θ)                          # define an unknown function
ur_ = sp.integrate(εrrg,(r))
ur_with_g = ur_ + g                             # g: unknown integral function of
display(Math(f"\\text{{Current $u_r$}}=\\small {latex(ur_with_g)}"))
```

$$\text{Current } u_r = g(\theta) + \frac{-r\left(c_1\nu + c_1 + 2c_2\nu - 2c_2\right) - \frac{c_0\nu + c_0}{r}}{E} + \frac{\left(-2c_1 r\nu + 2c_1 r\right)\log\left(r\right)}{E}$$

Next, using the second equation in Eq. (2.28), we have

$$\frac{\partial u_\theta}{\partial \theta} = r\varepsilon_{\theta\theta} - u_r \tag{3.35}$$

This is used to find the expression for u_θ, which introduces another unknown function:

```
h = sp.Function('h')(r)                         # h: integral unknown function
uθ_with_h = sp.integrate(r*εθθg - ur_with_g, (θ)) + h
display(Math(f"\\text{{Current displacement $u_{{θ}}$}} = \
                                        {latex(uθ_with_h.simplify())}"))
```

$$\text{Current displacement } u_\theta = h(r) - \int g(\theta)\, d\theta + \frac{4c_1 r\theta}{E}$$

Now, using the third equation in Eq. (2.31) with $\varepsilon_{r\theta} = 0$, we have

$$\frac{\partial u_r}{\partial \theta} + r\frac{\partial u_\theta}{\partial r} - u_\theta = 0 \tag{3.36}$$

This expression can be obtained using the following snippet:

```
expr3 = (ur_with_g.diff(θ) + r*uθ_with_h.diff(r) - uθ_with_h).simplify()
display(Math(f" \\text{{ The third expression}}: {latex(expr3)}= 0 "))
```

$$\text{The third expression: } r\frac{d}{dr}h(r) - h(r) + \frac{d}{d\theta}g(\theta) + \int g(\theta)\, d\theta = 0$$

The foregoing equation must be zero. We re-group them as follows:

$$\underbrace{r\frac{d}{dr}h(r) - h(r)}_{H(r)} = \underbrace{-\frac{d}{d\theta}g(\theta) - \int g(\theta)\, d\theta}_{G(\theta)} \tag{3.37}$$

It is seen that $H(r)$ is a function of only r, and $G(\theta)$ is a function of only θ. Equation (3.37) will not hold unless $H(r)$ and $G(\theta)$ are equal to a common constant, say C. Thus, we denote $H(r) = C$ and $G(\theta) = C$. This gives us the following two separated equations:

$$\begin{aligned} r\frac{d}{dr}h(r) - h(r) &= C \\ -\frac{d}{d\theta}g(\theta) - \int g(\theta)\, d\theta &= C \end{aligned} \tag{3.38}$$

It is observed that the first equation in Eq. (3.38) is a 1D differential equation, which can be easily solved for $h(r)$ using `sp.dsolve()`. The second equation contains an integral. We need to convert it into a differential equation by differentiating both sides once, which gives

$$-\frac{d^2}{d\theta^2}g(\theta) - g(\theta) = 0 \tag{3.39}$$

Now, this equation is a differential equation and can be solved for $g(\theta)$ using `sp.dsolve()`. Note that this differentiation eliminates the constant C from the equation. Therefore, the satisfaction of Eq. (3.36) may fail by a constant. We thus need to re-enforce it after $g(\theta)$ is found.

The following snippet solves for both $h(r)$ and $g(\theta)$:

```
C = symbols('C')                                              # constant
# find g(θ) using dsolve():
sln_g = sp.dsolve(-g.diff(θ,2) - g, g)
display(Math(f" \\text{{ Function $g(θ)$}}={latex(sln_g)}"))

# find h(r) using dsolve():
sln_h = sp.dsolve(r*h.diff(r) - h - C, h)  #+ 2*c1*r**2/E
display(Math(f" \\text{{ Function }} {latex(sln_h)}"))
```

Function $g(\theta) = g(\theta) = C_1 \sin(\theta) + C_2 \cos(\theta)$

Function $h(r) = -C + C_1 r$

We have successfully found these two unknown functions, but with four additional constants. The general form of the displacements can now be obtained as follows:

```
C1, C2, Ch = symbols('C1, C2, C_h')                    # integral constants

g = C1*sin(θ) + C2*cos(θ)                                   # re-define g(θ)
urg = (ur_ + g).simplify()
display(Math(f"\\text{{Current $u_r$}}=\\scriptsize \
   {latex(urg.collect(2*c1*r).collect(c1).collect(2*c2).collect(c0))}"))

h = -C + Ch*r                                               # re-define h(r)
uθg = sp.expand_log(sp.integrate( r*εθθg-urg, (θ)) + h).collect(c1*r**2)
display(Math(f"\\text{{Current $u_θ$:}}"))
display(Math(f"\\text{{}}\\scriptsize {latex(sum(uθg.args[:3]))}"))
display(Math(f"\\hspace{{.5cm}}+\\scriptsize {latex(uθg.args[-1])}"))
```

$$\text{Current } u_r = \frac{\begin{array}{c} Er\left(C_1\sin\left(\theta\right) + C_2\cos\left(\theta\right)\right) + c_0\left(-\nu - 1\right) \\ -2c_1r^2\left(\nu - 1\right)\log\left(r\right) + r^2\left(c_1\left(-\nu - 1\right) + 2c_2\left(1 - \nu\right)\right) \end{array}}{Er}$$

$$\text{Current } u_\theta\text{: } -C + C_hr + \frac{\begin{array}{c} C_1Er\cos\left(\theta\right) - C_2Er\sin\left(\theta\right) + c_0\theta\nu + c_0\theta \\ +c_1r^2\left(2\theta\nu\log\left(r\right) + \theta\nu - 2\theta\log\left(r\right) + \theta\right) \\ +2c_2r^2\theta\nu - 2c_2r^2\theta \end{array}}{Er}$$
$$+\frac{\theta\left(-c_0 + r^2\left(2c_1\log\left(r\right) + 3c_1 + 2c_2\right) - \nu\left(c_0 + r^2\left(2c_1\log\left(r\right) + c_1 + 2c_2\right)\right)\right)}{Er}$$

3.3.9 *Determining displacements using DBCs*

These displacements contain a total of four constants: C, C_1, C_2, and C_h. These constants must be determined using the displacement boundary conditions (DBCs) for the given problem. For our case, assume the DBCs are specified at point A ($r = a$, $\theta = 0$, see Fig. 3.5):

$$u_r = 0; \quad u_\theta = 0; \quad \frac{\partial u_r}{\partial \theta} = 0 \tag{3.40}$$

This gives us three conditions, which can be used to determine some of these four constants. The snippet is as follows:

```
# Using DBCs to determine some of the constants:
dic_rt = {r:a, θ:0}
expr1 = urg.subs(dic_rt)
expr2 = uθg.subs(dic_rt)
expr3 = urg.diff(θ).subs(dic_rt)
constants = sp.solve([expr1, expr2, expr3], [C1, C2, C, Ch])

C1_ = constants[C1]
C2_ = constants[C2]
C_  = constants[C]
display(Math(f" \\text{{ Constants found}}: "))
display(Math(f" \\text{{ $C_1$}}={latex(C1_)}"))
display(Math(f" \\text{{ $C_2$}}=\
                {latex(C2_.simplify().collect(c1).collect(2*c2))}"))
display(Math(f" \\text{{ $C$}}={latex(C_ )}"))
```

Constants found:

$C_1 = 0$

$$C_2 = \frac{a^2\left(c_1\left(\nu + \log\left(a^{2\nu-2}\right) + 1\right) + 2c_2\left(\nu - 1\right)\right) + c_0\left(\nu + 1\right)}{Ea}$$

$C = C_ha$

We see that the nonzero constant C_2 contains the constants c_0, c_1, and c_2, which were found earlier. We can now determine C_2 by substitution:

```
C2_ = C2_.subs(sln_cs).simplify()
display(Math(f" \\text{{ $C_2$ }}={latex(C2_)}"))
```

$$C_2 = \frac{4M\left(-a^2+b^2+\log\left(a^{-2b^2}b^{2b^2}\right)\right)}{C_{ab}Ea^3}$$

```
# Update the displacements using the constants found:

dic_C = {C: C_, C1: C1_, C2: C2_} #{C0: C0_, C1: C1_, C:0}
ur = sp.expand_log(urg.subs(dic_C).subs(sln_cs).simplify()).simplify()
uθ = uθg.subs(dic_C).subs(sln_cs).simplify()
#display(Math(f"\\text{{ General $u_r$}}=\\small {latex(ur)}"))
#display(Math(f"\\text{{ General $u_θ$}}=\\small {latex(uθ)}"))#Lengthy

# Check strain-displacement relations:
((εrr - ur.diff(r)).subs(dic_C).subs(sln_cs).simplify(),
(r*εθθ - ur- uθ.diff(θ)).subs(dic_C).subs(sln_cs).simplify(),
(ur.diff(θ) + r*uθ.diff(r)- uθ).subs(dic_C).subs(sln_cs).simplify())
```

$(0,\ 0,\ C_h a)$

It is observed that the third equation is not satisfied and depends only on the constant C_h. This occurs because the differentiation used to obtain Eq. (3.39) eliminated one constant. We now enforce Eq. (3.36) by setting $C_h = 0$, which also leads to $C = 0$. With all constants determined, we can now obtain the final solution for the displacements. The corresponding snippet is as follows:

```
dic_C = {C: 0, Ch: 0, C1: 0}
ur = sp.expand_log(urg.subs(dic_C).simplify()).simplify().collect(c1)\
                                          .collect(2*c2).collect(c0)
uθ = uθg.subs(dic_C).simplify()
display(Math(f" \\text{{ Final $u_r$}}=\\scriptsize {latex(ur)}"))
display(Math(f" \\text{{ Final $u_θ$}}=\\scriptsize {latex(uθ)}"))
```

$$\text{Final } u_r = \frac{C_2 Er\cos(\theta)+c_0(-\nu-1)-2c_1r^2(\nu-1)\log(r)+r^2(c_1(-\nu-1)+2c_2(1-\nu))}{Er}$$

$$\text{Final } u_\theta = -C_2\sin(\theta)+\frac{4c_1r\theta}{E}$$

The final expressions for the displacements are given as

$$\begin{aligned} u_r &= C_2 \cos\theta - \frac{c_0(1+\nu)}{Er} + \frac{2c_1(1-\nu)}{E} r \log r + \frac{[2c_2(1-\nu) - c_1(1+\nu)]\, r}{E} \\ u_\theta &= -C_2 \sin\theta + \frac{4c_1}{E} r\theta \end{aligned} \tag{3.41}$$

where

$$C_2 = \frac{a}{E}\left\{c_1\left[\nu + 2(\nu-1)\log a + 1\right] + 2c_2(\nu-1)\right\} + \frac{(\nu+1)}{Ea} c_0 \tag{3.42}$$

with constants c_0, c_1, and c_2 given in Eq. (3.34).

It is evident that, although the stresses are axisymmetric, the displacement components are not. The radial component u_r is mirror symmetric with respect to the plane $\theta = 0°$, while the tangential component u_θ is anti-mirror symmetric.

For validation, the following snippet computes the strains and stresses from the obtained displacements. These can then be compared with the stresses used initially to derive the displacements. Agreement between them indicates that the calculated displacements are likely correct:

```
# Final check on the displacement solutions:
R = sp.Matrix([r, θ])
dic_C2 = {C2: C2_}
ε = gr.displacement2StrainPolar([ur.subs(dic_C2).subs(sln_cs),
                                 uθ.subs(dic_C2).subs(sln_cs)], R)
σrr_, σθθ_, σrθ_ = gr.strain2stress2D(E, ν, ε, Ptype= 'stress')

display(Math(f" \\text{{Agrees with $σ_{{rr}}$ from Ref.? }}\
                          {latex((σrr_-σrr).simplify()==0)}"))
display(Math(f" \\text{{Agrees with $σ_{{θθ}}$ from Ref.? }}\
                          {latex((σθθ_-σθθ).simplify()==0)}"))
display(Math(f" \\text{{Agrees with $σ_{{rθ}}$ from Ref.? }}\
                          {latex((σrθ_-σrθ).simplify()==0)}"))
```

Agrees with σ_{rr} from Ref.? True

Agrees with $\sigma_{\theta\theta}$ from Ref.? True

Agrees with $\sigma_{r\theta}$ from Ref.? True

We have now successfully obtained the complete set of solutions for the stress axisymmetric problem.

3.4 Problems with mirror symmetry

3.4.1 *Curved beams under pure shearing at two ends*

Setting and analyzing the problem

Consider a thin semicircular curved beam of uniform thickness with an inner radius a and an outer radius b. It is subjected to a pair of horizontal shear forces P acting on the cross-sections at its two ends, as shown in Fig. 3.6. Assume that the beam undergoes very small deformations. The polar coordinates (r, θ), shown in Fig. 3.6, are defined with the origin at the center of the circles. The axis $\theta = 0°$ lies on the plane of symmetry of the semicircular beam, allowing the beam's geometric symmetry to be exploited in the analysis.

The geometry of the beam and the applied force P are symmetric with respect to $\theta = 0°$. We adopt the stress method, expressing the stress function ϕ as

$$\phi = f(r)\cos\theta \tag{3.43}$$

where $f(r)$ is an unknown function of r. The choice of $\cos\theta$ is motivated by three main reasons: (1) it is symmetric with respect to θ, (2) the beam is semicircular, and (3) it allows the stress function to be effectively reduced to a 1D problem, as follows.

Since the body force in the plate is zero, the assumed function ϕ must satisfy $\nabla^4\phi = 0$, which leads to

$$\left(\frac{d^4 f(r)}{dr^4} + \frac{2}{r}\frac{d^3 f(r)}{dr^3} - \frac{3}{r^2}\frac{d^2 f(r)}{dr^2} + \frac{3}{r^3}\frac{df(r)}{dr} - \frac{3f(r)}{r^4}\right)\cos\theta = 0 \tag{3.44}$$

or

$$\left(\frac{d^4 f(r)}{dr^4} + \frac{2}{r}\frac{d^3 f(r)}{dr^3} - \frac{3}{r^2}\frac{d^2 f(r)}{dr^2} + \frac{3}{r^3}\frac{df(r)}{dr} - \frac{3f(r)}{r^4}\right) = 0 \tag{3.45}$$

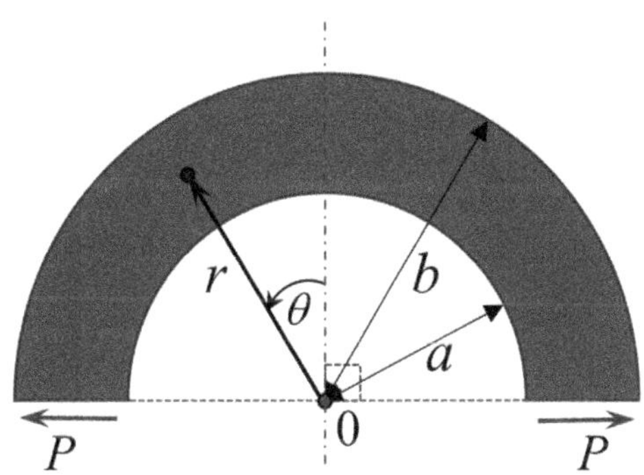

Figure 3.6. A thin semicircular curved beam of uniform thickness subjected to a pair of horizontal shear forces acting on the cross-sections at its two ends.

This is a 1D fourth-order differential equation (DE) for $f(r)$. The preceding equations and the general solution for ϕ can be found using the following code.

```
# Define variables for this problem:
r = sp.symbols('r', positive=True)   # symbolic variable for radial axis
θ = sp.symbols('θ', real=True)                           # for tangential axis

f = sp.Function('f')(r)                                  # 1D function of r only
ϕ = f*cos(θ)                   # sin(θ) can be use for antisymmetric problems

# The Laplace of ϕ(r,θ):
dϕ2 = ϕ.diff(r,2) + ϕ.diff(r)/r + ϕ.diff(θ,2)/r/r
display(Math(f" \\text{{ $∇^2ϕ$}}= \\scriptsize {latex(dϕ2)}"))

# Biharmonic of ϕ(r,θ):
dϕ4 = (dϕ2.diff(r,2) + dϕ2.diff(r)/r + dϕ2.diff(θ,2)/r/r ).expand()
display(Math(f" \\text{{ $∇^4ϕ$}}=\\scriptsize {latex(dϕ4)}"))

DE = dϕ4.coeff(cos(θ))                                    # get the 1D equation
display(Math(f" \\text{{ 1D differential equation for $f(r)$}}:\
                                        \\scriptsize {latex(DE)}=0"))
f_sln = sp.dsolve(DE, f)   # solve the 1D differential equation for f(r)
display(Math(f" \\text{{ General solution: }}\small {latex(f_sln)}"))
```

$$\nabla^2\phi = \cos(\theta)\frac{d^2}{dr^2}f(r) + \frac{\cos(\theta)\frac{d}{dr}f(r)}{r} - \frac{f(r)\cos(\theta)}{r^2}$$

$$\nabla^4\phi = \cos(\theta)\frac{d^4}{dr^4}f(r) + \frac{2\cos(\theta)\frac{d^3}{dr^3}f(r)}{r} - \frac{3\cos(\theta)\frac{d^2}{dr^2}f(r)}{r^2} + \frac{3\cos(\theta)\frac{d}{dr}f(r)}{r^3} - \frac{3f(r)\cos(\theta)}{r^4}$$

1D differential equation for $f(r)$: $\frac{d^4}{dr^4}f(r) + \frac{2\frac{d^3}{dr^3}f(r)}{r} - \frac{3\frac{d^2}{dr^2}f(r)}{r^2} + \frac{3\frac{d}{dr}f(r)}{r^3} - \frac{3f(r)}{r^4} = 0$

General solution: $f(r) = \frac{C_1}{r} + C_2 r + C_3 r^3 + C_4 r\log(r)$

We rewrite the solution of $f(r)$ as follows for later reference:

$$f = \frac{c_0}{r} + c_1 r + c_2 r^3 + c_3 r\log(r) \tag{3.46}$$

We can now form the general expression for the stress function:

```
ϕ = sp.Function('ϕ')(r, θ)
c0, c1, c2, c3 = symbols('c0, c1, c2, c3')          #integration constant
f = c0/r + c1*r + c2*r**3 + c3*r*log(r)
ϕ = f*sp.cos(θ)
display(Math(f" \\text{{ Stress function, ϕ}} = {latex(ϕ)}"))
```

Stress function, $\phi = \left(\frac{c_0}{r} + c_1 r + c_2 r^3 + c_3 r \log(r)\right)\cos(\theta)$

This general form of the stress function satisfies the compatibility condition in Eq. (3.45) and contains four arbitrary constants.

3.4.2 *General stress solution*

Using the stress function, we can now compute the stresses through the relation given in Eq. (2.59), assuming the absence of body forces:

```
# General expressions for stresses:

σrrg = (diff(diff(ϕ,θ),θ)/r/r+diff(ϕ,r)/r).expand().collect(cos(θ))
display(Math(f" \\text{{ The general solution of $σ_{{rr}}$}}=\
                                                {latex(σrrg)}"))
σθθg = (diff(diff(ϕ,r),r)).expand().collect(cos(θ))
display(Math(f" \\text{{ The general solution of $σ_{{θθ}}$}}=\
                                                {latex(σθθg)}"))
σrθg = (-diff(diff(ϕ,θ)/r,r)).expand().collect(sin(θ))
display(Math(f" \\text{{ The general solution of $σ_{{rθ}}$}}=\
                                                {latex(σrθg)}"))
```

The general solution of $\sigma_{rr} = \left(-\frac{2c_0}{r^3} + 2c_2 r + \frac{c_3}{r}\right)\cos(\theta)$

The general solution of $\sigma_{\theta\theta} = \left(\frac{2c_0}{r^3} + 6c_2 r + \frac{c_3}{r}\right)\cos(\theta)$

The general solution of $\sigma_{r\theta} = \left(-\frac{2c_0}{r^3} + 2c_2 r + \frac{c_3}{r}\right)\sin(\theta)$

It is found that:

1. These stresses contain only three unknown constants: c_0, c_2, and c_3. The constant c_1 does not affect the stresses, so only three constants need to be determined.
2. The normal stresses σ_{rr} and $\sigma_{\theta\theta}$ are mirror symmetric with respect to the plane $\theta = 0°$, while the shear stress $\sigma_{r\theta}$ is anti-mirror symmetric.

3. At $\theta = 90°$ or π, the normal stresses σ_{rr} and $\sigma_{\theta\theta}$ vanish. Only the shear stress $\sigma_{r\theta}$ is nonzero there. This indicates that on these boundary surfaces, the stress state is pure shear, allowing the force boundary conditions to be satisfied. Meanwhile, at $\theta = 0°$, the shear stress $\sigma_{r\theta}$ is zero, and the normal stresses σ_{rr} and $\sigma_{\theta\theta}$ are nonzero, meaning that σ_{rr} and $\sigma_{\theta\theta}$ are principal stresses at this location.
4. The terms in parentheses for σ_{rr} and $\sigma_{r\theta}$ are identical.

The following snippet verifies whether these stresses satisfy the equilibrium equations:

```
# Check the satisfaction of the equilibrium equations:
br = 0; bθ = 0
σvg = sp.Matrix([σrrg, σθθg, σrθg])
equi1=(σvg[0].diff(r)+σvg[2].diff(θ)/r+(σvg[0]-σvg[1])/r + br)\
                                              .simplify().factor()
equi2=(σvg[2].diff(r)+σvg[1].diff(θ)/r+2*σvg[2]/r + bθ)\
                                              .simplify().factor()
equi1, equi2  # should be (0, 0)
```

(0, 0)

These equilibrium equations are satisfied.

3.4.3 *Enforcing stress boundary conditions*

On the inner and outer curved surfaces of the beam, we apply the following stress boundary conditions (SBCs).

$$\begin{aligned} \sigma_{rr} = \sigma_{r\theta} = 0 \quad \text{at} \quad r = a \\ \sigma_{rr} = \sigma_{r\theta} = 0 \quad \text{at} \quad r = b \end{aligned} \tag{3.47}$$

The following snippet enforces these stress boundary conditions to determine some of the unknown constants.

First, for stress component σ_{rr}:

```
# Solve SBCs for some of the unknown constants:
a, b = symbols('a, b', positive=True)                          # radii
cs = sp.Matrix([c0, c2, c3])
sln_SBC = sp.solve([σrrg.subs(r,a), σrrg.subs(r,b)], cs)

# Simplify these constants:
sln_SBC = {key: value.subs(sln_SBC).simplify() for key, \
                                          value in sln_SBC.items()}
sln_cs = sln_SBC      # keep the constants in a unified dictionary sln_cs
print(f" Current len of sln_cs={len(sln_cs)}; Total cs={len(cs)}")
display(sln_cs)
```

```
Current len of sln_cs=2; Total cs=3
```

$$\left\{ c_0\colon \frac{a^2b^2c_3}{2\left(a^2+b^2\right)},\ c_2\colon -\frac{c_3}{2a^2+2b^2} \right\}$$

We find that the constants c_0 and c_2 are expressed in terms of c_3. Let us now update the stresses and check the shear stress on the inner and outer surfaces at $r = a$ and $r = b$:

```
(σrθg.subs(r,a).subs(sln_cs).simplify(),
 σrθg.subs(r,b).subs(sln_cs).simplify())                    # should be (0, 0)
```

(0, 0)

It is seen that the boundary conditions for the shear stress are already satisfied. This occurs because the terms in parentheses of the general expressions for σ_{rr} and $\sigma_{r\theta}$ are identical. Therefore, if σ_{rr} satisfies the zero-stress boundary condition, $\sigma_{r\theta}$ automatically satisfies it as well.

There remains one constant, c_3, to be determined. We shall now apply the integral equilibrium conditions, or force boundary conditions.

3.4.4 *Enforcing integral equilibrium conditions*

At the cross-section of a beam end ($\theta = \pm\pi/2$), as shown in Fig. 3.6, the following integral equilibrium conditions must be satisfied:

$$\begin{aligned} \sum M_z = 0\colon &\int_a^b \sigma_{\theta\theta}\left(\frac{a+b}{2} - r\right)dr = 0 \\ \sum F_r = 0\colon &\int_a^b \sigma_{r\theta} dr - P = 0 \end{aligned} \tag{3.48}$$

Note that the moment is measured with respect to the centroid axis of the beam cross-section. The following snippet enforces these conditions. First, for $\sum M_z$:

```
# Compute stresses with known constants and then compute the moment:
Mz = sp.integrate(σθθg.subs(θ, sp.pi/2).subs(sln_cs)*((a+b)/2 - r), \
                                                  (r, a, b)).simplify()
display(Math(f" \\text{{$∑M_z$}}=\\small {latex(Mz)}"))
```

$\sum M_z = 0$

It is found that the first equation in Eq. (3.33) is satisfied regardless of c_3. We shall now enforce the second equation in Eq. (3.33) to determine c_3 and then simplify it:

```
# Compute the sum of the force F_r on a cross-section via integral:
P = symbols('P', real=True)
equI=sp.integrate(σrθg.subs(sln_cs).subs(θ,sp.pi/2),(r,a,b)).simplify()-P
#display(equI)
sln_equI = sp.solve([equI], cs)                    # solve for the constant

# simplify the expression for the constant:
sln_equI = {key: value.subs(sln_equI).simplify() for key, \
                                    value in sln_equI.items()}
display(Math(f"\\text{{Constant found:}}=\\small {latex(sln_equI)}"))

Cab = sp.symbols('C_ab')   # defined for further simplifying expressions
c3_numer = (sp.numer(sln_equI[c3])/P).expand().simplify()
c3_denom = sp.denom(sln_equI[c3]).expand().simplify()   #sp.expand_log()

Cab_= (b**2/a**2-1) + (1+b**2/a**2)*(log(a)-log(b))  # simplify notation
display(Math(f"\\text{{Simplified denominator, }}C_{{ab}}={latex(Cab_)}"))

dffr = sp.expand_log((Cab_*a**2-c3_denom)).simplify()              # check
display(Math(f"\\text{{Is simplification correct:}}={latex(dffr==0)}"))
```

Constant found: $= \left\{ c_3 : \frac{P\left(-a^2 - b^2\right)}{a^2\log(a) - a^2\log(b) - a^2 + b^2\log(a) - b^2\log(b) + b^2} \right\}$

Simplified denominator, $C_{ab} = \left(1 + \frac{b^2}{a^2}\right)(\log(a) - \log(b)) - 1 + \frac{b^2}{a^2}$

Is simplification correct: = True

In the cell above, we applied some simplifications due to the length of the expressions. Now, these constants can be expressed in concise forms as follows:

```
sln_equI[c3] = P*c3_numer/(Cab*a**2)
sln_cs = {key: value.subs(sln_equI).simplify() for key, \
                                    value in sln_cs.items()}

sln_cs.update(sln_equI)                         # solve for these constants
#display(Math(f"\\text{{ Constants found:}} {latex(sln_cs)}"))
```

We have found all three constants. The formulas for these constants are rewritten as follows for future reference:

$$c_0 = -\frac{Pb^2}{2C_{ab}}; \quad c_2 = \frac{P}{2C_{ab}a^2}; \quad c_3 = -\frac{P\left(a^2 + b^2\right)}{C_{ab}a^2} \tag{3.49}$$

where

$$C_{ab} = \left(1 + \frac{b^2}{a^2}\right)(\log(a) - \log(b)) - 1 + \frac{b^2}{a^2}$$

We redefine these constant expressions in Python for later use:

```
c0_ =-P*b**2/(2*Cab)
c2_ = P/(2*Cab*a**2)
c3_ =-P*(a**2 + b**2)/(Cab*a**2)
```

3.4.5 *Final formulas for stresses*

Using the constants found, we obtain the final formulas for the stresses:

```
σrr = σrrg.subs(sln_cs)
σθθ = σθθg.subs(sln_cs)
σrθ = σrθg.subs(sln_cs)
display(Math(f" \\text{{ Final solution for stresses:}}"))
display(Math(f" \\text{{ $σ_{{rr}}$}}=\\scriptsize {latex(σrr)}"))
display(Math(f" \\text{{ $σ_{{θθ}}$}}=\\scriptsize {latex(σθθ)}"))
display(Math(f" \\text{{ $σ_{{rθ}}$}}=\\scriptsize {latex(σrθ)}"))
```

Final solution for stresses:

$$\sigma_{rr} = \left(\frac{Pb^2}{C_{ab}r^3} + \frac{Pr}{C_{ab}a^2} + \frac{P\left(-a^2 - b^2\right)}{C_{ab}a^2 r}\right)\cos(\theta)$$

$$\sigma_{\theta\theta} = \left(-\frac{Pb^2}{C_{ab}r^3} + \frac{3Pr}{C_{ab}a^2} + \frac{P\left(-a^2 - b^2\right)}{C_{ab}a^2 r}\right)\cos(\theta)$$

$$\sigma_{r\theta} = \left(\frac{Pb^2}{C_{ab}r^3} + \frac{Pr}{C_{ab}a^2} + \frac{P\left(-a^2 - b^2\right)}{C_{ab}a^2 r}\right)\sin(\theta)$$

3.4.6 *Final formulas for strains*

After obtaining all the stresses, we compute the strains using the constitutive equation (3.25), assuming the beam is very thin and that the problem is one of plane stress:

```
E, ν = symbols('E, ν', positive=True)        # given material constants

εrr = ((σrr - ν*σθθ)/E).simplify()
εθθ = ((σθθ - ν*σrr)/E).simplify()
γrθ = σrθ/E*(2*(1+ν))
display(Math(f" \\text{{ Final solution for strains}}:"))
display(Math(f" \\text{{$ε_{{rr}}$=}}\\scriptsize {latex(εrr)}"))
display(Math(f" \\text{{$ε_{{θθ}}$=}}\\scriptsize {latex(εθθ)}"))
display(Math(f" \\text{{$γ_{{rθ}}$=}}\\scriptsize {latex(γrθ)}"))
```

Final solution for strains:

$$\varepsilon_{rr} = \frac{P\left(a^2b^2 + r^4 - r^2\left(a^2+b^2\right) - \nu\left(-a^2b^2 + 3r^4 - r^2\left(a^2+b^2\right)\right)\right)\cos\left(\theta\right)}{C_{ab}Ea^2r^3}$$

$$\varepsilon_{\theta\theta} = \frac{P\left(-a^2b^2 + 3r^4 - r^2\left(a^2+b^2\right) - \nu\left(a^2b^2 + r^4 - r^2\left(a^2+b^2\right)\right)\right)\cos\left(\theta\right)}{C_{ab}Ea^2r^3}$$

$$\gamma_{r\theta} = \frac{\left(2\nu+2\right)\left(\frac{Pb^2}{C_{ab}r^3} + \frac{Pr}{C_{ab}a^2} + \frac{P\left(-a^2-b^2\right)}{C_{ab}a^2r}\right)\sin\left(\theta\right)}{E}$$

The following snippet confirms that all stresses and strains satisfy the necessary conditions and are therefore correct:

```
# Verification of equilibrium constants:

εrrg = ((σrrg - ν*σθθg)/E).simplify()     # compute strain using stresses
εθθg = ((σθθg - ν*σrrg)/E).simplify()
γrθg = σrθg/E*(2*(1+ν))
display(Math(f" \\text{{ General solution for strains}}:"))
display(Math(f" \\text{{$ε_{{rr}}$ = }}\\scriptsize {latex(εrrg)}"))
display(Math(f" \\text{{$ε_{{θθ}}$ = }}\\scriptsize {latex(εθθg)}"))
display(Math(f" \\text{{$γ_{{rθ}}$ = }}\\scriptsize {latex(γrθg)}"))

# Equilibrium Equations (zero body force):
br = 0; bθ = 0
equi1=(σrrg.diff(r)+σrθg.diff(θ)/r+(σrrg-σθθg)/r+br).simplify()
equi2=(σrθg.diff(r)+σθθg.diff(θ)/r+2*σrθg/r+bθ).simplify()
equis= [equi1, equi2]
display(Math(f" \\text{{Are the equilibrium equation satisfied? }}\
                    {latex(abs(equis[0]+abs(equis[0])).simplify()==0)}"))

# The Laplace of the stress trace:
σ_sum = (σrrg+σθθg)
cmptbl = σ_sum.diff(r,2)+σ_sum.diff(r)/r + σ_sum.diff(θ,2)/r/r
display(Math(f" \\text{{Is the compatibility equation satisfied? }}\
                                        {latex(cmptbl.simplify()==0)}"))
```

General solution for strains:

$$\varepsilon_{rr} = \frac{\left(-2c_0 + 2c_2r^4 + c_3r^2 - \nu\left(2c_0 + 6c_2r^4 + c_3r^2\right)\right)\cos\left(\theta\right)}{Er^3}$$

$$\varepsilon_{\theta\theta} = \frac{\left(2c_0 + 6c_2r^4 + c_3r^2 - \nu\left(-2c_0 + 2c_2r^4 + c_3r^2\right)\right)\cos\left(\theta\right)}{Er^3}$$

$$\gamma_{r\theta} = \frac{\left(2\nu + 2\right)\left(-\frac{2c_0}{r^3} + 2c_2r + \frac{c_3}{r}\right)\sin\left(\theta\right)}{E}$$

Are the equilibrium equation satisfied? True

Is the compatibility equation satisfied? True

It is seen that all necessary conditions for stresses are satisfied regardless of the values of the constants c_0, c_2, and c_3. These strains can now be used to compute the displacements. We use the general constants instead of the specific constants found earlier because the expressions are simpler for subsequent derivations. Additionally, the derived formulas are more general. The known constants can always be substituted when needed.

3.4.7 *Displacements in curved beam under shear forces*

As discussed earlier, the displacement is not axisymmetric. Therefore, the strain–displacement relation in Eq. (3.24) cannot be used. We must employ the more general formulas in Eq. (2.28), in which $\frac{\partial u_r}{\partial r} = \varepsilon_{rr}$. Hence, u_r can be obtained by integrating once using the strain ε_{rr} found earlier. The procedure is similar to that used in the previous section. This integration introduces an unknown function:

```
g = sp.Function('g')(θ)                    # define an unknown function
ur_ = sp.integrate(εrrg,(r))
ur_with_g = ur_ + g                       # g is an integral function of
display(Math(f"\\text{{Current displacement}}:"))
display(Math(f"\\text{{ $u_r$}}=\\scriptsize {latex(ur_with_g)}"))
```

Current displacement:

$$u_r = g\left(\theta\right) + \frac{-c_3\left(\nu - 1\right)\log\left(r\right)\cos\left(\theta\right) - r^2\left(3c_2\nu\cos\left(\theta\right) - c_2\cos\left(\theta\right)\right) - \frac{-c_0\nu\cos(\theta) - c_0\cos(\theta)}{r^2}}{E}$$

Next, using the second equation in Eq. (2.28), we have

$$\frac{\partial u_\theta}{\partial \theta} = r\varepsilon_{\theta\theta} - u_r \tag{3.50}$$

This is used to find the expression for u_θ, which introduces another unknown function:

```
h = sp.Function('h')(r)                          # h: integral unknown function
uθ_with_h = sp.integrate(r*εθθg - ur_with_g, (θ)) + h
display(Math(f"\\text{{Current displacement}}:"))
display(Math(f"\\text{{ $u_{{θ}}$}}=\\scriptsize \
 {latex(uθ_with_h.simplify().collect(sin(θ)).collect(r).collect(E))}"))
```

Current displacement:

$$u_\theta = \frac{Er^2\left(h(r) - \int g(\theta)\,d\theta\right) + \left(c_0\nu + c_0 + c_3\log\left(r^{r^2(\nu-1)}\right) + r^4\left(c_2\nu + 5c_2\right) + r^2\left(-c_3\nu + c_3\right)\right)\sin(\theta)}{Er^2}$$

Now, using the third equation in Eq. (2.28), we have

$$r\gamma_{r\theta} = \frac{\partial u_r}{\partial \theta} + r\frac{\partial u_\theta}{\partial r} - u_\theta \tag{3.51}$$

This expression can be obtained using the following snippet:

```
expr3=(r*γrθg - ur_with_g.diff(θ) - r*uθ_with_h.diff(r) + uθ_with_h)\
                                                    .simplify().expand()
display(Math(f" \\text{{ The third expression}}: {latex(expr3)}= 0 "))
```

The third expression: $-r\frac{d}{dr}h(r) + h(r) - \frac{d}{d\theta}g(\theta) - \int g(\theta)\,d\theta + \frac{4c_3\sin(\theta)}{E} = 0$

The foregoing equation must be zero. We re-group these terms in it as follows:

$$\underbrace{r\frac{d}{dr}h(r) - h(r)}_{H(r)} = \underbrace{-\frac{d}{d\theta}g(\theta) - \int g(\theta)\,d\theta + \frac{4c_3\sin(\theta)}{E}}_{G(\theta)} \tag{3.52}$$

It is seen that $H(r)$ is a function of only r, and $G(\theta)$ is a function of only θ. Equation (3.52) will hold only if $H(r)$ and $G(\theta)$ are both equal to a common constant, say C. We thus denote $H(r) = C$ and $G(\theta) = C$. This gives us the following two equations:

$$\begin{aligned} r\frac{d}{dr}h(r) - h(r) &= C \\ -\frac{d}{d\theta}g(\theta) - \int g(\theta)\,d\theta + \frac{4c_3\sin(\theta)}{E} &= C \end{aligned} \tag{3.53}$$

It is observed that the first equation in Eq. (3.53) is a 1D differential equation, which can be solved easily for $h(r)$ using `sp.dsolve()`.

The second equation is also 1D but contains an integral. We need to convert it to a differential equation by differentiating both sides of the equation once, which gives

$$-\frac{d^2}{d\theta^2}g(\theta) - g(\theta) + \frac{4c_3\cos(\theta)}{E} = 0 \tag{3.54}$$

This equation is now a 1D differential equation and can be solved for $g(\theta)$ using `sp.dsolve()`. Note that this differentiation eliminated the constant C from the equation. Therefore, the satisfaction of Eq. (3.51) may fail by a constant. We thus need to re-enforce it later after $g(\theta)$ is found.

The following snippet solves for both $h(r)$ and $g(\theta)$:

```
C = symbols('C')                                                # constant
# find g(θ) using dsolve():
sln_g = sp.dsolve(-g.diff(θ,2) - g + 4*c3*cos(θ)/E, g)
display(Math(f" \\text{{ Function $g(θ)$}}={latex(sln_g)}"))

# find h(r) using dsolve():
sln_h = sp.dsolve(r*h.diff(r) - h - C, h)
display(Math(f" \\text{{ Function }} {latex(sln_h)}"))
```

Function $g(\theta) = g(\theta) = C_2\cos(\theta) + \left(C_1 + \frac{2c_3\theta}{E}\right)\sin(\theta)$

Function $h(r) = -C + C_1 r$

We have successfully found these two unknown functions, but there are still four additional constants. The general form of the displacements can be obtained as follows:

```
C1, C2, Ch = symbols('C1, C2, C_h')                    #integral constants

g = C2*cos(θ) + (C1 + 2*c3*θ/E)*sin(θ)                  # re-define g(θ)
urg = (ur_ + g).simplify()
display(Math(f"\\text{{Current $u_r$}}=\\scriptsize \
                        {latex(urg.collect(c1).collect(c2).collect(c0))}"))

h = -C + Ch*r                                           # re-define h(r)
uθg = (sp.integrate(r*εθθ - urg, (θ)) + h).simplify().collect(sin(θ))
display(Math(f"\\text{{Current $u_θ$}}=\\scriptsize \
        {latex(uθg.collect(r**2*cos(θ)).collect(c0).collect(r**4*c2))}"))
```

$$\text{Current } u_r = \frac{C_2 E r^2 \cos(\theta) + r^2 (C_1 E + 2c_3\theta)\sin(\theta) + (c_0(\nu+1) + r^2(c_2 r^2 (1-3\nu) - c_3(\nu-1)\log(r)))\cos(\theta)}{E r^2}$$

$$\text{Current } u_\theta = \frac{\begin{gathered} C_{ab} E a^2 r^2 (-C + C_h r) + C_{ab} a^2 \left(c_3 \log\left(r^{r^2(\nu-1)\sin(\theta)}\right) + r^2 (C_1 E + 2c_3\theta)\cos(\theta) + \left(-C_2 E r^2 + c_0(-\nu-1) + c_2 r^4 (3\nu - 1) - 2c_3 r^2\right)\sin(\theta)\right) \\ - P\left(a^2 b^2 - 3r^4 + r^2(a^2+b^2) + \nu\left(a^2 b^2 + r^4 - r^2(a^2+b^2)\right)\right)\sin(\theta) \end{gathered}}{C_{ab} E a^2 r^2}$$

3.4.8 *Determine displacements using DBCs*

These displacements involve four constants: C, C_1, C_2, and C_h. These constants need to be determined using the displacement boundary conditions (DBCs) for the given problem. For our case, assume the DBCs are at $r = a$ and $\theta = 0$ (see Fig. 3.6):

$$u_r = 0; \quad u_\theta = 0; \quad \frac{\partial u_r}{\partial \theta} = 0 \tag{3.55}$$

This provides three conditions, which can be used to determine some of these four constants. The snippet is as follows:

```
# Using DBCs to determine some of the constants:
expr1 = uθg.subs({r:a, θ:0})
expr2 = urg.subs({r:a, θ:0})
expr3 = urg.diff(θ).subs({r:a, θ:0})

#display(expr1, expr2, expr3)
sln_DBC = sp.solve([expr1, expr2, expr3], [C, C1, C2, Ch])
display(sln_DBC)
```

$$\left\{C : C_h a,\ C_1 : 0,\ C_2 : \frac{3a^2 c_2 \nu}{E} - \frac{a^2 c_2}{E} + \frac{c_3 \nu \log(a)}{E} - \frac{c_3 \log(a)}{E} - \frac{c_0 \nu}{E a^2} - \frac{c_0}{E a^2}\right\}$$

We see that the nonzero constant C_2 contains the constants c_0, c_2, and c_3, which were found earlier. We can now determine C_2 by substitution:

```
C2_ = sln_DBC[C2].subs(sln_cs).simplify()
display(Math(f"\\text{{Constant $C_2$}}=\\scriptsize {latex(C2_ )}"))
```

$$\text{Constant } C_2 = \frac{P\left(a^2\left(3\nu-1\right)+b^2\nu+b^2-2\nu\left(a^2+b^2\right)\log\left(a\right) +2\left(a^2+b^2\right)\log\left(a\right)\right)}{2C_{ab}Ea^2}$$

```
# Check strain–displacement relations:
εu_expr1=(εrr - urg.diff(r)).subs(sln_DBC).subs(sln_cs).simplify()
εu_expr2=(r*εθθ - urg-uθg.diff(θ)).subs(sln_DBC).subs(sln_cs).simplify()
εu_expr3=(r*γrθg - urg.diff(θ) - r*uθg.diff(r) + uθg).subs(sln_DBC)\
                                                .subs(sln_cs).simplify()
(εu_expr1, εu_expr2, εu_expr3)
```

$(0, \ 0, \ -C_h a)$

It is seen that the third equation is not satisfied, and it depends only on the constant C_h. This is because the differentiation performed to obtain Eq. (3.54) removed one constant. We shall now re-enforce Eq. (3.36). The snippet is as follows:

```
sln_Ch = sp.solve([εu_expr3], [Ch])
sln_Cs = {key: value.subs(sln_Ch).simplify() for key, \
                                                value in sln_DBC.items()
sln_Cs.update(sln_Ch)
display(Math(f"\\text{{Constants found:}} "))
display(Math(f" \\scriptsize {latex(sln_Cs)}"))
```

Constants found:

$$\left\{ C : 0, \ C_1 : 0, \ C_2 : \frac{a^2\left(3a^2c_2\nu - a^2c_2 + c_3\nu\log\left(a\right) -c_3\log\left(a\right)\right) - c_0\left(\nu+1\right)}{Ea^2}, \ C_h : 0 \right\}$$

3.4.9 *Final displacement solution*

Finally, all the constants are determined, and we can obtain the final solution for these displacements:

```
dic_C = {C: 0, Ch: 0, C1: 0}
ur = sp.expand_log(urg.subs(dic_C).subs(sln_cs).simplify()).simplify()\
                                .collect(sin(θ)).collect(cos(θ)).collect(r)
uθ = sp.expand_log(uθg.subs(dic_C).subs(sln_cs).simplify()).simplify()\
                                .collect(sin(θ)).collect(cos(θ)).collect(r)
#display(Math(f" \\text{{ Final $u_r$}}=\\small {latex(ur)}"))
#display(Math(f" \\text{{ Final $u_θ$}}=\\small {latex(uθ)}")) #lengthy

# redefine the displacement solution explicitly for later use.
a2 = a**2; b2 = b**2; r2 = r**2;
ur_ = C2*cos(θ) + (-2*P*θ*(a2 + b2)*sin(θ))/(Cab*E*a2) -P*(a2*b2*(ν+1)
    + r2*(r2*(3*ν-1) - 2*(a2+b2)*(ν-1)*log(r)))*cos(θ)/(2*Cab*E*a2*r2)

uθ_ =-C2*sin(θ) + P*(-4*(a2+b2)*θ*cos(θ) + (4*a2+4*b2 + r2*(3*ν-1) \
    - 2*(a2+b2)*(ν-1)*log(r))*sin(θ))/(2*Cab*E*a2) + (P*a2*b2*(ν+1)/2 \
    - P*(a2*b2 - 3*r**4 + r2*(a2+b2) + ν*(a2*b2 + r**4 \
    - r2*(a2+b2))))*sin(θ)/(Cab*E*a2*r2)
(ur_-ur).simplify(), (uθ_-uθ).simplify()   # checking explicit formulas
```

(0, 0)

The final expressions for the displacements are rewritten in a more concise and readable form:

$$
\begin{aligned}
u_r &= C_2 \cos\theta - \frac{2P\theta\left(a^2+b^2\right)\sin\theta}{C_{ab}Ea^2} \\
&\quad - \frac{P\left\{a^2b^2(\nu+1) + r^2\left[r^2(3\nu-1) - \left(2a^2+2b^2\right)(\nu-1)\log r\right]\right\}}{2C_{ab}Ea^2r^2}\cos\theta \\
u_\theta &= -C_2\sin\theta + \frac{P}{2C_{ab}Ea^2}\left\{-4(a^2+b^2)\theta\cos\theta + \left[r^2(3\nu-1)\right.\right. \\
&\quad \left.\left. + (a^2+b^2)(4-2(\nu-1)\log r\right]\sin\theta\right\} \\
&\quad + \frac{P}{2C_{ab}Ea^2r^2}\left\{a^2b^2(\nu+1) - 2\left[a^2b^2(\nu+1) + r^4(\nu-3)\right.\right. \\
&\quad \left.\left. + r^2\left(a^2+b^2\right)(1-\nu)\right]\right\}\sin\theta
\end{aligned}
\tag{3.56}
$$

where

$$
C_2 = \frac{a^2c_2(3\nu-1) + c_3(\nu-1)\log a}{E} - \frac{c_0(\nu+1)}{Ea^2} \tag{3.57}
$$

with constants c_0, c_2, and c_3 given in Eq. (3.49).

It is clear that the displacement component u_r is mirror symmetric with the plane of $\theta = 0°$, while the component u_θ is anti-mirror symmetric.

As a validation, the following snippet computes the strains and stresses using the obtained displacements:

```
# Final check on the displacement solutions:
R = sp.Matrix([r, θ])
ε = gr.displacement2StrainPolar([urg.subs(sln_Cs).subs(sln_cs),
                                 uθg.subs(sln_Cs).subs(sln_cs)], R)
#ε = [εrr, εθθ, γrθ/2]
#ε = [εrrg.subs(sln_cs), εθθg.subs(sln_cs), γrθg.subs(sln_cs)/2]
σrr_, σθθ_, σrθ_ = gr.strain2stress2D(E, ν, ε, Ptype= 'stress')

display(Math(f" \\text{{Agrees with $σ_{{rr}}$ from Ref.? }}\
                        {latex((σrr_-σrr).simplify()==0)}"))
display(Math(f" \\text{{Agrees with $σ_{{θθ}}$ from Ref.? }}\
                        {latex((σθθ_-σθθ).simplify()==0)}"))
display(Math(f" \\text{{Agrees with $σ_{{rθ}}$ from Ref.? }}\
                        {latex((σrθ_-σrθ).simplify()==0)}"))
```

Agrees with σ_{rr} from Ref.? True
Agrees with $\sigma_{\theta\theta}$ from Ref.? True
Agrees with $\sigma_{r\theta}$ from Ref.? True

It is seen that all necessary conditions for stresses are satisfied. We have successfully solved the problem of a curved circular beam under pure shear forces at its two ends.

3.5 Review questions

1. Consider a plane strain, axisymmetric problem without body force, typically a cylindrical long pipe.
   (1) Derive the full set of solutions (displacements, strains, and stresses).
   (2) Use Eq. (3.13), apply the substitution given in Eq. (3.14), and then compare it with the solution obtained in step 1.
2. Consider Case I of a disk that does not have a hole in the center, but is subjected to a body force $b_r = qr$, where q is a constant. Find the full solutions, considering the body force.
3. Consider a disk with an inner radius a and outer radius b. Its inner circle is fully attached to a rigid shaft spinning at a constant angular velocity ω. Find the solution for stresses and displacements in the disk following the procedure given in Section 3.1, but considering the body force.
4. Consider a 2D circular disk with an inner radius of a and an outer radius of b. It is subjected to internal pressure p_i over the inner surface and external

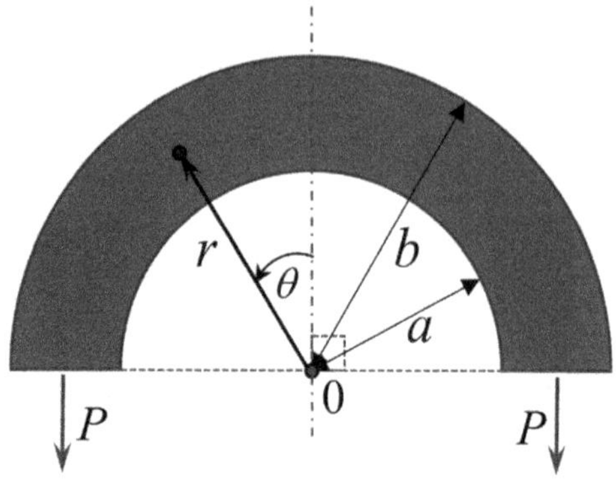

Figure 3.7. A thin, semicircular curved beam of uniform thickness loaded by a pair of vertical forces acting on the cross-sections at its two ends.

pressure p_o on the outer surface, as shown in Fig. 3.1. We assume zero body force in the system.

(1) Determine the general formulas for the displacements within the disk, using the stress solution formulas (which are axisymmetric) given in Chapter 3, without assuming the axisymmetric condition for the displacements.
(2) Derive the stress formulas using the general formulas of the displacements obtained in item 1, and validate them against the stress formulas used in item 1.
(3) Compare the displacement solution (which are axisymmetric) given in Chapter 3 with the formulas derived in item 1, assuming axisymmetry for the displacements.
(4) Discuss the results obtained, highlighting the differences and implications of the assumptions made.

5. Consider a thin, semicircular curved beam of uniform thickness with an inner radius of a and an outer radius of b. It experiences a pair of vertical forces P acting on the cross-sections at its two ends, as shown in Fig. 3.7.

Find the stress solutions for the beam. You may use the stress function $\phi = f(r)\sin\theta$.

3.6 Remarks

This chapter has discussed stress methods for solving problems with various types of symmetry. We conclude with a few remarks:

1. An effective methodology has been presented, including detailed step-by-step procedures and Python codes for different types of problems.

2. Sets of formulas for computing displacements and stresses have been provided for various classes of problems.
3. The symmetry property is used to define the stress function, which is an effective step in applying a stress method and helps reduce the number of unknowns.
4. Procedurally, the approach is very similar to the stress methods presented in Ref. [1] for problems in the Cartesian coordinate system.
5. For the first time, displacement solutions have been obtained for curved beams exhibiting stress symmetry or mirror symmetry.

In the following chapter, the symmetry property will be employed to assume expressions for displacements, which constitutes a crucial first step in applying a displacement method.

References

[1] G. R. Liu, *Solid Mechanics with Python: Cartesian Coordinates*, World Scientific, New Jersey, 2025.
[2] G. R. Liu, *Mechanics of Materials: Formulations and Solutions with Python*, World Scientific, New Jersey, 2024.
[3] A. P. Boresi, K. Chong and J. D. Lee, *Elasticity in Engineering Mechanics*, John Wiley & Sons, New York, 2010.

Chapter 4

Displacement Methods for Problems with Symmetry

Let us import the necessary external modules or dependencies for later use in this chapter:

```
# Place cursor in this cell, and press Ctrl+Enter to import dependencies.
import sys                              # for accessing the computer system
sys.path.append('../grbin/')              # Add in the path to your system

from commonImports import *         # Import dependencies from '../grbin/'
import grcodes as gr                      # Import the module of the author
importlib.reload(gr)                    # When grcodes is modified, reload it

init_printing(use_unicode=True)           # For latex-like quality printing
np.set_printoptions(precision=4,suppress=True)    # Digits in print-outs
```

In the previous chapter, we discussed stress methods for solving problems with certain symmetry properties. If a stress function can be found, the stresses can then be easily obtained. In computing the displacement, however, indefinite integrals are required, which introduces unknown functions into the solution. Determining these functions involves careful, mostly manual operations.

If the displacement can be found first, computing the strains — and hence the stresses — becomes straightforward, requiring only differentiation, which can always be done. This displacement method has been demonstrated in detail in Ref. [1] for problems defined in the Cartesian coordinate system. In this chapter, we present displacement methods for similar types of problems using polar coordinates. The discussion begins with the axisymmetric problem, which has neat formulas and many applications. The solution procedure largely follows that outlined in the last section of Chapter 2.

4.1 Axisymmetric problems

4.1.1 *Basic equations*

When a structural component is axisymmetric, it exhibits several features, as stated in Section 3.1.1, including: all variables are independent of θ; zero body force component b_θ; zero displacement component u_θ; and zero shear stress $\sigma_{r\theta}$. Therefore, the equilibrium equation becomes one-dimensional (1D):

$$\frac{\partial \sigma_{rr}}{\partial r} + \frac{\sigma_{rr} - \sigma_{\theta\theta}}{r} + b_r = 0 \tag{4.1}$$

where b_r is the body force applied in the r-direction. In this chapter, we assume the body force is purely linear in r:

$$b_r = qr \tag{4.2}$$

where q is a given constant.

The unknown stresses in an axisymmetric problem are σ_{rr} and $\sigma_{\theta\theta}$. The constitutive equation is given as (see Chapter 5 of Ref. [2])

$$\begin{aligned}
\sigma_{rr} &= \frac{E}{(1-\nu^2)}(\varepsilon_{rr} + \nu\varepsilon_{\theta\theta}) \\
\sigma_{\theta\theta} &= \frac{E}{(1-\nu^2)}(\varepsilon_{\theta\theta} + \nu\varepsilon_{rr}) \\
\sigma_{r\theta} &= \frac{E}{(1+\nu)}\varepsilon_{r\theta}
\end{aligned} \tag{4.3}$$

which has been coded in the Python function `gr.strain2stress2D()`.

The strain–displacement relation becomes

$$\varepsilon_{rr} = \frac{\partial u_r}{\partial r}; \quad \varepsilon_{\theta\theta} = \frac{u_r}{r}; \quad \varepsilon_{r\theta} = 0 \tag{4.4}$$

4.1.2 *General expression of displacement*

As discussed in detail in the last section of Chapter 2, a displacement method begins with an assumption of the displacement field. For our axisymmetric problem, the radial displacement u_r is assumed to be a linear combination of monomial bases of r, as follows:

$$u_r = c_0 + c_1 \frac{1}{r^3} + c_2 \frac{1}{r} + c_3 r + c_4 r^3 + c_5 r^5 \tag{4.5}$$

Here, both positive and negative powers of r are included [3]. Even powers are deliberately excluded because, in any axisymmetric setting, the displacement u_r can only be an odd function of r. In general, all lower-order monomials are included unless there is a specific reason to omit them.

Once the displacement expression is assumed, the strains can be computed using Eq. (4.4) through differentiation, the stresses using Eq. (4.3) through algebraic operations, and the equilibrium equation using Eq. (4.1), which also involves only differentiation and algebra.

With this strategy clearly laid out, the following code snippets demonstrate the process step by step:

```
# Define variables for displacement method for axisymmetric problems:
E, ν = symbols('E, ν', positive=True)            # given material constants
r = sp.symbols('r', positive=True)   # symbolic variable for radial axis

# Define given constants for body force br:
q =sp.symbols('q', real=True)
br = q*r                                # Linear body force in the r-direction

# Define unknown constants for displacement ur:
c0, c1, c2, c3, c4, c5 = sp.symbols('c0, c1, c2, c3, c4, c5',real=True)
cs_list = [c0, c1, c2, c3, c4, c5]
cs = sp.Matrix(cs_list)
print(f"Total number of unknown constants = {len(cs)}")

ur = sp.symbols('u_r', real=True)
εrr, εθθ = sp.symbols('ε_rr, ε_θθ ',real=True)
σrr, σθθ = sp.symbols('σ_rr, σ_θθ ',real=True)

# Define the general form of displacement ur:
urg = c0 + c1/r**3 + c2/r + c3*r + c4*r**3 + c5*r**5
order = 8                      # total order = positive order + negative order
display(Math(f" \\text{{General solution for displacement $u_r$ = }}\
                                                            {latex(urg)}"))
```

```
Total number of unknown constants = 6
```

General solution for displacement $u_r = c_0 + \frac{c_1}{r^3} + \frac{c_2}{r} + c_3 r + c_4 r^3 + c_5 r^5$

In this general approximation of the displacement using elementary basis functions, there are a total of six unknown constants.

4.1.3 *General formulas for stresses*

The strains are first computed using the strain–displacement relation Eq. (4.4), and then the stresses are obtained through the constitutive equation. All these steps involve only differentiation and algebraic operations, which can always be carried out easily for elementary functions in Python:

```
# Compute strains using the displacement:
εrr = urg.diff(r)
εθθ = urg/r
εvg = sp.Matrix([εrr, εθθ, 0])                # define strains as a vector

# Compute stress vector using strains, using constitutive equation:
σvg = sp.Matrix(gr.strain2stress2D(E, ν, εvg, Ptype= 'stress'))
σvg = σvg.expand()

display(Math(f" \\text{{General solution for stresses:}} "))
display(Math(f" \\text{{$σ_{{rr}}$=}} \
                \\scriptsize {latex(σvg[0].collect(r).simplify())}"))
display(Math(f" \\text{{$σ_{{θθ}}$=}} \
                \\scriptsize {latex(σvg[1].collect(r).simplify())}"))
```

General solution for stresses:

$$\sigma_{rr} = \frac{E\left(-c_0 r^3 \nu + c_1(3-\nu) + c_2 r^2(1-\nu) - r^4\left(c_3\nu + c_3 + c_4 r^2(\nu+3) + c_5 r^4(\nu+5)\right)\right)}{r^4(\nu^2-1)}$$

$$\sigma_{\theta\theta} = \frac{E\left(-c_0 r^3 + c_1(3\nu-1) + c_2 r^2(\nu-1) - r^4\left(c_3\nu + c_3 + c_4 r^2(3\nu+1) + c_5 r^4(5\nu+1)\right)\right)}{r^4(\nu^2-1)}$$

We have obtained the general formulas for stresses for any axisymmetric problem. The general solution contains all the unknown constants.

4.1.4 *Enforcing differential equilibrium equations*

To determine some of these constants, we use the equilibrium equation Eq. (4.1). Since this equation must hold at every point in the domain, all coefficients of the various powers of r in the equation must be zero. This requirement leads to a set of algebraic equations for the constants.

To perform this systematically, we use the following Python function, which groups the coefficients, forms the system of equations, and solves for the unknown constants. This function will be applied multiple times throughout this chapter to update the constants:

```
def constantUpdate(eqns, r, order, cs, print=False):
    '''Compute and update constants `cs` for a given 1D sympy `eqns`.
    return: sln_eqn - solution of cs in dictionary.
    '''
    exprs = []
    for eqn in eqns:
        for i in range(order):
            exprs.append(eqn.coeff(r,i))

    exprs = gr.remove_zero_exprs(exprs)      # remove exprs that are 0
    if prnt: display(exprs)
    sln_eqns = sp.solve(exprs, cs)         # solve for unknown constants
    return sln_eqns                              # can be very lengthy
```

We define the equilibrium equation in Python and simplify it to obtain a rational expression. We then focus on the numerator, which must vanish. Typically, the numerator is a polynomial in r, so all coefficients corresponding to each power of r must be set to zero. This procedure yields a system of equations for the unknown constants.

All of this is performed by invoking the function `constantUpdate()`, passing the numerator expression as the argument. The corresponding code snippet is as follows:

```
# Define equilibrium equation that needs to be satisfied:
equi = (σvg[0].diff(r) + (σvg[0] - σvg[1])/r + br).simplify()
equi = sp.numer(equi).expand()    # numerator of the equilibrium equation
display(Math(f"\\text{{Numerator of equili. equation=}}{latex(equi)}"))

sln_cs = {}                 # initialize solution of unknown constants, cs
sln_equi = constantUpdate([equi], r, order+1, cs)

# Update unknown constants:
sln_cs = {**sln_equi}
sln_cs = {key: value.subs(sln_cs) for key, value in sln_cs.items()}
print(f" Current len of sln_cs={len(sln_cs)}; Total cs={len(cs)}")

σv = sp.simplify(σvg.subs(sln_cs))                      # update stresses
display(Math(f"\\text{{Updated stresses = }}{latex(σv)}"))
```

Numerator of equili. equation $= Ec_0r^3 - 8Ec_1 - 8Ec_4r^6 - 24Ec_5r^8 + qr^6\nu^2 - qr^6$

```
Total number of expressions = 9
Number of nonzero expressions = 4
Current len of sln_cs=4; Total cs=6
```

$$\text{Updated stresses} = \begin{bmatrix} \frac{E\left(-c_2\nu+c_2-c_3r^2\nu-c_3r^2\right)+\frac{qr^4\left(-3\nu^2-\nu\left(\nu^2-1\right)+3\right)}{8}}{r^2\left(\nu^2-1\right)} \\ \frac{E\left(c_2\nu-c_2-c_3r^2\nu-c_3r^2\right)+\frac{qr^4\left(-\nu^2-3\nu\left(\nu^2-1\right)+1\right)}{8}}{r^2\left(\nu^2-1\right)} \\ 0 \end{bmatrix}$$

We have successfully determined four of the unknown constants and used them to update the stresses. These constants can also be used to update the displacements (or strains) using `sp.subs()`:

```
# update the displacement:
ur = urg.subs(sln_cs).collect(q)
#display(Math(f"\\text{{$u_r$ satisfying equilibrium = }}{latex(ur)}"))
print(ur)
# define ur explicitly:
urg_ = c2/r + c3*r + q*r**3*(ν**2 - 1)/(8*E)
```

```
c2/r + c3*r + q*r**3*(ν**2 - 1)/(8*E)
```

It is observed that the constants corresponding to the highest-order, lowest-order, and constant terms are automatically set to zero when the equilibrium equation is enforced.

4.1.5 *General displacement for axisymmetric problems*

We have now derived a fundamental general displacement equation applicable to all axisymmetric problems, which inherently satisfies the equilibrium equation:

$$u_r = \frac{c_2}{r} + c_3 r - \frac{qr^3\left(1-\nu^2\right)}{8E} \tag{4.6}$$

This equation contains two unknown constants, c_2 and c_3, indicating that any axisymmetric problem requires only two unknown constants in the assumed displacement. The relevant monomial bases are $\frac{1}{r}$, r, and r^3. The presence of the r^3 term arises from the linearly distributed body force. If the body force is zero ($q = 0$), only the monomial bases $\frac{1}{r}$ and r remain.

In future applications, Eq. (4.6) should be used directly when assuming the displacement for axisymmetric problems, eliminating the need to verify the equilibrium equation, provided the body force is linear in r.

The constants c_2 and c_3 must be determined using the boundary conditions specific to each problem. Every axisymmetric problem inherently provides two boundary conditions, ensuring a unique solution. To illustrate this, we revisit examples studied in the previous chapter, allowing a direct comparison of the solutions.

4.1.6 *Example: Spinning solid disk*

Consider a thin solid disk of uniform thickness with an outer radius of b, on which a constant pressure p_b is applied. It is fixed on a stiff solid shaft of small radius, which rotates together with the shaft at a constant angular velocity ω, as shown in Fig. 4.1. Polar coordinates (r, θ) are defined with the origin at the centroid of the disk (shaft). The body force on the disk domain is the centrifugal force, given by

$$b_r = \rho\omega^2 r \tag{4.7}$$

4.1.6.1 *Determination of the constants*

Assume the problem is axisymmetric. We aim to derive a complete solution for the displacement (referenced to the shaft), stresses, and strains. The general formulas derived above can be applied directly, with the boundary conditions of this problem used to determine the two unknown constants:

Condition 1: At the center of the disk ($r = 0$), the displacement must be zero (referenced to the shaft). Therefore, c_2 in Eq. (4.6) must be zero.

Condition 2: At the outer edge of the disk ($r = b$), the radial stress must satisfy $\sigma_{rr} = -p_b$.

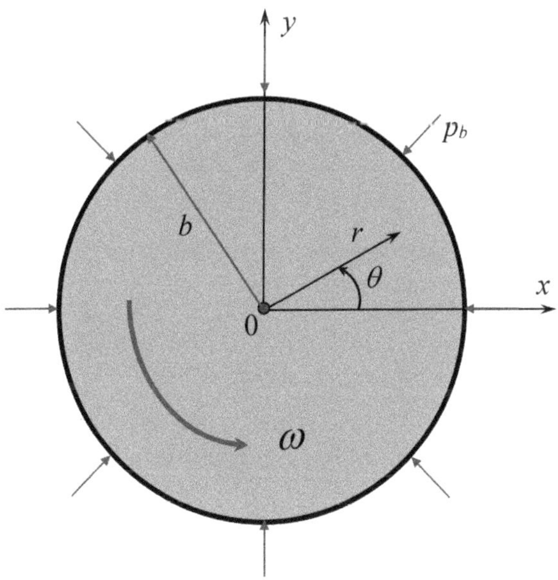

Figure 4.1. A thin solid disk of uniform thickness. It rotates at a constant angular velocity along with a stiff solid shaft of small radius.

The Python snippets for imposing these two conditions are as follows:

```
b = sp.symbols('b', positive=True)     # symbolic variable, outer radius
pb = sp.symbols('p_b', real=True)          # given stress on boundaries
ρ, ω= sp.symbols('ρ, ω', positive=True)   # for centrifugal body force

sln_c2 = {c2:0}                                        # use condition 1
sln_cs = {**sln_equi, **sln_c2}                       # update constants
σv = sp.simplify(σvg.subs(sln_cs))                     # update stresses

# Define stress boundary conditions (SBCs) at r=b (condition 2):
SBCrb = (σv[0].subs(r,b) + pb).simplify()  # stress boundary conditions
SBCrb = sp.numer(SBCrb).expand()

sln_SBCs = constantUpdate([SBCrb], r, order, cs)
display(sln_SBCs)
```

```
Total number of expressions = 8
Number of nonzero expressions = 1
```

$$\left\{c_3: \frac{-b^2 q\nu^2 - 2b^2 q\nu + 3b^2 q + 8p_b\nu - 8p_b}{8E}\right\}$$

4.1.6.2 *Final stress solution*

We have found the constant c_3. Now, we update and display the stress solutions:

```
sln_cs = {**sln_equi, **sln_c2, **sln_SBCs}             # update constants
σv = sp.simplify(σvg.subs(sln_cs))                     # update stresses
σv = σv.applyfunc(lambda expr:sp.collect(expr,q*r**2/8)\
                        .collect(b**2*q/8).collect(q/8)).subs(q,ρ*ω**2)

display(Math(f" \\text{{Final stress vector $σ_v^T$:}} "))
display(Math(f" \\small {latex(σv.T)}"))
```

Final stress vector σ_v^T:

$$\left[-p_b + \frac{\rho\omega^2\left(b^2(\nu+3)+r^2(-\nu-3)\right)}{8} \quad -p_b + \frac{\rho\omega^2\left(b^2(\nu+3)+r^2(-3\nu-1)\right)}{8} \quad 0\right]$$

This is the final solution for all stresses. If we set $\omega = 0$, the results reduce to those of Case I, in which the solid disk is subjected to a constant stress, as discussed in Section 3.1.

For a spinning disk rotating at a constant angular velocity ω, with $q = \rho\omega^2$, the stress solutions are expressed as

$$\begin{aligned}\sigma_{rr} &= -p_b + \frac{\rho\omega^2\left[b^2(\nu+3) - r^2(\nu+3)\right]}{8}\\ \sigma_{\theta\theta} &= -p_b + \frac{\rho\omega^2\left[b^2(\nu+3) - r^2(3\nu+1)\right]}{8}\\ \sigma_{r\theta} &= 0\end{aligned} \tag{4.8}$$

4.1.6.3 *Final solution for displacement and strains*

The displacement and strains are obtained as follows through substitution:

```
ur_final = urg.subs(sln_cs).collect(b).collect(8*pb).collect(q)\
                                                 .subs(q,ρ*ω**2)
display(Math(f" \\text{{Final displacement $u_r$ = }}\
                                         \\small {latex(ur_final)}"))
display(Math(f" \\text{{When ω=0, $u_r$ = }}\
                              \\small {latex(ur_final.subs(ω, 0))}"))

εv = sp.simplify(εvg.subs(sln_cs))                     # update strains
εv = εv.applyfunc(lambda expr:sp.collect(expr,q*r**2/8)\
                        .collect(b**2*q/8).collect(q/8)).subs(q,ρ*ω**2)
display(Math(f" \\text{{Final strain vector $ε_v$ = }} {latex(εv)}"))
display(Math(f" \\text{{When $ω=0$, $ε_v^T$ = }}\
                              {latex(εv.subs(ω, 0).simplify().T)}"))
```

Final displacement

$$u_r = \frac{r^3\rho\omega^2\left(\nu^2-1\right)}{8E} + \frac{r\left(b^2\rho\omega^2\left(-\nu^2-2\nu+3\right)+8p_b\left(\nu-1\right)\right)}{8E}$$

When $\omega = 0$, $u_r = \dfrac{p_b r\left(\nu-1\right)}{E}$

Final strain vector $\varepsilon_v = \begin{bmatrix}\frac{-b^2\nu^2\rho\omega^2-2b^2\nu\rho\omega^2+3b^2\rho\omega^2+8p_b\nu-8p_b+3r^2\rho\omega^2\left(\nu^2-1\right)}{8E}\\ \frac{-b^2\nu^2\rho\omega^2-2b^2\nu\rho\omega^2+3b^2\rho\omega^2+8p_b\nu-8p_b+r^2\rho\omega^2\left(\nu^2-1\right)}{8E}\\ 0\end{bmatrix}$

When $\omega = 0$, $\varepsilon_v^T = \begin{bmatrix}\frac{p_b(\nu-1)}{E} & \frac{p_b(\nu-1)}{E} & 0\end{bmatrix}$

We have now obtained the complete solution set. It is worth noting that when there is no body force (e.g., $\omega = 0$), the linear displacement solution coincides with that presented in Section 3.1, which was derived using the stress method.

4.1.7 *Example: Spinning hollow disk with edge pressure*

Consider a hollow disk with an inner radius a, subjected to a pressure p_b on its outer surface at $r = b$. This problem was previously analyzed in Section 3.1 using the stress method. The setup is identical to the previous example, except that the shaft now has a finite radius a. Our goal is to derive the complete solution set for this configuration.

In this case, we cannot set $c_2 = 0$ because the hollow disk does not extend to $r = 0$. However, there are now two boundary locations, $r = a$ and $r = b$, which provide two independent conditions. The following code snippet establishes these boundary conditions and solves for the unknown constants.

4.1.7.1 *General displacement*

We will carry out this analysis independently of the previous results, so the following code is fully self-contained. Since the problem remains axisymmetric, the displacement field can again be expressed in terms of only two constants:

```
# Define variables for displacement method for axisymmetric problems:
E, ν = symbols('E, ν', positive=True)          # given material constants
r, a, b = sp.symbols('r, a, b', positive=True)  # variables for r-axis

q =sp.symbols('q', real=True)                          # for body force br
pb = sp.symbols('p_b', real=True)              # given stress on boundaries
ρ, ω= sp.symbols('ρ, ω', positive=True)   # for centrifugal body force
br = q*r                              # linear body force in the r-direction

# Define unknown constants for displacement ur:
c0, c1 = sp.symbols('c0, c1', real=True)
cs_list = [c0, c1]
cs = sp.Matrix(cs_list)
print(f"Total number of unknown constants = {len(cs)}")

ur = sp.symbols('u_r', real=True)
σrr, σθθ, εrr, εθθ = sp.symbols('σrr, σθθ, εrr, εθθ',real=True)

# Define the general displacement ur (obtained earlier):
order = 2                   # total order = positive order + negative order
urg = c0/r + c1*r - q*r**3*(1-ν**2)/(8*E)        # equilibrium satisfied
display(Math(f" \\text{{General solution for displacement $u_r$ = }}\
                                                        {latex(urg)}"))
εrr = urg.diff(r)       # Compute general strains using the displacement
εθθ = urg/r
εvg = sp.Matrix([εrr, εθθ, 0])              # define strains as a vector

# Compute general stress vector using strain vector:
σvg = sp.Matrix(gr.strain2stress2D(E, ν, εvg, Ptype= 'stress'))
σvg = σvg.expand()
```

```
Total number of unknown constants = 2
```

General solution for displacement $u_r = \frac{c_0}{r} + c_1 r - \frac{qr^3\left(1-\nu^2\right)}{8E}$

4.1.7.2 *Determination of the constants*

To determine the unknown constants for this problem, we define the following boundary conditions:

- **Condition 1**: Since the disk is rigidly attached to the shaft at the inner radius $r = a$, the radial displacement u_r relative to the shaft must be zero along this boundary. Therefore,

$$u_r\Big|_{r=a} = 0 \tag{4.9}$$

- **Condition 2**: At the outer edge of the disk, where $r = b$, the radial stress σ_{rr} must be equal to the applied pressure $-p_b$:

$$\sigma_{rr}\Big|_{r=b} = -p_b \tag{4.10}$$

The following Python code snippet illustrates how these boundary conditions can be implemented to solve for the unknown constants:

```
BCa = urg.subs(r,a)                # displacement boundary condition at r=a
#BCa = σvg[0].subs(r,a)+pi        # if use stress boundary condition at r=a
BCa = sp.numer(BCa).expand()
BCb = σvg[0].subs(r,b) + pb
BCb = sp.numer(BCb).expand()

sln_BCs = constantUpdate([BCa, BCb], r, order, cs)
c0_ = sln_BCs[c0].collect(q).collect(a).collect(b).collect(pb)
c1_ = sln_BCs[c1].collect(q).collect(a).collect(b).collect(pb)
#display(Math(f" \\text{{Constants found: }}"))
#display(Math(f" \\text{{ $c_0$=}} \\small \
#  {latex(sln_BCs[c0].collect(q).collect(a).collect(b).collect(pb))}"))
#display(Math(f" \\text{{ $c_1$=}} \\small \
#  {latex(sln_BCs[c1].collect(q).collect(a).collect(b).collect(pb))}"))
```

```
Total number of expressions = 4
Number of nonzero expressions = 2
```

To handle the complexity of the resulting lengthy expressions, we simplify the final formulae by introducing new symbolic variables to represent recurring terms. This **book-keeping technique** reduces expression size, enhances readability, and facilitates subsequent symbolic or numerical calculations.

The following Python code snippet demonstrates how certain lengthy expressions are replaced with new symbolic variables to streamline the final displacement and stress computations:

```
v1, v2, v3, v4, v5, v6= sp.symbols('v1, v2, v3, v4, v5, v6')
v1_ = c0_.args[1].args[0].args[1]. args[0].args[-1]    # get long exprs
v2_ = c0_.args[1].args[0].args[1]. args[1].args[-1]
v3_ = c1_.args[1].args[0].args[1]. args[0].args[-1]
v4_ = c1_.args[1].args[0].args[1]. args[1].args[-1]

dic_v = {v1_:v1, v2_:v2, v3_:v3, v4_:v4, (-3*v**3-v**2+3*v+1):v5}
dicI_v = {value: key for key, value in dic_v.items()}    # inverse dict

display(Math(f" \\text{{Simplified constants:}} "))
display(Math(f"\\text{{ $c_0$=}}\\small {latex(c0_.subs(dic_v))}; \
               \\; \\text{{ $c_1$=}}\\small {latex(c1_.subs(dic_v))}"))
display(Math(f" \\text{{where,}}"))
dicI_v_2 = gr.split_dict_n(dicI_v, 2)
display(Math(f" \\text{{}} {latex(dicI_v_2[0])} "))
display(Math(f" \\text{{}} {latex(dicI_v_2[1])} "))
```

Simplified constants:

$$c_0 = \frac{a^2 b^2 p_b \left(8\nu^2 - 8\right) + q\left(a^4 b^2 \nu_2 + a^2 b^4 \nu_1\right)}{a^2\left(8E\nu - 8E\right) + b^2\left(-8E\nu - 8E\right)};$$

$$c_1 = \frac{b^2 p_b \left(8 - 8\nu^2\right) + q\left(a^4 \nu_3 - b^4 \nu_1\right)}{a^2\left(8E\nu - 8E\right) + b^2\left(-8E\nu - 8E\right)}$$

where

$$\left\{\nu_1 : -\nu^3 - 3\nu^2 + \nu + 3,\ \nu_2 : \nu^3 + \nu^2 - \nu - 1\right\}$$

$$\left\{\nu_3 : -\nu^3 + \nu^2 + \nu - 1,\ \nu_4 : \nu^3 + 3\nu^2 - \nu - 3,\ \nu_5 : -3\nu^3 - \nu^2 + 3\nu + 1\right\}$$

4.1.7.3 *Solution for spinning disk under centrifugal force*

Having determined the two unknown constants, we can now compute the final solutions for the displacement, stresses, and strains. These resulting expressions are often quite lengthy. To improve clarity, we present the results separately — those arising from the body force and those from the pressure load.

The following Python code snippet displays the results corresponding to the **body force only**, obtained by setting the pressure terms to zero:

```
# Update constants, displacements, stresses & strains (zero pressure):
sln_cs = {c0:c0_.subs(dic_ν), c1:c1_.subs(dic_ν)}

dic_={pb:0, q:ρ*ω**2}
ur_final = urg.subs(sln_cs).subs(dic_).simplify()\
                        .collect(q).collect(a).collect(b).collect(r)
ur_ra = ur_final.subs(r, a).subs(dicI_ν).simplify()
ur_rb = ur_final.subs(r, b).subs(dicI_ν).simplify()

σv_final = sp.simplify(σvg.subs(sln_cs).subs(dic_))
εv_final = sp.simplify(εvg.subs(sln_cs).subs(dic_))
σv_final = σv_final.applyfunc(lambda expr: sp.collect(expr,q)\
                      .collect(a).collect(b).collect(r).subs(dic_ν)\
                .collect(a**2*ν2 + b**2*ν1).collect(a**4*ν3 - b**4*ν1))
εv_final = εv_final.applyfunc(lambda expr: sp.collect(expr,q)\
                              .collect(a).collect(b).collect(r))

display(Math(f" \\text{{Final displacement $u_r$:}} "))
display(Math(f" \hspace{{1.5cm}}  \scriptsize {latex(ur_final)}"))

display(Math(f" \\text{{At $r=a$, $u_r$ = }} {latex(ur_ra)}"))
display(Math(f" \\text{{At $r=b$, $u_r$=}} \scriptsize {latex(ur_rb)}"))

display(Math(f" \\text{{Final stress $σ$ =}}{latex(σv_final)}"))
display(Math(f" \\text{{Final strains $ε$ =}}{latex(εv_final)}"))
```

Final displacement u_r:

$$\frac{\rho\omega^2\left(a^2b^2\left(a^2\nu_2+b^2\nu_1\right)+r^4\left(\nu^2-1\right)\left(a^2\left(\nu-1\right)+b^2\left(-\nu-1\right)\right)+r^2\left(a^4\nu_3-b^4\nu_1\right)\right)}{8Er\left(a^2\left(\nu-1\right)+b^2\left(-\nu-1\right)\right)}$$

At $r=a$, $u_r=0$

At $r=b$, $u_r=\dfrac{b\rho\omega^2\left(a^4\nu^2-a^4-2a^2b^2\nu^2+2a^2b^2+b^4\nu^2-b^4\right)}{4E\left(a^2\nu-a^2-b^2\nu-b^2\right)}$

Final stress

$$\sigma=\begin{bmatrix}\frac{\rho\omega^2\left(a^2b^2(1-\nu)\left(a^2\nu_2+b^2\nu_1\right)+r^4\nu_1\left(a^2(\nu-1)+b^2(-\nu-1)\right)+r^2(-\nu-1)\left(a^4\nu_3-b^4\nu_1\right)\right)}{8r^2(\nu^2-1)(a^2(\nu-1)+b^2(-\nu-1))}\\ \frac{\rho\omega^2\left(a^2b^2(\nu-1)\left(a^2\nu_2+b^2\nu_1\right)+r^4\nu_5\left(a^2(\nu-1)+b^2(-\nu-1)\right)+r^2(-\nu-1)\left(a^4\nu_3-b^4\nu_1\right)\right)}{8r^2(\nu^2-1)(a^2(\nu-1)+b^2(-\nu-1))}\\ 0\end{bmatrix}$$

Final strains

$$\varepsilon=\begin{bmatrix}\frac{\rho\omega^2\left(-a^2b^2\left(a^2\nu_2+b^2\nu_1\right)+3r^4\left(\nu^2-1\right)\left(a^2(\nu-1)+b^2(-\nu-1)\right)+r^2\left(a^4\nu_3-b^4\nu_1\right)\right)}{8Er^2(a^2(\nu-1)+b^2(-\nu-1))}\\ \frac{\rho\omega^2\left(a^2b^2\left(a^2\nu_2+b^2\nu_1\right)+r^4\left(\nu^2-1\right)\left(a^2(\nu-1)+b^2(-\nu-1)\right)+r^2\left(a^4\nu_3-b^4\nu_1\right)\right)}{8Er^2(a^2(\nu-1)+b^2(-\nu-1))}\\ 0\end{bmatrix}$$

This represents the final solution for a hollow disk spinning at a constant angular velocity ω. The result is identical to the one obtained in the previous chapter using the stress function method (Section 3.2.7 with $\omega_t = 0$ and hence axisymmetric). This equivalence can be verified using the following Python code snippet:

```
ur_strssMthd=ρ*ω**2*(a**2-r**2)*(ν**2-1)*(a**2*(b**2*(ν+1)+r**2*(1-ν))\
            -b**4*(ν+3)+b**2*r**2*(ν+1))/(8*E*r*(a**2*(ν-1)-b**2*(ν+1)))
display(Math(f" \\text{{Agrees with $u_r$ from the stress method? }}\
         {latex((ur_final.subs(dicI_ν)-ur_strssMthd).simplify()==0)}"))
```

Agrees with u_r from the stress method? True

It is observed that under a linearly distributed body force (centrifugal force), the displacement varies with r^3, which is **one order higher** than that in a one-dimensional bar subjected to a linear body force. This higher-order dependence arises because the **mass of a sectoral strip** in the disk increases linearly with r. It is also noted that the **stresses vary one order lower** than the displacement.

The hoop stresses at $r = a$ and $r = b$ can be obtained as follows:

```
σrr_ra = σv_final.subs(r,a).simplify()[0].subs(dicI_ν).factor()\
                                                .collect(a).collect(b)
σrr_rb = σv_final.subs(r,b).simplify()[0].subs(dicI_ν).factor()\
                                                .collect(a).collect(b)
display(Math(f" \\text{{At $r=a$, $σ_{{rr}}$=}}\small {latex(σrr_ra)}"))
display(Math(f" \\text{{At $r=b$, $σ_{{rr}}$=}}\small {latex(σrr_rb)}"))

σθθ_ra = σv_final.subs(r,a).simplify()[1].subs(dicI_ν).factor()\
                                                .collect(a).collect(b)
σθθ_rb = σv_final.subs(r,b).simplify()[1].subs(dicI_ν).factor()\
                                                .collect(a).collect(b)
display(Math(f" \\text{{At $r=a$, $σ_{{θθ}}$=}}\small {latex(σθθ_ra)}"))
display(Math(f" \\text{{At $r=b$, $σ_{{θθ}}$=}}\small {latex(σθθ_rb)}"))
```

At $r = a$, $\sigma_{rr} = -\dfrac{\rho\omega^2 (a-b)(a+b)\left(a^2(\nu-1) + b^2(-\nu-3)\right)}{4\left(a^2(\nu-1) + b^2(-\nu-1)\right)}$

At $r = b$, $\sigma_{rr} = 0$

At $r = a$, $\sigma_{\theta\theta} = -\dfrac{\nu\rho\omega^2 (a-b)(a+b)\left(a^2(\nu-1) + b^2(-\nu-3)\right)}{4\left(a^2(\nu-1) + b^2(-\nu-1)\right)}$

At $r = b$, $\sigma_{\theta\theta} = \dfrac{\rho\omega^2 (a-b)^2 (a+b)^2 (\nu-1)(\nu+1)}{4\left(a^2(\nu-1) + b^2(-\nu-1)\right)}$

It is observed that both the radial and hoop stresses remain positive throughout the disk, indicating that every point is under tensile stress due to the centrifugal force, as expected. Since the outer edge of the disk is free, the boundary condition $\sigma_{rr} = 0$ is satisfied.

4.1.7.4 *Plot of stress distribution*

The following code snippet plots the variation of both radial and hoop stresses along the radial coordinate of the disk:

```
# Convert SymPy expressions to Python functions:
dic_ = {a:1, b:5, ρ:1, ω:1, ν:0.3, pb:0}
func1 = sp.lambdify(r, σv_final.subs(dicI_ν)[0].subs(dic_), 'numpy')
func2 = sp.lambdify(r, σv_final.subs(dicI_ν)[1].subs(dic_), 'numpy')

labels=[r'$σ_{rr},a=1,b=5,ρ=1,ω=1,ν=0.3, p_{b} =0$',
        r'$σ_{θθ},a=1,b=5,ρ=1,ω=1,ν=0.3, p_{b} =0$']

r_data = np.linspace(1, 5, 400)                  #  r values for the plot
x_data = [r_data, r_data]
y_data = [func1(r_data), func2(r_data)]                # functions values

#gr.plot_fs(x_data,y_data,labels,'r','stresses',\
#                                        'imagesSMp/SpinningDiskDisp')
```

The distributions of the radial and hoop stresses in a thin disk of uniform thickness, spinning together with a shaft at a constant angular velocity, are shown in Fig. 4.2. The outer edge surface of the disk is free of traction.

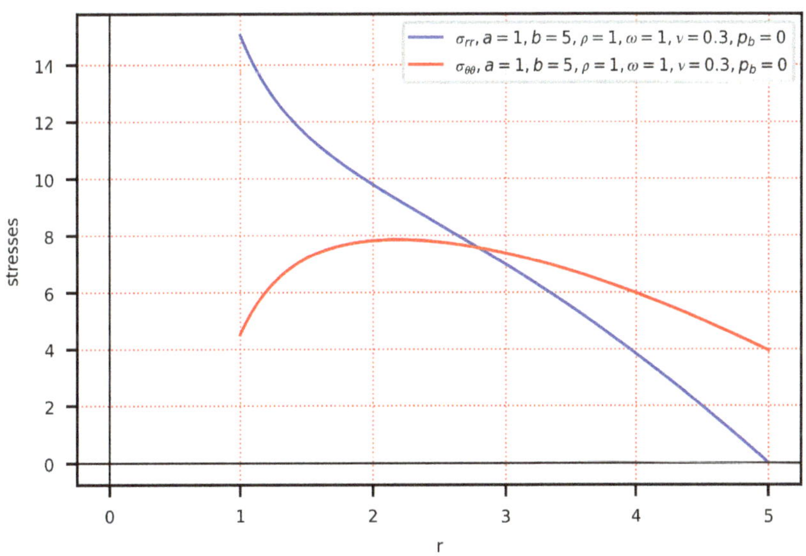

Figure 4.2. Stress distribution in a spinning thin disk of uniform thickness.

The plot confirms that both the radial and hoop stresses remain positive, indicating tension throughout the disk. The maximum radial stress occurs at $r = a$, where the disk exhibits a concave boundary prone to stress concentration. This stress decreases with increasing r, reaching zero at the free edge. The maximum hoop stress appears within the disk, approximately at a distance of $\frac{b-a}{3}$ from $r = a$.

4.1.7.5 *Solution for disk on shaft under edge pressure*

The following code snippet presents the results due to edge pressure, with the body force neglected ($q = 0$):

```
# Final solution for displacements, stresses & strains (zero body force):

sln_cs = {**sln_BCs}
sln_cs = {key: value.subs(sln_cs) for key, value in sln_cs.items()}
print(f" Current len of sln_cs = {len(sln_cs)}; Total cs = {len(cs)}")

ur_final = urg.subs(sln_cs).subs({q:0}).simplify().collect(a)\
                              .collect(b).collect(pb).collect(ν**2-1)

σv_final = sp.simplify(σvg.subs(sln_cs).subs({q:0}))
εv_final = sp.simplify(εvg.subs(sln_cs).subs({q:0}))
σv_final = σv_final.applyfunc(lambda expr: sp.collect(expr,pb)\
                                    .collect(a).collect(b).collect(r))
εv_final = εv_final.applyfunc(lambda expr: sp.collect(expr,pb)\
                               .collect(a).collect(b).collect(ν**2-1))
display(Math(f"\\text{{Displacement solution for $u_r$=}}\
                                    \\small {latex(ur_final)}"))
display(Math(f"\\text{{Stress solution, $σ_v^T$=}}{latex(σv_final.T)}"))
display(Math(f"\\text{{Strain solution, $ε_v^T$=}}{latex(εv_final.T)}"))
```

```
Current len of sln_cs = 2; Total cs = 2
```

$$\text{Displacement solution for } u_r = \frac{b^2 p_b \left(-a^2 + r^2\right) \left(\nu^2 - 1\right)}{Er \left(a^2 \left(1 - \nu\right) + b^2 \left(\nu + 1\right)\right)}$$

$$\text{Stress solution, } \sigma_v^T = \begin{bmatrix} \frac{b^2 p_b \left(a^2(\nu-1)+r^2(-\nu-1)\right)}{r^2(a^2(1-\nu)+b^2(\nu+1))} & \frac{b^2 p_b \left(a^2(1-\nu)+r^2(-\nu-1)\right)}{r^2(a^2(1-\nu)+b^2(\nu+1))} & 0 \end{bmatrix}$$

$$\text{Strain solution, } \varepsilon_v^T = \begin{bmatrix} \frac{b^2 p_b \left(a^2+r^2\right)\left(\nu^2-1\right)}{Er^2(a^2(1-\nu)+b^2(\nu+1))} & \frac{b^2 p_b \left(-a^2+r^2\right)\left(\nu^2-1\right)}{Er^2(a^2(1-\nu)+b^2(\nu+1))} & 0 \end{bmatrix}$$

This is the final solution for a disk fixed at $r = a$ and subjected to a pressure p_b at $r = b$. In this case, the problem is purely axisymmetric, and the shear stress vanishes identically. Readers may visualize the results using the previously provided plotting code.

4.2 Anti-axisymmetric problems

Another class of frequently encountered problems is anti-axisymmetric problems, a representative example being disks spinning with angular acceleration.

4.2.1 *Problem setup*

Consider the same spinning thin disk studied in the previous section. The disk has a uniform thickness, an inner radius a, and an outer radius b. Its angular velocity ω is now time-dependent, with an angular acceleration $\dot{\omega}$. This gives rise to a body force with two components:

$$b_r = \rho r \omega^2, \qquad b_\theta = -\rho r \dot{\omega} \tag{4.11}$$

The body force b_θ acts in the circumferential (θ) direction. It is independent of θ but varies with r. The displacement method will be employed to derive the solution.

Problem Analysis:

For this case, the equilibrium equations form a coupled system of two equations:

$$\begin{aligned} \frac{\partial \sigma_{rr}}{\partial r} + \frac{\sigma_{rr} - \sigma_{\theta\theta}}{r} + b_r = 0 \\ \frac{\partial \sigma_{r\theta}}{\partial r} + \frac{2\sigma_{r\theta}}{r} + b_\theta = 0 \end{aligned} \tag{4.12}$$

Note that these two partial differential equations are decoupled, allowing them to be solved independently. This effectively separates the problem into two subproblems. The first subproblem is identical to the one studied in the previous section, for which we already have a complete solution.

Therefore, our focus will be on the second subproblem, governed by the second differential equation in Eq. (4.12).

The unknown in this problem is the shear stress $\sigma_{r\theta}$, which is related to the shear strain $\varepsilon_{r\theta}$ through the shear modulus. To determine $\sigma_{r\theta}$, we must first compute $\varepsilon_{r\theta}$.

4.2.2 *Strain–displacement relation for mixed axisymmetric problems*

Due to the presence of the shear stress $\sigma_{r\theta}$, this is a mixed axisymmetric problem, as discussed in Section 4.2 and illustrated in Fig. 3.4. The shear stress $\sigma_{r\theta}$ is anti-axisymmetric, and the displacement u_θ is also anti-axisymmetric with respect to θ, yet both are independent of θ.

Accordingly, the strain–displacement relation Eq. (4.4) simplifies to

$$\begin{aligned} \varepsilon_{rr} &= \frac{\partial u_r}{\partial r}; \quad \varepsilon_{\theta\theta} = \frac{u_r}{r} \\ 2\varepsilon_{r\theta} &= \frac{\partial u_\theta}{\partial r} - \frac{u_\theta}{r} \end{aligned} \tag{4.13}$$

The first two equations were used in the previous section for the axisymmetric problem. From the third equation in Eq. (4.13), to find $\varepsilon_{r\theta}$, we must find u_θ.

4.2.3 *General expressions for displacement and stress*

As shown in Fig. 3.4, the displacement component u_θ is also antisymmetric with respect to r for a spinning disk. It increases as we move away from the origin and decreases as we approach the origin. Therefore, u_θ can be assumed in a similar manner to u_r, in the following form using monomial basis functions of r:

$$u_\theta = d_0\frac{1}{r} + d_1 r + d_2 r^3 \tag{4.14}$$

where d_0, d_1, and d_2 are three unknown constants. Based on our study in the previous chapter, we know that d_2 should be determined using the equilibrium equation: the second equation in Eq. (4.12). The code snippet to accomplish this is as follows:

```
# Variables for displacement method for anti-axisymmetric problems:
E, ν, G= symbols('E, ν, G', positive=True)  # given material constants
r, a, b = sp.symbols('r, a, b', positive=True)   # variables for r-axis

q =sp.symbols('q', real=True)                          # for body force br
τb = sp.symbols('τ_b', real=True)         # given shear stress at r = b
ρ, ωt= sp.symbols('ρ, ω_t', positive=True)  # ωt: angular acceleration
bθ = q*r                              # Linear body force in the r-direction

# Define unknown constants for displacement uθ:
d0, d1, d2 = sp.symbols('d0, d1, d2', real=True)
cs_list = [d0, d1, d2]
cs = sp.Matrix(cs_list)
print(f"Total number of unknown constants = {len(cs)}")

uθ = sp.symbols('u_θ', real=True)
σrθ, εrθ = sp.symbols('σ_rθ, ε_rθ',real=True)

# Define the general displacement uθ:
order = 4                    # total order = positive order + negative order
uθg = d0/r + d1*r + d2*r**3                  # equilibrium not yet satisfied
display(Math(f" \\text{{General solution for displacement $u_θ$ = }}\
                                                  {latex(uθg)}"))

εrθg = (uθg.diff(r) - uθg/r)/2          # general strains in displacement

# Compute general stress vector using strain vector:
σrθg = G*(2*εrθg)                                       # G = E/(2*(1+ν))
display(Math(f" \\text{{General solution for shear stress $σ_{{rθ}}$}}\
                                            =\small {latex(σrθg)}"))
```

```
Total number of unknown constants = 3
```

General solution for displacement $u_\theta = \frac{d_0}{r} + d_1 r + d_2 r^3$

General solution for shear stress

$$\sigma_{r\theta} = G\left(-\frac{d_0}{r^2} + d_1 + 3d_2r^2 - \frac{\frac{d_0}{r} + d_1r + d_2r^3}{r}\right)$$

Next, let us enforce the equilibrium equation:

```
# Define equilibrium equation that needs to be satisfied:
equi = (σrθg.diff(r) + 2*σrθg/r + bθ).simplify().factor()
equi = sp.numer(equi).expand()

sln_cs = {}                # initialize solution of unknown constants, cs
sln_equi = constantUpdate([equi], r, order+1, cs)
display(sln_equi)

# Update unknown constants:
sln_cs = {**sln_equi}
sln_cs = {key: value.subs(sln_cs) for key, value in sln_cs.items()}
print(f" Current len of sln_cs={len(sln_cs)}; Total cs={len(cs)}")

uθ = sp.simplify(uθg.subs(sln_cs))                      # update displacement
display(Math(f"\\text{{Current displacement $u_{{θ}}$=}}{latex(uθ)}"))

σrθ = sp.simplify(σrθg.subs(sln_cs))                        # update stresses
display(Math(f"\\text{{Current stress $σ_{{rθ}}$ = }} {latex(σrθ)}"))
```

```
Total number of expressions = 5
Number of nonzero expressions = 1
```

$$\left\{d_2 : -\frac{q}{8G}\right\}$$

```
Current len of sln_cs=1; Total cs=3
```

Current displacement $u_\theta = \dfrac{d_0}{r} + d_1 r - \dfrac{qr^3}{8G}$

Current stress $\sigma_{r\theta} = \dfrac{-8Gd_0 - qr^4}{4r^2}$

It is observed that the constant associated with the highest-order term has been determined using the equilibrium equation.

4.2.4 *General displacement for anti-axisymmetric problems*

We have obtained a fundamental equation for the general displacement u_θ, valid for all anti-axisymmetric problems. It satisfies the equilibrium equation:

$$u_\theta = \frac{d_0}{r} + d_1 r - \frac{qr^3}{16G} \tag{4.15}$$

This expression contains two unknown constants, d_0 and d_1. Hence, any anti-axisymmetric problem requires only two constants in the assumed displacement. The relevant monomial bases are $\frac{1}{r}$, r, and r^3. The r^3 term arises from the linearly distributed body force b_θ. If the body force is zero ($q = 0$), only the monomial bases $\frac{1}{r}$ and r are needed.

In future applications, when assuming displacement for anti-axisymmetric problems, Eq. (4.15) should be used directly instead of Eq. (4.14). In such cases, it is unnecessary to check the equilibrium equation explicitly, provided the body force is linear in r.

The constants d_0 and d_1 are determined from the boundary conditions of the specific problem. Since any anti-axisymmetric problem provides two boundary conditions, the solution can always be uniquely determined. For our current problem, the Python implementation is as follows:

```
BCa = uθg.subs(r,a)              # displacement boundary condition at r=a
BCa = sp.numer(BCa).expand()
BCb = σrθg.subs(r,b) + τb
BCb = sp.numer(BCb).expand()

sln_BCs = constantUpdate([BCa, BCb], r, order, cs)
sln_cs = {**sln_equi, **sln_BCs}                      # update constants
sln_cs = {key: value.subs(sln_cs) for key, value in sln_cs.items()}
print(f" Current len of sln_cs={len(sln_cs)}; Total cs={len(cs)}")
display(Math(f" \\text{{Constants found: }} {latex(sln_cs)}"))
```

```
Total number of expressions = 8
Number of nonzero expressions = 2
Current len of sln_cs=3; Total cs=3
```

Constants found:

$$\left\{d_0: -\frac{b^4q}{8G}+\frac{b^2\tau_b}{2G},\ d_1: -\frac{b^2\tau_b}{2Ga^2}-\frac{q\left(-a^4-b^4\right)}{8Ga^2},\ d_2: -\frac{q}{8G}\right\}$$

4.2.5 *Final solutions for displacement and stress*

The final solutions for the displacement and stress are obtained by substituting the constants back into the general expressions:

```
# Final solution for displacements, stresses & strains (τ_b=0):

dic_={τb:0, q:-ρ*ωt}
uθ_final = uθg.subs(sln_cs).subs(dic_).simplify()\
                        .collect(q).collect(a).collect(b).collect(r)
σrθ_final = sp.simplify(σrθg.subs(sln_cs).subs(dic_))
εrθ_final = sp.simplify(εrθg.subs(sln_cs).subs(dic_))

display(Math(f"\\text{{Final displacement $u_θ$ =}}{latex(uθ_final)}"))
display(Math(f"\\text{{Final stress $σ_{{rθ}}$ =}}{latex(σrθ_final)}"))
display(Math(f"\\text{{Final strains $ε_{{rθ}}$=}}{latex(εrθ_final)}"))
```

Final displacement $u_\theta = \dfrac{\rho\omega_t\left(a^2\left(b^4+r^4\right)+r^2\left(-a^4-b^4\right)\right)}{8Ga^2r}$

Final stress $\sigma_{r\theta} = \dfrac{\rho\omega_t\left(-b^4+r^4\right)}{4r^2}$

Final strains $\varepsilon_{r\theta} = \dfrac{\rho\omega_t\left(-b^4+r^4\right)}{8Gr^2}$

This is the final solution for a hollow disk spinning at an accelerated angular velocity $\dot{\omega}$. It is the same as was obtained in the previous chapter using the stress method (Section 3.2.7). Let us confirm this using the following snippet:

```
G_ = E/(2*(1+ν))
uθ_strssMthd=ρ*ωt*(r**2-a**2)*(ν+1)*(a*r-b**2)*(a*r+b**2)/(4*E*a**2*r)
display(Math(f" \\text{{Agrees with $u_θ$ from the stress method? }}\
        {latex((uθ_final.subs(G, G_)-uθ_strssMthd).simplify()==0)}"))
```

Agrees with u_θ from the stress method? True

It is observed that under a linearly distributed body force (due to angular acceleration), the displacement increases proportional to r^3, which is one order higher than in a 1D bar under a linear body force. This behavior is analogous to the centrifugal force problem. The reason is that the mass of a sector-strip in the disk increases linearly with r. Consequently, the stresses are always one order lower than the corresponding displacement.

4.2.6 *Plot of displacement and stress distributions*

The following code snippet plots the distribution of the displacement in the θ-direction and the corresponding shear stress $\sigma_{r\theta}$ along the radial axis of the disk:

```
# Convert SymPy expressions to Numpy functions:
dic_ = {a:1, b:5, ρ:1, ωt:1, G:1, τb:0}
func1 = sp.lambdify(r, σrθ_final.subs(dic_), 'numpy')
func2 = sp.lambdify(r,  uθ_final.subs(dic_), 'numpy')

labels=[r'$σ_{rθ},a=1,b=5,ρ=1,ω_{t}:1, G:1, τb:0 =0$',
        r'$u_{θ}, a=1,b=5,ρ=1,ω_{t}:1, G:1, τb:0 =0$']

r_data = np.linspace(1, 5, 400)                   #  r values for the plot
x_data = [r_data, r_data]
y_data = [func1(r_data), func2(r_data)]                  # function values

#gr.plot_fs(x_data,y_data,labels,'r','displacement, stress',\
#                                              'imagesSMp/Disk_utsrt')
```

The distributions of the displacement in the θ-direction and shear stress in a thin disk of uniform thickness, spinning with a shaft under a constant angular acceleration, are shown in Fig. 4.3. The outer edge of the disk is free of stress.

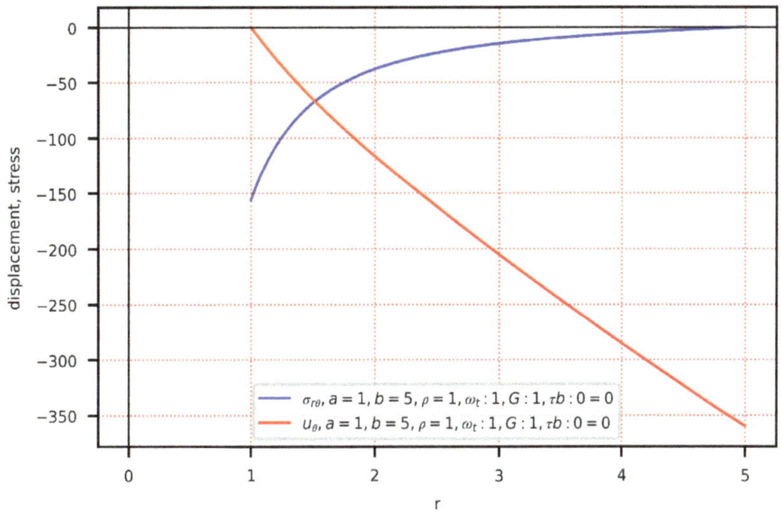

Figure 4.3. Displacement and shear stress distributions in a thin disk of uniform thickness spinning with a shaft under constant angular acceleration.

Both the displacement in the θ-direction and the shear stress are negative, reflecting tension or compression due to the inertial body forces induced by angular acceleration. The maximum shear stress occurs at the inner radius $r = a$, where the disk has a concave boundary and is prone to stress concentration. It decreases quadratically with r, reaching zero at the outer edge. The maximum displacement occurs at the disk edge.

4.2.7 *Solution for disk on shaft under edge shear*

The following snippet presents the results for a disk subjected to shear traction at its edge, with the body force set to zero:

```
# Final solution for displacements, stresses & strains (zero body force):

sln_cs = {**sln_equi, **sln_BCs}
uθ_final = uθg.subs(sln_cs).subs({q:0}).simplify().collect(a)\
                                .collect(b).collect(pb).collect(ν**2-1)

σrθ_final = sp.simplify(σrθg.subs(sln_cs).subs({q:0}))
εrθ_final = sp.simplify(εrθg.subs(sln_cs).subs({q:0}))
#σv_final = σv_final.applyfunc(lambda expr: sp.collect(expr,pb)\
#                                   .collect(a).collect(b).collect(r))

display(Math(f"\\text{{Displacement solution for $u_θ$=}}\
                                            \\small {latex(uθ_final)}"))
display(Math(f"\\text{{Stress solution, $σ_{{rθ}}$=}}{latex(σrθ_final)}))
display(Math(f"\\text{{Strain solution, $ε_{{rθ}}$=}}{latex(εrθ_final)}"))
```

Displacement solution for $u_\theta = \dfrac{b^2\tau_b\left(a^2 - r^2\right)}{2Ga^2r}$

Stress solution, $\sigma_{r\theta} = -\dfrac{b^2\tau_b}{r^2}$

Strain solution, $\varepsilon_{r\theta} = -\dfrac{b^2\tau_b}{2Gr^2}$

The maximum shear stress occurs at $r = a$, at the inner edge of the disk. The applied shear traction at the disk edge is amplified by a factor of $\frac{b^2}{a^2}$. Consequently, for a disk mounted on a small shaft, the shear stress on the shaft can become very large.

4.3 Half-space subject to a point downward force

In this section, we introduce a displacement method for solving axisymmetric problems in cylindrical coordinates. We illustrate this method using a half-space under a vertical point force.

Consider a three-dimensional (3D) half-space made of uniform, isotropic, linear elastic material. It is subjected to a point vertical force applied on its surface, as shown in Fig. 4.4. Our goal is to obtain analytic formulas for the displacements and stresses at any point in the 3D space.

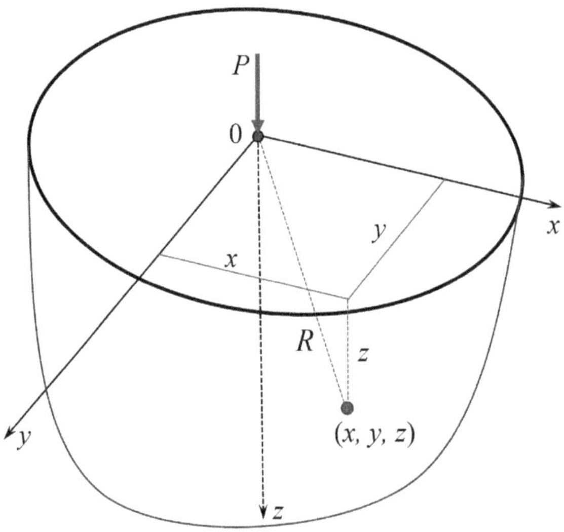

Figure 4.4. Three-dimensional half-space subjected to a vertical point force on its surface in the z-direction.

This problem is axisymmetric about the z-axis, meaning that any rotation about the z-axis (with corresponding rotation of the x- and y-axes) does not change the solution. Therefore, cylindrical coordinates can be used to simplify the analysis. The problem conditions are as follows:

1. Stress boundary conditions on the top surface:

$$\sigma_{zr} = \sigma_{zz} = 0 \quad \text{at } z = 0 \text{ and } r \neq 0 \tag{4.16}$$

2. Integral equilibrium conditions in the domain:

   Since the external force is a point force, the stress solution is singular at the application point. Moreover, as the force acts on the top boundary, the differential equilibrium equations alone cannot capture its effect. Thus, integral equilibrium conditions must be applied on a plane away from the singularity.

   For convenience, we choose a plane parallel to the surface at $z > 0$. The integration of the stress component σ_{zz} over this plane must satisfy the equilibrium conditions corresponding to the applied vertical force P:

$$\sum F_z = 0 : \int_{-\infty}^{\infty}\int_{-\infty}^{\infty} \sigma_{zz} dx dy + P = 0 \tag{4.17}$$

4.3.1 *Displacement functions in cylindrical coordinate system*

Similar to the stress function used in the stress method, this displacement method uses the so-called Love displacement function $\xi(r, z)$. The displacement components can be written in terms of $\xi(r, z)$ as [4]

$$\begin{aligned} u_r &= -\frac{1}{2G}\frac{\partial^2 \xi}{\partial r \partial z} \\ u_z &= \frac{1}{2G}\left[2(1-\nu)\nabla^2 - \frac{\partial^2}{\partial z^2}\right]\xi \end{aligned} \tag{4.18}$$

The equilibrium equation can be written using Eq. (4.19) as (zero body force)

$$\begin{aligned} (G+\lambda)\frac{\partial e}{\partial r} + G\left(\nabla^2 u_r - \frac{u_r}{r^2}\right) = 0 \\ (G+\lambda)\frac{\partial e}{\partial z} + G\nabla^2 u_z = 0 \end{aligned} \tag{4.19}$$

where ∇^2 is the Laplace operator in the cylindrical coordinates for axisymmetric problems:

$$\nabla^2 = \frac{\partial^2}{\partial r^2} + \frac{\partial}{r\partial r} + \frac{\partial^2}{\partial z^2} \tag{4.20}$$

Substituting Eq. (4.18) into Eq. (4.19), we have

$$\nabla^4 \xi = 0 \tag{4.21}$$

This means that if we can find such a displacement function satisfying the biharmonic equation (4.21), we can then have solutions for the displacement using Eq. (4.18). The stresses can, in turn, be found using the stress–displacement relation in Eq. (2.51).

The following snippet confirms that the displacements given in Eq. (4.18) satisfy the equilibrium Eq. (4.19):

```
G, ν= sp.symbols("G, ν", nonnegative=True)
r, z = sp.symbols("r, z")
ur, uz = sp.symbols("u_r, u_z")

ξ = Function("ξ")(r, z)                              # displacement function
nab2ξ = ξ.diff(r,2)+ξ.diff(r,1)/r+ξ.diff(z,2)                          #∇^2ξ
nab4ξ = nab2ξ.diff(r,2)+nab2ξ.diff(r,1)/r+nab2ξ.diff(z,2)              #∇^4ξ

ur=-1/(2*G)*(ξ.diff(r,1)).diff(z,1)                            # write ur in ξ
uz= 1/(2*G)*(2*(1-ν)*nab2ξ-ξ.diff(z,2))                        # write uz in ξ

nab_ur = ur.diff(r,2)+ur.diff(r,1)/r+ur.diff(z,2)                     #∇^2ur
e =ur.diff(r,1)+ur/r+uz.diff(z,1)
eqn1 = 1/(1-2*ν)*e.diff(r,1)+nab_ur-ur/r**2  # the first Equilibrium Eq.
display(Math(f" \\text{{Is the first equilibrium Eq.= 0? }}\
                                          {latex(eqn1.simplify()==0)}"))
```

Is the first equilibrium Eq. = 0? True

This means that the first equation in Eq. (4.19) is satisfied regardless of the form of the displacement-function $\xi(r, z)$. Let us substitute $\xi(r, z)$ into the second equation in Eq. (4.19):

```
nab_uz = uz.diff(r,2)+uz.diff(r,1)/r+uz.diff(z,2)                    #∇^2 uz
eqn2 = (1/(1-2*ν)*e.diff(z,1) +nab_uz).simplify() # second Equilibrium Eq.
eqn2 = (eqn2 - nab4ξ*(1-ν)/G).simplify()     # subtraction  ∇^4ξ*(1-ν)/G
display(Math(f" \\text{{Is second equilibrium Eq.= 0? }}{latex(eqn2==0)}"))
```

Is second equilibrium Eq. = 0? True

This implies that the second equation in Eq. (4.19) reduces to $\frac{1-\nu}{G}\nabla^4\xi$. Therefore, if $\nabla^4\xi = 0$, the second equation in Eq. (4.19) is automatically satisfied.

Next, we attempt to find a Love displacement function that satisfies the biharmonic equation (4.21).

4.3.2 *Love displacement functions*

For the problem of a half-space under a vertical point force, the displacement function can be written as [4]

$$\xi(r, z) = c_0 R + c_1 (R - z\ln(R + z)) \tag{4.22}$$

where $R = \sqrt{r^2 + z^2}$. These terms are selected based on Eqs. (5.44) and (5.45) in Ref. [1]. Substituting Eq. (4.22) into Eq. (4.21) confirms that the biharmonic equation is satisfied.

Note that since these functions involve square roots, symbolic computations become significantly more complex, and the resulting expressions are much longer than those for polynomial functions. This can be implemented in Python or symbolic computation software using the following snippet.

4.3.3 *Use of displacement functions*

```
G, ν, z = sp.symbols("G, ν, z", nonnegative=True)     # define variables
r0, r, R, P = sp.symbols("r_0, r, R, P", nonnegative=True)
c0, c1 = sp.symbols("c0, c1")

ur, uz = sp.symbols("u_r, u_z")
σrr, σθθ, σzz, σrz = sp.symbols("σ_rr, σ_θθ, σ_zz, σ_rz")
R = sp.sqrt(r**2+z**2)

ξ = c0*R + c1*(R-z*sp.log(R+z))                        # Displacement function
nab2ξ = (ξ.diff(r,2)+ξ.diff(r,1)/r+ξ.diff(z,2))
nab2ξ_= nab2ξ.simplify()
nab4ξ = nab2ξ_.diff(r,2)+nab2ξ_.diff(r,1)/r+nab2ξ_.diff(z,2)
display(Math(f" \\text{{Is $∇^4ξ=0$? }}{latex(nab4ξ.simplify()==0)}"))
```

Is $\nabla^4\xi = 0$? True

It is observed that the biharmonic equation is satisfied by the displacement function given in Eq. (4.22) regardless of the values of c_0 and c_1. We thus use $\xi(r, z)$ to compute the displacement components using Eq. (4.18) and then the stresses using Eq. (2.51):

```
ur =-1/(2*G)*(ξ.diff(r,1)).diff(z,1)              # compute displacements
uz = 1/(2*G)*(2*(1-ν)*nab2ξ-ξ.diff(z,2))
e  = (ur.diff(r,1)+ur/r+uz.diff(z,1))

σzz = (2*G*(ν/(1-2*ν)*e+uz.diff(z,1)))                # compute stresses
σrz = (G*(ur.diff(z,1)+uz.diff(r,1))).expand()
display(Math(f" \\text{{Sample terms in general stress $σ_{{rz}}$:}}"))
display(Math(f"\small {latex(σrz.args[0])}+...+{latex(σrz.args[10])}+.."))
```

Sample terms in general stress σ_{rz}:

$$
-\frac{4c_0 r}{r^2\sqrt{r^2+z^2}+z^2\sqrt{r^2+z^2}}+\cdots
$$

$$
+\frac{4c_1 rz}{r^4+2r^2z^2+r^2z\sqrt{r^2+z^2}+z^4+z^3\sqrt{r^2+z^2}}+\cdots
$$

The expressions for these stresses are very lengthy (about eight pages), and hence only two sample terms are displayed. The point is that these stresses contain unknown constants c_0 and c_1.

Next, let us use the stress boundary condition Eq. (4.16) on the surface of the half-space to solve for these constants:

```
# Use stress boundary condition. Zero shear stress on surface:
sln = sp.solve([σrz.subs(z,0)], [c0, c1])  # solve for unknown constants
sln                                         # should be: {c0: c1/(2*ν - 1)}
```

$$
\left\{c_0\colon \frac{c_1}{2\nu-1}\right\}
$$

We found the relationship between these two constants. To fully determine them, we will now use the integral equilibrium condition Eq. (4.17). This requires computing an infinite integral of a complicated integrand, which may take about 20 minutes to run on the author's laptop. The reason for this lengthy computation is that it is done symbolically to produce an exact formula. Readers may uncomment the code lines in the following cell to carry out the integration to compute `szz_int`:

```
## Use the integral equilibrium condition. First, evaluate the integral:
#σzz_c1 = σzz.subs({c0:sln[c0], z:1}).factor(c1)
#szz_int=(integrate(2*sp.pi*σzz_c1*r,(r,r0,oo)).subs(r0,0)).simplify()
#szz_int     # It takes ~20m to run, result should be: -2*pi*c1/(2*ν-1)
```

```
# Use the integral equilibrium condition solve for the constants:
szz_int = -2*pi*c1/(2*ν-1)                  # redefine the integral value
sln_c1 = sp.solve(szz_int+P, c1)[0]
sln_c0 = sln_c1/(2*ν-1)
cs = {c0:sln_c0,c1:sln_c1}
cs                  # should be: {c0:P/(2*sp.pi), c1:P*(2*ν-1)/(2*sp.pi)}
```

$$\left\{c_0: \frac{P}{2\pi},\ c_1: \frac{P\left(2\nu-1\right)}{2\pi}\right\}$$

We have found both constants, which can be substituted into the displacement expressions to obtain the final solution for the displacements:

```
ur_final_ = ur.subs(cs).simplify()
ur_final_                # P*r*((2ν-1)*R**2 +z*(z+R))/(4*pi*G*R**3*(R+z))
```

$$\frac{Pr\left(2r^2\nu - r^2 + 2z^2\nu + z\sqrt{r^2+z^2}\right)}{4\pi G\left(r^4 + 2r^2z^2 + r^2z\sqrt{r^2+z^2} + z^4 + z^3\sqrt{r^2+z^2}\right)}$$

This result can be manually simplified and rewritten as follows:

```
ur_final = P*r*((2*ν-1))/(4*sp.pi*G*R*(R+z)) +P*r*z/(4*sp.pi*G*R**3)
#print((ur_final_-ur_final).simplify())      # to check simplied formula
display(Math(f" \\text{{ Final displacement $u_{{r}}$}}-\
                                                \small {latex(ur_final)}"))
```

$$\text{Final displacement } u_r = \frac{Prz}{4\pi G\left(r^2+z^2\right)^{\frac{3}{2}}} + \frac{Pr\left(2\nu-1\right)}{4\pi G\sqrt{r^2+z^2}\left(z+\sqrt{r^2+z^2}\right)}$$

This is the exact solution for u_x in Cartesian coordinates, given in Eq. (5.52) in Ref. [1] related to P_z. The displacement u_z is found using:

```
uz_final_ = uz.subs(cs).simplify()
uz_final_.factor().collect(r**2).collect(z**2)
```

$$-\frac{P\left(r^2\left(2\nu-2\right)+z^2\left(2\nu-3\right)\right)\left(r^2+2z^2+2z\sqrt{r^2+z^2}\right)}{4\pi G\left(r^2+z^2\right)\left(r^2\left(2z+\sqrt{r^2+z^2}\right)+2z^3+2z^2\sqrt{r^2+z^2}\right)}$$

This result can be manually simplified and rewritten as follows:

```
uz_final = P*2*(1-ν)/(4*sp.pi*G*R) + P*z**2/(4*sp.pi*G*R**3)
#print((uz_final_-uz_final).simplify())    # to check simplified formula
uz_final
```

$$\frac{Pz^2}{4\pi G\left(r^2+z^2\right)^{\frac{3}{2}}}+\frac{P\left(1-\nu\right)}{2\pi G\sqrt{r^2+z^2}}$$

This is the exact solution term for u_z in Eq. (5.52) in Ref. [1] related to P_z.

With the simplified form of displacement formulas, we can now compute all these stresses in their final forms. We will not present the detailed process but provide the code below and list the final simplified formulas:

```
e = (ur_final.diff(r,1)+(ur_final/r).simplify()+uz_final.diff(z,1))

σrr_final = (2*G*(ν/(1-2*ν)*e+ur_final.diff(r,1))).doit()
σθθ_final = (2*G*(ν/(1-2*ν)*e+(ur_final/r).simplify())).doit()
σzz_final = (2*G*(ν/(1-2*ν)*e+uz_final.diff(z,1))).doit()
σrz_final = (G*(ur_final.diff(z,1)+uz_final.diff(r,1))).doit()
#σrr_final
```

The full set of solutions is given as follows [4]:

$$\begin{aligned}
u_r &= \frac{P}{4\pi G}\left\{\frac{rz}{R^3}-\frac{(1-2\nu)r}{R(R+z)}\right\}\\
u_z &= \frac{P}{4\pi G}\left\{\frac{2(1-\nu)}{R}+\frac{z^2}{R^3}\right\}\\
\sigma_{rr} &= \frac{P}{2\pi}\left\{\frac{1-2\nu}{R(R+z)}-\frac{3r^2z}{R^5}\right\}\\
\sigma_{\theta\theta} &= \frac{(1-2\nu)P}{2\pi}\left\{\frac{z}{R^3}-\frac{1}{R(R+z)}\right\}\\
\sigma_{zz} &= -\frac{3Pz^3}{2\pi R^5};\quad \sigma_{rz} = -\frac{3Prz^2}{2\pi R^5}
\end{aligned}\tag{4.23}$$

where $R=\sqrt{x^2+y^2+z^2}$ and $r=\sqrt{x^2+y^2}$. One may use SymPy code to verify the correctness of the handwritten formulas against the computed results.

Similar to the 3D cases studied in Chapter 5 of Ref. [1], the raw outputs of the stress expressions can be complicated and lengthy due to the complexity of the displacement functions. Therefore, careful and laborious simplification — using both SymPy and manual techniques — is required to obtain concise formulas, such as those given in Eq. (4.23). Alternatively, one may rely on numerical computations, as demonstrated in Chapter 5 of Ref. [1].

4.3.4 *Useful observations*

It can be seen from Eq. (4.23) that:

1. The solutions are singular at $R = 0$, indicating that the values approach infinity as $R \to 0$. This implies that critical locations, where maximum values are likely to occur, are around $r = 0$.
2. The solutions on the surface of the half-space at $z = 0$ become

$$
\begin{aligned}
u_r &= -\frac{(1-2\nu)P}{4\pi G r} \quad \text{move towards } P(>0) \\
u_z &= \frac{(1-\nu)P}{2\pi G r} \quad \text{move with } P(>0)(\text{subsidence}) \\
\sigma_{rr} &= \frac{(1-2\nu)P}{2\pi r^2} \quad \text{tension for } P>0 \\
\sigma_{\theta\theta} &= -\frac{(1-2\nu)P}{2\pi r^2} \quad \text{compression for } P>0 \\
\sigma_{zz} &= \sigma_{rz} = 0 \quad \text{enforced BCs on the surface}
\end{aligned}
\tag{4.24}
$$

The displacements vary as $1/r$, while the stresses vary as $1/r^2$. The stresses are therefore one order more singular and also decay one order faster, since they are obtained by differentiating the displacements.

4.3.5 *Plots of displacements varying with depth*

It is of interest to examine the distribution of displacements as they change with depth. The following code performs this analysis, with parameters set as $P_z = 1$, $G = 1$, and $\nu = 0.3$. Since both u_z and u_r are singular at $z = 0$, this point is avoided when plotting. Additionally, u_r is zero at $r = 0$, a small value such as $r = 0.1$ is used:

```
### plot displacements along the vertical axis z:
dic_ud001 = {P:1, G:1, ν:0.3, r:0.001}
dic_ud01  = {P:1, G:1, ν:0.3, r:0.01}
dic_ud1    = {P:1, G:1, ν:0.3, r:0.1}
dic_uz    = {P:1, G:1, ν:0.3, r:0.0 }
funcs = [sp.lambdify(z, ur_final.subs(dic_ud001), 'numpy'),
         sp.lambdify(z, ur_final.subs(dic_ud01), 'numpy'),
         sp.lambdify(z, ur_final.subs(dic_ud1), 'numpy'),
         sp.lambdify(z, uz_final.subs(dic_uz), 'numpy')]

labels=[r'$u_{r}$; '+f'${dic_ud001}$',r'$u_{r}$; '+f'${dic_ud01}$',
        r'$u_{r}$; '+f'${dic_ud1}$'  ,r'$u_{z}$; '+f'${dic_uz}$']

z_data_ur  = np.linspace(.0001, .1, 800)          #  r values for ur
z_data_uz  = np.linspace(.008,  .1, 800)  # r values for uz (singular)
x_data = [z_data_ur, z_data_ur, z_data_ur, z_data_uz]
y_data = [funcs[i](x_data[i]) for i in range(len(x_data))]     # funcs

#gr.plot_fs(x_data,y_data,labels,'z','Displacements',\
#                                          'imagesSMp/HS_uruz_Pz')
```

The distribution of displacements along the depth (z-direction) is shown in Fig. 4.5.

We observe that u_z decays rapidly, following a rational function $1/z$. In contrast, u_r varies significantly near the surface, although its magnitude is much smaller than that of u_z. For $P > 0$, particles initially move toward the

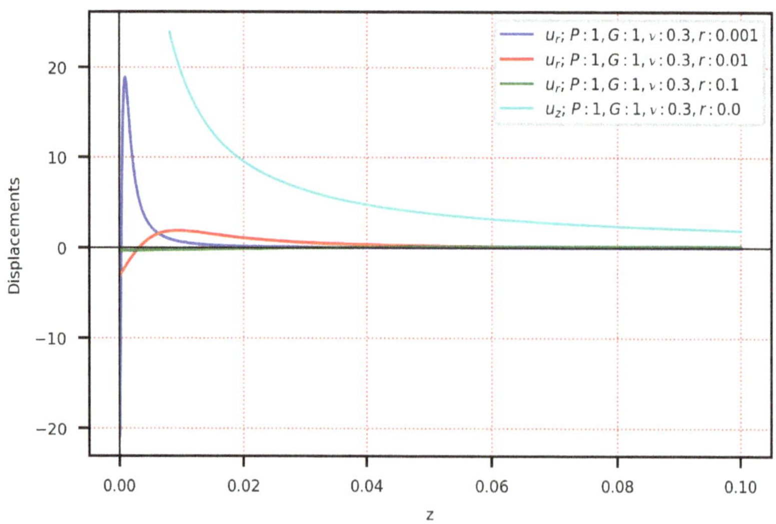

Figure 4.5. Distribution of displacements along the z-direction in the half-space subjected to a unit vertical point force at $r = 0, z = 0$ on the top surface.

point force and quickly reverse direction as the depth increases. Additionally, u_r decays rapidly with increasing r.

4.3.6 *Plots of stresses varying with depth*

It is also of interest to examine how the stresses vary with depth. The following code computes this distribution using the same parameter settings. Due to the singularity at the origin, we use a small value for r, such as $r = 0.01$:

```
# plot stresses along the vertical axis z:
dic = {G:1, ν:0.3, r:0.01, P:1}
funcs = [sp.lambdify(z, σrr_final.subs(dic), 'numpy'),
         sp.lambdify(z, σθθ_final.subs(dic), 'numpy'),
         sp.lambdify(z, σzz_final.subs(dic), 'numpy')]

labels=[r'$σ_{rr}$; '+f'${dic}$',r'$σ_{θθ}$; '+f'${dic}$',
        r'$σ_{zz}$; '+f'${dic}$']

z_data = np.linspace(0.0001, 0.1, 400)          # z values for the plot
x_data = [z_data, z_data, z_data]
y_data = [funcs[i](x_data[i]) for i in range(len(x_data))]   # stresses

#gr.plot_fs(x_data, y_data, labels,'z','stresses','imagesSMp/HS_Sij_Pz')
```

The distribution of the stress components along the depth (z-direction) is shown in Fig. 4.6.

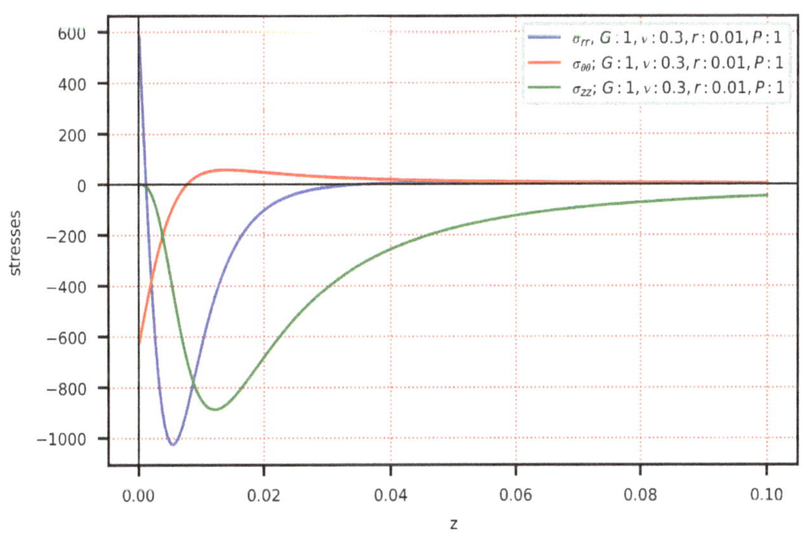

Figure 4.6. Distribution of the stress components along the z-direction in the half-space subjected to a unit vertical point force at $r = 0, z = 0$ on the top surface.

We observe that these stresses decay rapidly, following a rational function $1/z^2$. In contrast, σ_{rr} and $\sigma_{\theta\theta}$ vary significantly near the surface, including sign changes. The amplitude of σ_{rr} can exceed that of σ_{zz}.

```
# plot stresses along the vertical axis z:
dic_rd001 = {G:1, ν:0.3, r:0.005, P:1}
dic_rd01  = {G:1, ν:0.3, r:0.01,  P:1}
dic_rd1   = {G:1, ν:0.3, r:0.1,   P:1}
dic = {G:1, ν:0.3, r:0.01, P:1}
funcs = [sp.lambdify(z, σrr_final.subs(dic_rd001), 'numpy'),
         sp.lambdify(z, σrr_final.subs(dic_rd01), 'numpy'),
         sp.lambdify(z, σrr_final.subs(dic_rd1), 'numpy'),
         sp.lambdify(z, σzz_final.subs(dic), 'numpy')]

labels=[r'$σ_{rr}$; '+f'${dic_rd001}$',r'$σ_{rr}$; '+f'${dic_rd01}$',
        r'$σ_{rr}$; '+f'${dic_rd1}$', r'$σ_{zz}$; '+f'${dic}$']

z_data = np.linspace(0.0001, 0.1, 400)          # z values for the plot
x_data = [z_data, z_data, z_data, z_data]
y_data = [funcs[i](x_data[i]) for i in range(len(x_data))]

#gr.plot_fs(x_data, y_data, labels,'z','stresses','imagesSMp/HS_Srr_Pz')
```

Distribution of stress σ_{rr} with different r values along the depth z-direction is shown in Fig. 4.7, along with σ_{zz} as a reference.

We found that σ_{rr} and $\sigma_{\theta\theta}$ vary drastically, with sign changes near the surface. The amplitude of σ_{rr} can be much larger than that of σ_{zz}.

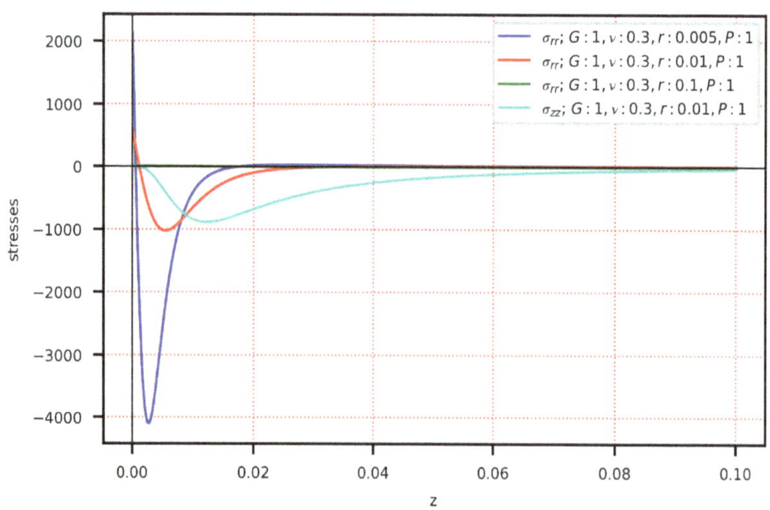

Figure 4.7. Distribution of stress σ_{rr} with different r values along the z-direction in the half-space subjected to a unit vertical point force at $r = 0, z = 0$ on the top surface.

```
# plot stresses along the vertical axis z:
dic_rd001 = {G:1, ν:0.3, r:0.005, P:1}
dic_rd01  = {G:1, ν:0.3, r:0.01,  P:1}
dic_rd1   = {G:1, ν:0.3, r:0.1,   P:1}
dic = {G:1, ν:0.3, r:0.01, P:1}
funcs = [sp.lambdify(z, σθθ_final.subs(dic_rd001), 'numpy'),
         sp.lambdify(z, σθθ_final.subs(dic_rd01), 'numpy'),
         sp.lambdify(z, σθθ_final.subs(dic_rd1), 'numpy'),
         sp.lambdify(z, σzz_final.subs(dic), 'numpy')]

labels=[r'$σ_{θθ}$; '+f'${dic_rd001}$',r'$σ_{θθ}$; '+f'${dic_rd01}$',
        r'$σ_{θθ}$; '+f'${dic_rd1}$', r'$σ_{zz}$; '+f'${dic}$']

z_data = np.linspace(0.0001, 0.1, 400)          # z values for the plot
x_data = [z_data, z_data, z_data, z_data]
y_data = [funcs[i](x_data[i]) for i in range(len(x_data))]   # stresses

#gr.plot_fs(x_data, y_data, labels,'z','stresses','imagesSMp/HS_Stt_Pz')
```

The distribution of stress $\sigma_{\theta\theta}$ for different values of r along the depth (z-direction) is shown in Fig. 4.8, with σ_{zz} included as a reference.

We observe that $\sigma_{\theta\theta}$ varies significantly near the surface, including sign changes. Its amplitude can also be much larger than that of σ_{zz}.

In conclusion, the stress components near the point of load application exhibit pronounced variation. They display a second-order singularity, meaning the stress values increase at a rate of $1/r^2$ as $r \to 0$.

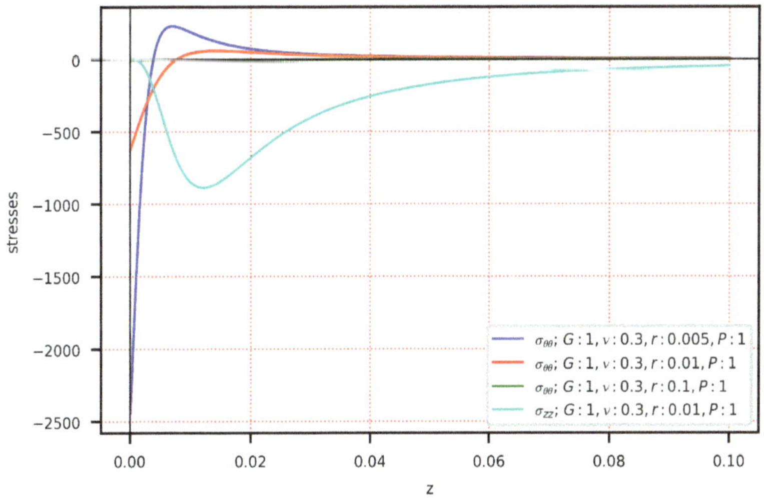

Figure 4.8. Distribution of stress $\sigma_{\theta\theta}$ for different r values along the z-direction in the half-space subjected to a unit vertical point force at $r = 0, z = 0$ on the top surface.

4.4 Half-space with circularly distributed force

In most real-life problems, the applied load is rarely a true point force; it is usually distributed over a finite area. Therefore, solutions for distributed loadings are often more practically useful.

Consider the same problem shown in Fig. 4.4, but now the point force P is replaced by a circularly distributed force p centered at $(R, z) = (0, 0)$, with a finite radius a. We aim to find analytic formulas for the surface subsidence (displacement u_z) and the stresses along the z-direction.

Since the solution for a point force P is already known, it can serve as the fundamental solution. The solution for a distributed force is then obtained by integration. Figure 4.9 illustrates the coordinate system used for integrating to compute the subsidence at point M on the surface, located outside the circular load.

4.4.1 *Result outside the loading circle*

Consider an infinitesimal area element $dA = sd\beta ds$. The corresponding differential point force is $dP = psd\beta ds$. Using the second equation in Eq. (4.24), the differential subsidence caused by dP is computed as

$$du_{z0} = du_z\big|_{z=0} = \frac{(1-\nu)dP}{2\pi Gs} = \frac{(1-\nu)psd\beta ds}{2\pi Gs} \tag{4.25}$$

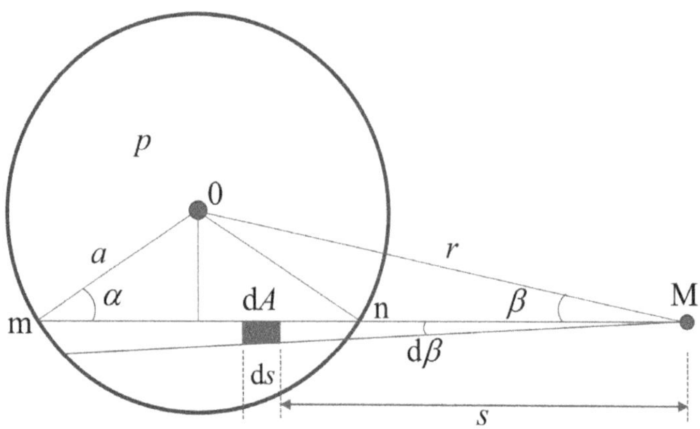

Figure 4.9. Top view of a circularly distributed force p on a half-space, showing the coordinate setup for elliptic integration to compute the subsidence at point M outside the loading circle on the surface.

The total subsidence at point M is then obtained by integrating over the circular area A:

$$u_{z0} = \int_A du_{z0} = \int_A \frac{(1-\nu)p}{2\pi G} d\beta ds = \frac{(1-\nu)}{2\pi G} \int_A p d\beta ds \tag{4.26}$$

Since the length $m - n$ is $2\sqrt{a^2 - r^2 \sin\beta}$, which corresponds to the integral over s, Eq. (4.26) becomes

$$u_{z0} = 2\frac{(1-\nu)p}{2\pi G} \int_0^{\beta_t} 2\sqrt{a^2 - r^2 \sin\beta}\, d\beta \tag{4.27}$$

where the factor of 2 in front of the integral is due to the symmetry of the integral domain and β_t is the maximum angle at which $M - m$ is tangent to the circle. Since

$$a \sin\alpha = r \sin\beta \quad \text{and} \quad (r\cos\beta)^2 + (a \sin\alpha)^2 = r^2 \tag{4.28}$$

we have

$$d\beta = \frac{a\cos\alpha d\alpha}{r\cos\beta} = \frac{a\cos\alpha d\alpha}{r\sqrt{1 - \frac{a^2}{r^2}\sin^2\alpha}} \tag{4.29}$$

Using Eq. (4.29), the integral over β can be converted to one over α, and the limits become from 0 to $\pi/2$. Equation (4.27) becomes

$$\begin{aligned} u_{z0} - 2\frac{(1-\nu)p}{\pi G} &\int_0^{\pi/2} \frac{a^2\cos^2\alpha}{r\sqrt{1 - \frac{a^2}{r^2}\sin^2\alpha}} d\alpha \\ = 2\frac{(1-\nu)pr}{\pi G} &\left\{ \underbrace{\int_0^{\pi/2} \sqrt{1 - m\sin^2\alpha}\, d\alpha}_{E(m)} - (1-m) \underbrace{\int_0^{\pi/2} \frac{d\alpha}{\sqrt{1 - m\sin^2\alpha}}}_{K(m)} \right\} \end{aligned} \tag{4.30}$$

where $m = \frac{a^2}{r^2}$, $K(m)$ is the complete elliptic integral of the first kind and $E(m)$ is the complete elliptic integral of the second kind.

Equation (4.30) provides the formula for computing the subsidence of a half-space under a circularly uniformly distributed force p at any point M located outside the circle. Although Eq. (4.30) can be implemented in SymPy, the evaluation is essentially numerical due to the complexities introduced by the square-root terms. The corresponding code is as follows:

```
# define symbolic variables
α, β, p = symbols('α, β, p')
a, ν, r, z, G, ν, m = symbols('a, ν, r, z, G, ν, m', positive=True)
a_value=1.0; r_value=1.000001    # M is near the circle, but a bit away
m_value = (a_value/r_value)**2

int_sln=2*(1-ν)*p*r/(sp.pi*G)*(sp.elliptic_e(m)-(1-m)/sp.elliptic_k(m))
print(f'The integral solution: {int_sln.subs(m,m_value)}')
```

```
The integral solution: 1.00000719581907*p*r*(2 - 2*ν)/(pi*G)
```

Note that at $r = a$ (on the edge of the circle), Eq. (4.30) can be simplified as

$$u_{z0} = 2\frac{(1-\nu)p}{\pi G}\int_0^{\pi/2} \frac{a^2\cos^2\alpha d\alpha}{\sqrt{r^2 - a^2 - a^2\cos^2\alpha}} d\alpha$$
$$= 2\frac{(1-\nu)pa}{\pi G}\int_0^{\pi/2} \cos\alpha d\alpha = 2\frac{(1-\nu)pa}{\pi G} \quad (4.31)$$

which supports the numerical results obtained in the previous code cell.

Most importantly, we observe that when the loading is distributed, the singularity in the displacement disappears.

4.4.2 *Result inside the loading circle*

When the observation point M is located inside the circle, the integral coordinates are shown in Fig. 4.10.

In this case, the procedure for performing the integral remains the same, but the geometric relations are slightly modified.

Equation (4.26) still holds, but the length l_{mn} is now

$$l_{mn} = 2a\cos\alpha = 2a\sqrt{1-\sin^2\alpha} = 2\sqrt{a^2 - r^2\sin^2\beta} \quad (4.32)$$

Here, we used the relation $a\sin\alpha = r\sin\beta$. Note that l_{mn} is integrated over s, so Eq. (4.26) becomes

$$u_{z0} = 2\frac{(1-\nu)p}{2\pi G}\int_0^{\pi/2} a\sin\alpha d\beta \quad (4.33)$$

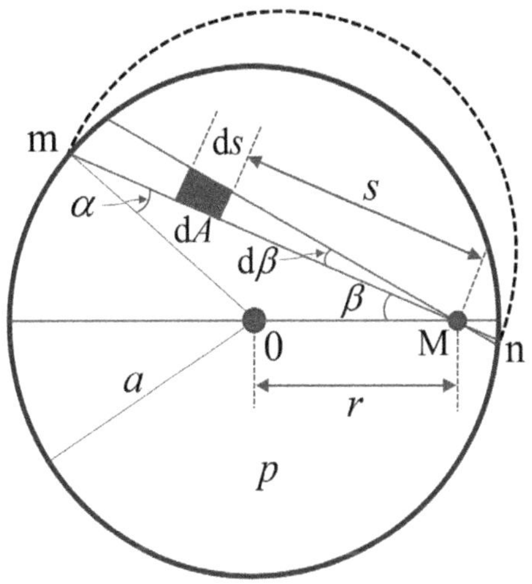

Figure 4.10. Top view of a circularly distributed force p on a half-space, showing the coordinate system for elliptic integration to compute the subsidence at point M inside the loading circle on the surface.

Observe that the integral limits for β now range from 0 to $\pi/2$. Using $a\sin\alpha - r\sin\beta$ again and $\cos\alpha = \sqrt{1-\sin^2\alpha}$, we obtain

$$u_{z0} = 2\frac{(1-\nu)pa}{\pi G}\underbrace{\int_0^{\pi/2}\sqrt{1-\frac{1}{m}\sin^2\beta}\,d\beta}_{E(1/m)} \tag{4.34}$$

This time, only the second-kind elliptic integral is involved.

At the center of the loading circle, $r = 0$, Eq. (4.34) gives

$$u_{z0} = 2\frac{(1-\nu)pa}{\pi G}\int_0^{\pi/2} d\beta = \frac{(1-\nu)pa}{G} \tag{4.35}$$

This represents the maximum subsidence.

The following code can be used to compute the subsidence at point M, both at the center and inside, very close to the edge of the loading circle:

```
a_value=1.0; r_value=0.999999  # M near the circle edge, but a bit away
m_inv = (r_value/a_value)**2                           # inverse of m

int_sln = 2*(1-ν)*p*a/(sp.pi*G)*(sp.elliptic_e(m))
display(Math(f" \\text{{Subsidence on the load circle edge}}=\
                              \\small {latex(int_sln.subs(m,m_inv))}"))
display(Math(f" \\text{{Max subsidence at the load circle center}}=\
                                  \\small {latex(int_sln.subs(m,0))}"))
```

$$\text{Subsidence on the load circle edge} = \frac{1.00000744747772ap\,(2-2\nu)}{\pi G}$$

$$\text{Max subsidence at the load circle center} = \frac{ap\,(2-2\nu)}{2G}$$

The result is very close to that obtained earlier for the point at the edge of the loading circle. The maximum subsidence at the center of the loading circle agrees with Eq. (4.35) and is $\pi/2$ times the subsidence at the edge of the loading circle. Moreover, the maximum subsidence is proportional to the diameter of the loading circle.

4.4.3 *Stresses caused by distributed force*

The same superposition technique, via integration, can also be used to obtain stress solutions. To compute the normal stress component σ_{zz} along the z-axis, we use the fifth equation in Eq. (4.23), set $dP = 2\pi r dr p$, and integrate over r:

$$\begin{aligned}\sigma_{zz} &= -\frac{3z^3 dP}{2\pi R^5} = -\int_0^a \frac{6\pi p z^3 r dr}{2\pi(r^2+z^2)^{5/2}}\\ &= -3pz^3\int_0^a \frac{rdr}{(r^2+z^2)^{5/2}} = -p\left[1-\frac{z^3}{(a^2+z^2)^{3/2}}\right]\end{aligned} \tag{4.36}$$

```
σzz = integrate(-3*p*z**3*r/sp.sqrt(r**2+z**2)**5, (r,0,a)).simplify()
σzz
```

$$\frac{pz^3}{(a^2+z^2)^{\frac{3}{2}}} - p$$

When computing the stress components σ_{rr} and $\sigma_{\theta\theta}$ along the z-axis, special care must be taken in handling the superposition. This is because contributions from dP at different locations within the loading circle affect these stresses in a slightly more complex manner.

Figure 4.11 illustrates the stress state at an observation point on the z-axis.

The contributions from the E and W cells are given by the third and fourth equations in Eq. (4.23):

$$\begin{aligned}d\sigma_{rr}^{EW} &= 2\frac{rdrd\theta p}{2\pi}\left\{\frac{1-2\nu}{R(R+z)} - \frac{3r^2 z}{R^5}\right\}\\ d\sigma_{\theta\theta}^{EW} &= 2\frac{(1-2\nu)rdrd\theta p}{2\pi}\left\{\frac{z}{R^3} - \frac{1}{R(R+z)}\right\}\end{aligned} \tag{4.37}$$

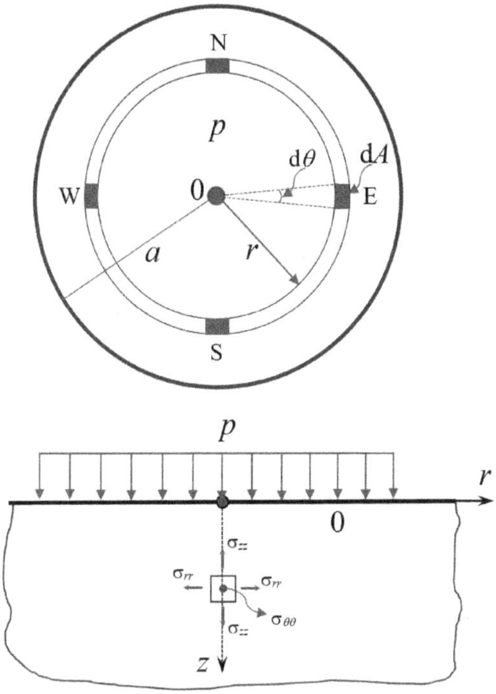

Figure 4.11. Stress state at an observation point on the z-axis in the half-space subjected to a uniform circularly distributed force p.

The contributions from the N and S cells are also given by the third and fourth equations in Eq. (4.23), but in a swapped order:

$$
\begin{aligned}
d\sigma_{rr}^{NS} &= 2\frac{(1-2\nu)rdrd\theta p}{2\pi}\left\{\frac{z}{R^3} - \frac{1}{R(R+z)}\right\} \\
d\sigma_{\theta\theta}^{NS} &= 2\frac{rdrd\theta p}{2\pi}\left\{\frac{1-2\nu}{R(R+z)} - \frac{3r^2 z}{R^5}\right\}
\end{aligned}
\tag{4.38}
$$

The superimposed contributions should be

$$
\begin{aligned}
d\sigma_{rr} &= d\sigma_{rr}^{EW} + d\sigma_{rr}^{NS} = \frac{rdrd\theta p}{\pi}\left\{(1-2\nu)\frac{z}{R^3} - \frac{3r^2 z}{R^5}\right\} \\
d\sigma_{\theta\theta} &= d\sigma_{\theta\theta}^{EW} + d\sigma_{\theta\theta}^{NS} = \frac{rdrd\theta p}{\pi}\left\{(1-2\nu)\frac{z}{R^3} - \frac{3r^2 z}{R^5}\right\}
\end{aligned}
\tag{4.39}
$$

$d\sigma_{rr}$ and $d\sigma_{\theta\theta}$ are found to be equal. This is due to axial symmetry along the z-axis, which makes them identical. The integration for the total circularly distributed force is given by

$$\begin{aligned}\sigma_{rr} = \sigma_{\theta\theta} &= \int_0^a \int_0^{\pi/2} d\sigma_{rr} \\ &= \frac{p}{2}\int_0^a \left\{ \frac{(1-2\nu)z}{(r^2+z^2)^{3/2}} - \frac{3r^2 z}{(r^2+z^2)^{5/2}} \right\} r dr d\theta \\ &= -\frac{p}{2}\left\{ (1+2\nu) + \frac{z^3}{(a^2+z^2)^{3/2}} - \frac{2(1+\nu)z}{(a^2+z^2)^{1/2}} \right\} \end{aligned} \tag{4.40}$$

Note that $\sigma_{\theta\theta}$ is equivalent to the hoop stress, which is equal to the normal stress σ_{rr}. The integration above can be computed using:

```
R   = sp.sqrt(r**2+z**2)
σθθ = integrate(p/2*((1-2*ν)*z*r/R**3-3*r**3*z/R**5), (r, 0, a)).\
                    subs(sp.sqrt(z**2),z).simplify()
σθθ.collect(a**2).collect(z**2)
```

$$\frac{p\left(z\left(a^2\left(2\nu+2\right)+z^2\left(2\nu+1\right)\right)-\left(a^2+z^2\right)^{\frac{3}{2}}\left(2\nu+1\right)\right)}{2\left(a^2+z^2\right)^{\frac{3}{2}}}$$

```
σθθ0= σθθ.subs(z,0)                                                # at z=0
σθθ0
```

$$-\frac{p\left(2\nu+1\right)}{2}$$

It is clear that the hoop stress (and the normal stress σ_{rr}) on the top surface is constant.

We also observe that when the loading is distributed, the singularity in the stresses disappears. However, discontinuities may still occur at the edges where the loading changes abruptly.

4.4.4 *Maximum shear stress*

From the last equation in Eq. (4.23), we note that σ_{rz} is zero along the z-axis ($r = 0$). Therefore, σ_{rr}, $\sigma_{\theta\theta}$, and σ_{zz} serve as the principal stresses at any point along the z-axis. The maximum shear stress occurs on a plane oriented at 45° to the z-axis. Using Mohr's circle theory [2], the maximum shear stress is given by

$$\tau_{\max} = \frac{1}{2}(\sigma_{\theta\theta} - \sigma_{zz}) = \frac{p}{4}\left\{(1-2\nu) + \frac{2(1+\nu)z}{(a^2+z^2)^{1/2}} - \frac{3z^3}{(a^2+z^2)^{3/2}}\right\} \quad (4.41)$$

The maximum shear stress $\tau_{\max}$ reaches its maximum value at

$$\tau_{\max,\max} = \frac{p}{4}\left\{(1-2\nu) + \frac{4}{9}(1+\nu)\sqrt{2(1+\nu)}\right\} \quad (4.42)$$

which is located at

$$z = a\sqrt{\frac{2(1+\nu)}{7-2\nu}} \quad (4.43)$$

All these formulas can be obtained using the following code snippet:

```
τ_max = (σθθ-σzz)/2
dτ_dz = sp.diff(τ_max, z)                                # derivative of τ_max

# Find zs for dτ_dz(zs)=0
stationary_zs=simplify(Matrix(sp.solve(dτ_dz,z))[1]) # choose second root
display(Math(f"\\text{{The stationary point}}=\
                                     \\small {latex(stationary_zs)}"))
                         # should be: sqrt(2)*a*sqrt(-1/(2*ν-7))*sqrt(ν+1)

τ_max_max = τ_max.subs(z, stationary_zs).simplify()  # find the maximum
display(Math(f"\\text{{Maximum of the maximum shear stress}}=\
                                         \\small {latex(τ_max_max)}"))
```

The stationary point $= \sqrt{2}a\sqrt{-\frac{1}{2\nu-7}}\sqrt{\nu+1}$

Maximum of the maximum shear stress

$$= \frac{p\left(4\nu\sqrt{2\nu+2} - 18\nu + 4\sqrt{2\nu+2} + 9\right)}{36}$$

```
print(f"When ν=1/3, the z-location of the max shear stress =" \
                              f"{stationary_zs.subs(ν,1/3).evalf()}")
print(f"When ν=1/3, the max shear stress = {τ_max_max.subs(ν,1/3)}")
```

```
When ν=1/3, the z-location of the max shear stress =0.648885684523O5*a
When ν=1/3, the max shear stress = 0.325258246200808*p
```

As an example, when $\nu = 1/3$, the maximum shear stress is $0.32526p$ at $z = 0.64889a$ on a plane inclined at 45° to the z-axis.

4.4.5 *Plots of stresses varying with depth*

```
# plot stresses along the vertical axis z:
dic = {a:1, ν:0.3333, p:1}
funcs = [sp.lambdify(z, σzz.subs(dic), 'numpy'),
         sp.lambdify(z, σθθ.subs(dic), 'numpy'),
        sp.lambdify(z, τ_max.subs(dic), 'numpy')]

labels=[r'$σ_{zz}$; '+f'${dic}$',r'$σ_{θθ}$; '+f'${dic}$',
        r'$τ_{max}$; '+f'${dic}$']

z_data  = np.linspace(0.00001, 5, 400)          # avoid zero-division
x_data = [z_data, z_data, z_data]
y_data = [funcs[i](x_data[i]) for i in range(len(x_data))]

#gr.plot_fs(x_data, y_data, labels, 'z', 'stresses',\
#                                  'imagesSMp/StressHalfSpace_p')
```

The distributions of stresses along the depth in the half-space subjected to a circularly distributed surface pressure are shown in Fig. 4.12.

As seen, all stresses decay rapidly with increasing depth. The location of the maximum shear stress is consistent with the observations discussed earlier. The magnitude of the stresses does not exceed the applied pressure.

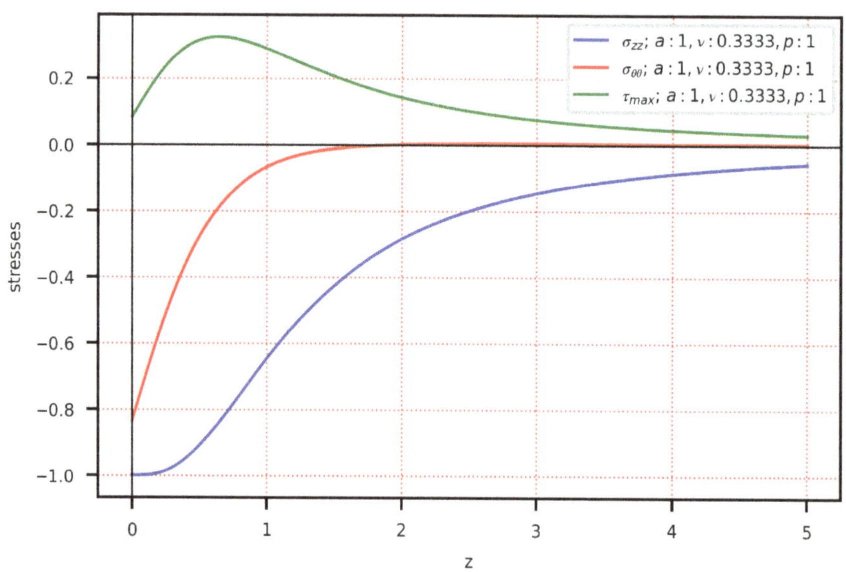

Figure 4.12. Stress distribution along the z-axis in the half-space subjected to a uniform circularly distributed force p.

4.5 Contact of two elastic solids

Contact between solids is a fundamental problem in solid mechanics, as contact forces generate deformations and stresses within the solids. Figure 4.13 illustrates an elastic solid subjected to a force with both normal and tangential components. The photoelasticity technique is used to experimentally visualize the stresses. High stress concentrations are observed, which can lead to crack initiation and failure in practical applications. This section discusses techniques and formulas for computing the displacements and stresses resulting from the contact of two elastic solids.

4.5.1 *Contact between two spherical balls*

4.5.1.1 *Setting of the problem*

Consider two elastic balls, ball-1 and ball-2, with radii R_1 and R_2, respectively. These balls are pressed into contact with each other, as illustrated in Fig. 4.14. Although the contact surface may be complex, its projection onto the $z = 0$ plane is circular. Assume that the deformations of the two balls are small and that the radius of the resulting contact surface is much smaller than the radii of the balls themselves. The objective is to determine the stresses and displacements caused by the contact force. This classical

Figure 4.13. Stresses in the contact area of an elastic solid subjected to a force with both normal and tangential components. These stresses are visualized using photoelasticity. Image from Wikimedia Commons by Reibungsphysik, in the public domain.

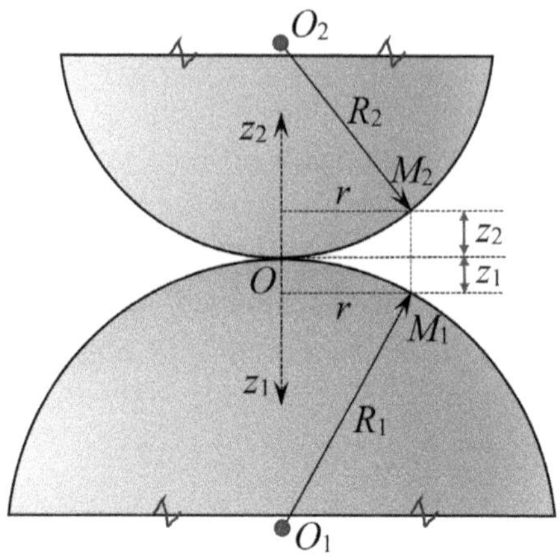

Figure 4.14. Cross-sectional view of two elastic balls pressed into contact. Ball-1 is at the bottom and ball-2 is on top.

problem was first solved by Hertz in the 1880s and is commonly referred to as the **Hertz contact problem**. The following discussion is based on the textbook reference [4].

4.5.1.2 *Initial geometric relations*

When the contact force is zero, the two balls touch at a single point, denoted as point O. The normal vectors of the balls at this point are denoted by z_1 and z_2, respectively. These vectors are aligned and point in opposite directions. Thus, the problem is axisymmetric with respect to the z-axis.

Consider a distance r from the common z-axis, which intersects the surfaces of the two balls at points M_1 and M_2, respectively. The following geometric relations hold:

$$\begin{aligned}(R_1 - z_1)^2 + r^2 &= R_1^2 \\ (R_2 - z_2)^2 + r^2 &= R_2^2\end{aligned} \tag{4.44}$$

These geometric relations are nonlinear and need to be linearized. We then rewrite these relations as

$$\begin{aligned}z_1 &= \frac{r^2}{2R_1 - z_1} \\ z_2 &= \frac{r^2}{2R_2 - z_2}\end{aligned} \tag{4.45}$$

When M_1 and M_2 are close to point O, the vectors z_1 and z_2 will be much smaller than $2R_1$ and $2R_2$, respectively. Therefore, the terms involving z_1 and z_2 in the denominators of Eq. (4.45) can be approximated by omitting them. This results in a simplified linear form:

$$z_1 = \frac{r^2}{2R_1}$$
$$z_2 = \frac{r^2}{2R_2} \tag{4.46}$$

The distance between M_1 and M_2 becomes

$$z_1 + z_2 = r^2\left(\frac{1}{2R_1} + \frac{1}{2R_2}\right) = r^2\frac{R_1 + R_2}{2R_1R_2} = qr^2 \tag{4.47}$$

where q is given by

$$q = \frac{R_1 + R_2}{2R_1R_2} \tag{4.48}$$

It is a measure of the relative radii of the two balls. This quantity is constant for the given balls and has dimensions of inverse length.

4.5.1.3 *Deformation analysis*

When the two balls are subjected to an equal and opposite pair of forces P, they are pressed into contact. This causes a local deformation around the contact surface, which is assumed to be very small.

Consider two points far from the contact area on each of the balls. These points experience a relative displacement in the z-direction, referred to as the **total displacement** and denoted by d. Let the vertical displacement of point M_1 along the z_1 direction be w_1 and the vertical displacement of point M_2 along the z_2 direction be w_2. As the balls come into contact, points M_1 and M_2 coincide. Therefore, the following in-contact geometric relation holds:

$$d - (w_1 + w_2) = z_1 + z_2 \tag{4.49}$$

or

$$w_1 + w_2 = d - (z_1 + z_2) \tag{4.50}$$

Using Eq. (4.47), we obtain

$$w_1 + w_2 = d - qr^2 \tag{4.51}$$

Due to the spherical geometry of the balls, the contact surface forms a portion of a spherical surface. When viewed along the z-axis, the contact area appears circular. Therefore, it can be represented as in Fig. 4.10. The point M in the figure corresponds to the contact point M_1 on ball-1 before deformation. Consequently, Eq. (4.26) can be used to evaluate the displacement w_1:

$$w_1 = \frac{(1-\nu_1)}{2\pi G_1}\int_A p d\beta ds \tag{4.52}$$

where ν_1 and G_1 are the Poisson's ratio and shear modulus of ball-1 and A denotes the area of the contact surface.

Similarly, for ball-2, we have

$$w_2 = \frac{(1-\nu_2)}{2\pi G_2}\int_A p d\beta ds \tag{4.53}$$

where ν_2 and G_2 are the Poisson's ratio and shear modulus of ball-2. The sum of w_1 and w_2 is

$$w_1 + w_2 = (k_1 + k_2)\int_A p d\beta ds \tag{4.54}$$

where

$$k_1 = \frac{1-\nu_1}{2\pi G_1} = \frac{1-\nu_1^2}{\pi E_1}; \quad k_2 = \frac{(1-\nu_2)}{2\pi G_2} = \frac{1-\nu_2^2}{\pi E_2} \tag{4.55}$$

Using Eq. (4.50), we arrive at

$$(k_1 + k_2)\int_A p d\beta ds = d - qr^2 \tag{4.56}$$

Note that p is the distributed pressure between the two balls, which is still unknown. The task is now to determine p such that Eq. (4.56) is satisfied.

Hertz showed that the pressure distribution p can be expressed as a semi-sphere, with the contact surface forming the base circle. The cross-sections of this pressure distribution are illustrated in Fig. 4.15.

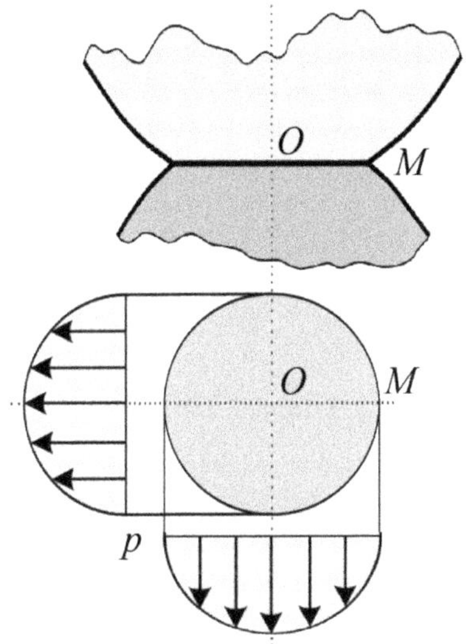

Figure 4.15. Schematic of the interfacial pressure on the contact area of two balls. Image modified based on Wikimedia Commons by MHz, in the public domain.

The formula for the contact pressure distribution is given by

$$p(r) = p_0 \left(1 - \frac{r^2}{a^2}\right)^{\frac{1}{2}} \tag{4.57}$$

where p_0 is the peak contact pressure at the center of the contact circle. This semi-spherical distribution of p satisfies Eq. (4.56). The proof is outlined as follows.

The ratio between the peak pressure and the peak height of the semi-sphere is p_0/a, where a is the radius of the spherical pressure distribution. The pressure along the m–n segment in Fig. 4.10 follows the semi-circle drawn with the dashed line. According to Hertz's assumption, we have

$$\int_s p ds = \frac{p_0}{a} A_p \tag{4.58}$$

where A_p is the area of the semi-circle of the pressure distribution along the m–n segment indicated in Fig. 4.10. Note that the value of p_0 is yet to be determined; it will be found later using the equilibrium condition.

Since the length of the m–n segment is $l_{mn} = 2\sqrt{a^2 - r^2 \sin^2 \beta}$, A_p can be calculated as

$$A_p = \frac{1}{2}\frac{\pi}{4} l_{mn}^2 = \frac{\pi}{2}(a^2 - r^2 \sin^2 \beta) \tag{4.59}$$

Substituting Eq. (4.59) into (4.56) gives

$$(k_1 + k_2)2\int_0^{\pi/2} \frac{p_0}{a}\frac{\pi}{2}(a^2 - r^2\sin^2\beta)d\beta = d - qr^2 \tag{4.60}$$

where the factor of 2 in front of the integral is due to the symmetry of the domain of integration. Equation (4.60) gives

$$(k_1 + k_2)\frac{\pi^2 p_0}{4a}(2a^2 - r^2) = d - qr^2 \tag{4.61}$$

This is a polynomial in r. To satisfy Eq. (4.61) for any r, both the coefficients of r^0 and r^2 must be zero. We thus have

$$\begin{aligned}(k_1 + k_2)\frac{\pi^2 a p_0}{2} &= d\\ (k_1 + k_2)\frac{\pi^2 p_0}{4a} &= q\end{aligned} \tag{4.62}$$

Next, we need to determine p_0. This can be done by enforcing that the total pressure on the contact area A equals the externally applied force P:

$$P = \int_A p dA = \int_A p_0\left(1 - \frac{r^2}{a^2}\right)^{\frac{1}{2}} dA \tag{4.63}$$

The computation is done as follows:

```
a, r = symbols('a, r', positive=True)
p0, θ = symbols('p0, θ', real=True)
P = p0/a*integrate(integrate(sqrt(a**2-r**2)*r, (r,0,a)),(θ,0,2*sp.pi))
print('Contact force P =',P)
#display(Math(f" \\text{{ Contact force $P$}}={Latex(P)}"))
```

```
Contact force P = 2*pi*a**2*p0/3
```

We found

$$P = \frac{2\pi a^2 p_0}{3} \tag{4.64}$$

This is, in fact, $\frac{p_0}{a}$ multiplied by the volume of the half-sphere representing the pressure distribution, $\frac{2}{3}\pi a^3$. Equation (4.64) gives the peak pressure as

$$p_0 = \frac{3P}{2\pi a^2} \tag{4.65}$$

which is 1.5 times the average pressure, $\frac{P}{\pi a^2}$.

Substituting Eqs. (4.65) and (4.48) into Eq. (4.62), we obtain a system of two equations. We can then solve this system for the two unknowns: the radius of the contact circle a and the total relative displacement d. The results are

$$\begin{aligned} a &= \sqrt[3]{\frac{3\pi P(k_1+k_2)R_1R_2}{4(R_1+R_2)}} \\ p_0 = \frac{3P}{2\pi a^2} &= \frac{3P}{2\pi}\sqrt[2/3]{\frac{4(R_1+R_2)}{3\pi P(k_1+k_2)R_1R_2}} \\ d &= \sqrt[3]{\frac{9\pi^2P^2(k_1+k_2)^2(R_1+R_2)}{16R_1R_2}} \end{aligned} \tag{4.66}$$

These solutions can be determined using the following Python snippet. It uses Eqs. (4.48) and (4.65), and solves Eq. (4.62) for a and d:

```
p0, P, d = symbols('p0, P, d')
ν1, G1, E1, ν2, G2, E2, E = symbols('ν1,G1,E1,ν2,G2,E2,E',positive=True)
a, R1, R2, k1, k2 = symbols('a, R1, R2, k1, k2 ', positive=True)

p0  = 3*P/(2*sp.pi *a**2)
q   = (R1+R2)/(2*R1*R2)
eq1 = (k1+k2)*sp.pi**2*a*p0/2-d
eq2 = (k1+k2)*sp.pi**2*p0/(4*a)- q
sln = sp.solve([eq1, eq2], [a, d])
sln[0]                                          # solution for a and d
```

$$\left(\frac{\sqrt[3]{6}\sqrt[3]{\pi}\sqrt[3]{R_1}\sqrt[3]{R_2}\sqrt[3]{k_1+k_2}\,(P^2)^{\frac{2}{3}}}{2P\sqrt[3]{R_1+R_2}},\ \frac{6^{\frac{2}{3}}\pi^{\frac{2}{3}}\sqrt[3]{R_1+R_2}\,(k_1+k_2)^{\frac{2}{3}}\sqrt[3]{P^2}}{4\sqrt[3]{R_1}\sqrt[3]{R_2}}\right)$$

4.5.2 *An elastic ball on a half-space*

Consider an elastic ball with radius R being pressed onto a half-space, as shown in Fig. 4.16(a).

The solution can be readily obtained using the result in Eq. (4.66) from the previous section by taking the limit $R_1 \to \infty$ and setting $R_2 = R$:

$$\begin{aligned} a &= \sqrt[3]{\frac{3\pi P(k_1+k_2)R}{4}} \\ p_0 = \frac{3P}{2\pi a^2} &= \frac{3P}{2\pi}\sqrt[2/3]{\frac{4}{3\pi P(k_1+k_2)R}} \\ d &= \sqrt[3]{\frac{9\pi^2P^2(k_1+k_2)^2}{16R}} \end{aligned} \tag{4.67}$$

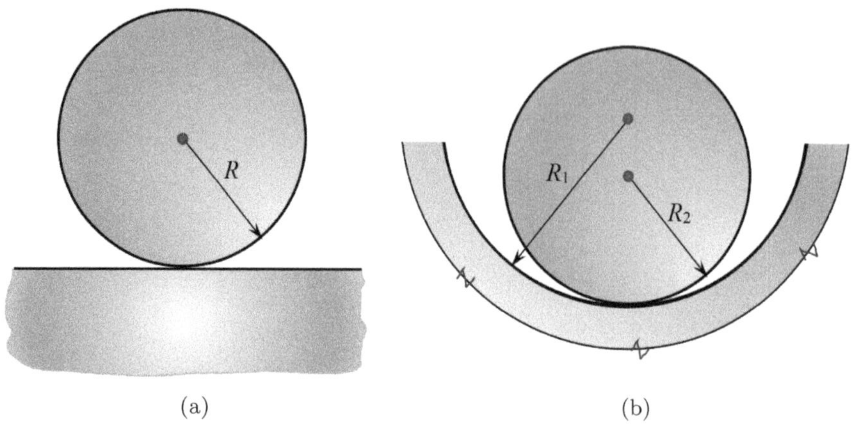

Figure 4.16. (a) An elastic ball on an elastic half-space. (b) An elastic ball on an elastic spherical bearing.

4.5.3 *An elastic ball on a bearing*

Consider an elastic ball with radius R_2 being pressed onto a bearing with radius R_1, as shown in Fig. 4.16(b). In this case, the solution can be readily obtained using Eq. (4.66) for the two-ball contact by simply reversing the sign of R_1. Therefore, we will not repeat the equations here.

4.5.4 *Contact of an ellipsoidal ball*

If the ball is ellipsoidal, the projection of the contact surface onto the $z = 0$ plane will be an ellipse. The key idea remains the Hertz assumption of a semi-ellipsoidal interfacial pressure, as shown in Fig. 4.17.

The solution procedure is similar to the spherical ball case but is more complex. We will not discuss it further. Interested readers may refer to textbooks such as Ref. [4].

4.5.5 *Some special cases*

Consider a special case in which the two spherical balls are made of the same material and have a Poisson's ratio of 0.3. In this scenario, a set of simple,

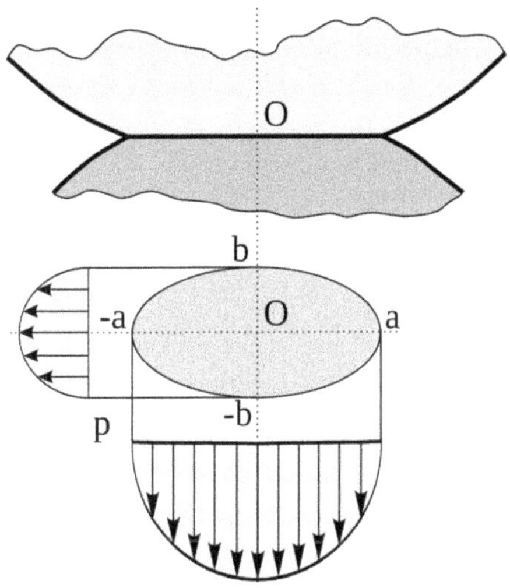

Figure 4.17. A schematic drawing of the interfacial pressure on the contact area of an ellipsoid on another solid. Image modified from Wikimedia Commons by MHz, as, public domain.

easy to use equations can be obtained:

$$\begin{aligned} a &= 1.109\sqrt[3]{\frac{PR_1R_2}{E(R_1+R_2)}} \\ p_0 &= 0.3880\sqrt[3]{\frac{PE^2(R_1+R_2)^2}{R_1^2R_2^2}} \\ d &= 1.231\sqrt[3]{\frac{P^2(R_1+R_2)}{E^2R_1R_2}} \end{aligned} \tag{4.68}$$

If ball-1 is a half-space, we have

$$\begin{aligned} a &= 1.109\sqrt[3]{\frac{PR}{E}} \\ p_0 &= 0.3880\sqrt[3]{\frac{PE^2}{R^2}} \\ d &= 1.231\sqrt[3]{\frac{P^2}{E^2R}} \end{aligned} \tag{4.69}$$

These equations can be obtained using the following code snippet and can also be used to compute other special cases:

```
a_sln = sln[0][0]        # solution for the radius of the contact circle
d_sln = sln[0][1]        # solution for far-points relative displacement
ν1_ = sp.Rational(1,3);  ν2_ = sp.Rational(1,3)
G1_ = E/(2*(1+ν1_))
G2_ = E/(2*(1+ν2_))

k1_ = (1-ν1_)/(2*sp.pi*G1_)
k2_ = (1-ν2_)/(2*sp.pi*G2_)

dic = {ν1:ν1_, ν2:ν2_,k1:k1_,k2:k2_}
n_digits = 4
a_dic = a_sln.subs(dic).evalf(n=n_digits)
d_dic = d_sln.subs(dic).evalf(n=n_digits)
p0_dic = (p0.subs(a,a_dic)).subs(dic).evalf(n=n_digits)
display(Math(f" \\text{{ Specific case, $p_{{0}}$}}={latex(p0_dic)}"))
display(Math(f" \\text{{ Specific case, $p_{{0}}$}}={latex(a_dic)}"))
display(Math(f" \\text{{ Specific case, $p_{{0}}$}}={latex(d_dic)}"))
```

$$\text{Specific case, } p_0 = \frac{0.3942E^{0.6667}P^3\left(R_1+R_2\right)^{0.6667}}{R_1^{0.6667}R_2^{0.6667}\left(P^2\right)^{1.333}}$$

$$\text{Specific case, } p_0 = \frac{1.101R_1^{0.3333}R_2^{0.3333}\left(P^2\right)^{0.6667}}{E^{0.3333}P\left(R_1+R_2\right)^{0.3333}}$$

$$\text{Specific case, } p_0 = \frac{1.211\left(R_1+R_2\right)^{0.3333}\left(P^2\right)^{0.3333}}{E^{0.6667}R_1^{0.3333}R_2^{0.3333}}$$

Contact problems are well studied, and many of them have been solved. With the foundation established in this section, readers should be able to study and apply these solutions when needed. For further reading, interested readers may refer to the Contact Mechanics Wikipedia page and the references therein.

4.6 Hollow ball subject to pressure loads

There are problems that are spherically symmetric. A typical example is a hollow ball subjected to pressure loads.

4.6.1 *Setting of the problem*

Consider a hollow ball with an inner radius a and an outer radius b. It is subjected to internal pressure p_i and external pressure p_o, as shown in Fig. 4.18.

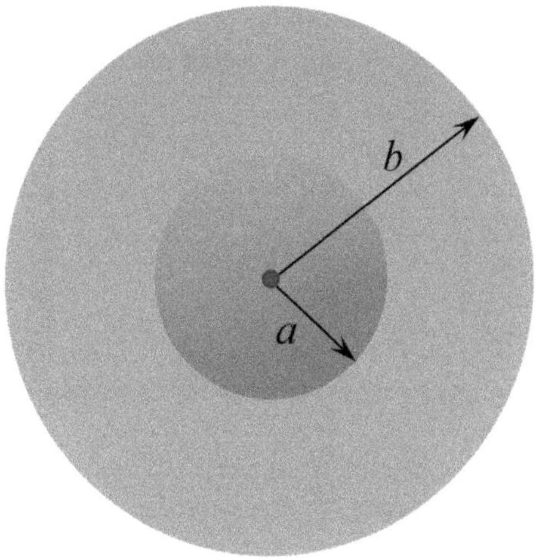

Figure 4.18. A hollow ball subjected to pressures on the inner and outer surfaces.

Assuming zero body force, determine the stresses and displacement within the ball. This is a classical and important spherically symmetric problem.

4.6.2 *Formulation*

This problem is clearly spherically symmetric. All variables are independent of the polar and azimuthal angles and depend only on the radial coordinate r. The only nonzero displacement component is the radial displacement u_r. The second and third equilibrium equations in Eq. (2.56) are automatically satisfied, while the first equation simplifies to

$$\frac{\partial \sigma_{rr}}{\partial r} + \frac{2}{r}(\sigma_{rr} - \sigma_{\theta\theta}) = 0 \tag{4.70}$$

Here, we use the fact that $\sigma_{\theta\theta} = \sigma_{\phi\phi}$ due to the spherical symmetry.

Using the first two equations in the strain–displacement relations in Eq. (4.4), we have

$$\varepsilon_{rr} = \frac{\partial u_r}{\partial r}; \quad \varepsilon_{\theta\theta} = \frac{u_r}{r} \tag{4.71}$$

Furthermore, using the constitutive equation for 3D solids [2], we obtain

$$\begin{aligned} \sigma_{rr} &= \frac{E}{(1+\nu)(1-2\nu)}\left[(1-\nu)\varepsilon_{rr} + 2\nu\varepsilon_{\theta\theta})\right] \\ \sigma_{\theta\theta} &= \frac{E}{(1+\nu)(1-2\nu)}\left[\varepsilon_{\theta\theta} + \nu\varepsilon_{rr})\right] \end{aligned} \tag{4.72}$$

By substituting Eq. (4.71) into Eq. (4.72), we first obtain expressions for the stresses. Substituting these stress expressions into Eq. (4.70) then yields the equilibrium equation for spherically symmetric problems:

$$\frac{d^2u_r}{dr^2} + \frac{2}{r}\frac{du_r}{dr} - \frac{2u_r}{r^2} = 0 \tag{4.73}$$

This is a one-dimensional differential equation (DE) for the only nonzero radial displacement u_r in a spherically symmetric problem. The following snippet demonstrates how to derive this differential equation:

```
r, a, b, E, ν = sp.symbols('r, a, b, E, ν', positive=True)  # variables
σrr, σθθ, εrr, εθθ = sp.symbols('σ_rr, σ_θθ, ε_rr, ε_θθ',real=True)
ur   = sp.Function('u_r' )(r)             # the nonzero radial displacement

λdv = E/((1+ν)*(1-2*ν))         # the coefficient in the constitutive eq.
εrr = ur.diff(r)                                                 # strains
εθθ = ur/r
σrr = λdv*((1-ν)*εrr + 2*ν*εθθ)                                  # stresses
σθθ = λdv*(εθθ + ν*εrr)
DE = ((σrr.diff(r) + 2/r*(σrr - σθθ))).simplify()               # ED in ur
DE = ((sp.numer(DE)/E).simplify().collect(2*ur)/(ν-1)).cancel((ν-1))
display(Math(f"\\text{{Differential equation for spherical problems:}}\
                                                 \\; \\; {latex(DE)}=0"))
```

Differential equation for spherical problems: $r^2\frac{d^2}{dr^2}u_r(r) + 2r\frac{d}{dr}u_r(r) - 2u_r(r) = 0$

4.6.3 *General solutions*

The DE given in Eq. (4.73) can be solved using the following snippet:

```
diff_eq = sp.Eq(DE, 0)                      # define differential equation
g_ur = sp.dsolve(diff_eq, ur).args[1] # solve the differential equation
display(Math(f" \\text{{ General solution for displacement $u_r$}} = \
                                                      {latex(g_ur)}"))
```

General solution for displacement $u_r = \frac{C_1}{r^2} + C_2 r$

```
# Redefine the general solution of ur:
c1, c2 = sp.symbols('c1, c2',real=True)
g_ur = c1/r**2 + c2*r
```

Using the general solution for the displacement, we can obtain the expressions for the strains using Eq. (4.71):

```
g_εrr = g_ur.diff(r)
g_εθθ = (g_ur/r).simplify()
display(Math(f" \\text{{ General solution for strain $ε_{{rr}}$}}:\
                                                    {latex(g_εrr)}"))
display(Math(f" \\text{{ General solution for strain $ε_{{θθ}}$}}:\
                                                    {latex(g_εθθ)}"))
```

General solution for strain ε_{rr} : $-\dfrac{2c_1}{r^3} + c_2$

General solution for strain $\varepsilon_{\theta\theta}$: $\dfrac{c_1}{r^3} + c_2$

Using the general solution for the strains, we can obtain the expressions for the stresses using Eq. (4.72):

```
g_σrr = (λdv*((1-ν)*g_εrr + 2*ν*g_εθθ)).simplify()
g_σθθ = (λdv*(g_εθθ + ν*g_εrr)).simplify()
display(Math(f" \\text{{ General solution for stress $σ_{{rr}}$}}:\
                                                    {latex(g_σrr)}"))
display(Math(f" \\text{{ General solution for stress $σ_{{θθ}}$}}:\
                                                    {latex(g_σθθ)}"))
```

General solution for stress σ_{rr} : $\dfrac{E\left(-4c_1\nu + 2c_1 - c_2r^3\nu - c_2r^3\right)}{r^3\left(2\nu^2 + \nu - 1\right)}$

General solution for stress $\sigma_{\theta\theta}$: $-\dfrac{E\left(c_1 + c_2r^3 - \nu\left(2c_1 - c_2r^3\right)\right)}{r^3\left(\nu + 1\right)\left(2\nu - 1\right)}$

4.6.4 *Determination of the constants*

The general solution contains two unknown constants, c_1 and c_2. To determine these constants, we apply the following stress boundary conditions:

For a hollow ball subjected to internal and external pressures, the boundary conditions are

$$\sigma_{rr}\Big|_{r=a} = -p_i; \quad \sigma_{rr}\Big|_{r=b} = -p_o \tag{4.74}$$

The following snippet solves for the constants c_1 and c_2:

```
pi, po = sp.symbols('p_i, p_o', real=True)
bc_eq1 = g_σrr.subs(r, a) + pi
bc_eq2 = g_σrr.subs(r, b) + po

c12_sln = sp.solve([bc_eq1, bc_eq2], [c1, c2])
c1_=c12_sln[c1].simplify().collect(pi).collect(po).collect(a).collect(b)
display(Math(f" \\text{{ Constant found, $c_{{1}}$}}={latex(c1_)}"))
c2_=c12_sln[c2].simplify().collect(pi).collect(po).collect(a).collect(b)
display(Math(f" \\text{{ Constant found, $c_{{2}}$}}={latex(c2_)}"))
```

$$\text{Constant found, } c_1 = \frac{a^3 b^3 \left(p_i \left(-\nu - 1\right) + p_o \left(\nu + 1\right)\right)}{2E \left(a^3 - b^3\right)}$$

$$\text{Constant found, } c_2 = \frac{a^3 p_i \left(2\nu - 1\right) + b^3 p_o \left(1 - 2\nu\right)}{E \left(a^3 - b^3\right)}$$

4.6.5 *Final solutions and findings*

Substituting the determined constants back into the general expressions yields the final solutions for the displacement and stresses in this specific problem:

```
ur_final = g_ur.subs(c12_sln).simplify().collect(pi).collect(po)
ur_final = ur_final.collect(a).collect(b).collect(1+ν).collect(1-2*ν)
display(Math(f" \\text{{ Final displacement $u_r$ }}=\
                                          \small {latex(ur_final)}"))
σrr_final = g_σrr.subs(c12_sln).simplify().collect(pi).collect(po)
σrr_final = σrr_final.collect(a).collect(b)
display(Math(f" \\text{{ Final stress $σ_{{rr}}$ }}=\
                                          \small {latex(σrr_final)}"))
σθθ_final = g_σθθ.subs(c12_sln).simplify().collect(pi).collect(po)
σθθ_final = σθθ_final.collect(a).collect(b)
display(Math(f" \\text{{ Final stress $σ_{{θθ}}$ }}=\
                                          \small {latex(σθθ_final)}"))
```

Final displacement

$$u_r = \frac{a^3 b^3 \left(-p_i + p_o\right)\left(\nu + 1\right) + 2r^3 \left(1 - 2\nu\right)\left(-a^3 p_i + b^3 p_o\right)}{2Er^2 \left(a^3 - b^3\right)}$$

$$\text{Final stress } \sigma_{rr} = \frac{a^3 p_i \left(b^3 - r^3\right) + b^3 p_o \left(-a^3 + r^3\right)}{r^3 \left(a^3 - b^3\right)}$$

$$\text{Final stress } \sigma_{\theta\theta} = \frac{a^3 p_i \left(-\frac{b^3}{2} - r^3\right) + b^3 p_o \left(\frac{a^3}{2} + r^3\right)}{r^3 \left(a^3 - b^3\right)}$$

Assume $p_0 = 0$, we have the following:

```
ur_pi = ur_final.subs(po,0).collect(pi).collect(a)
display(Math(f" \\text{{With $p_i$ only, $u_r$}}= {latex(ur_pi)}"))

σrr_pi = σrr_final.subs(po,0)
display(Math(f" \\text{{With $p_i$ only, $σ_{{rr}}$}}={latex(σrr_pi)}"))

σθθ_pi = σθθ_final.subs(po,0)
display(Math(f" \\text{{With $p_i$ only, $σ_{{θθ}}$}}={latex(σθθ_pi)}"))
```

$$\text{With } p_i \text{ only, } u_r = \frac{a^3 p_i \left(-b^3 (\nu + 1) - 2r^3 (1 - 2\nu)\right)}{2Er^2 (a^3 - b^3)}$$

$$\text{With } p_i \text{ only, } \sigma_{rr} = \frac{a^3 p_i \left(b^3 - r^3\right)}{r^3 (a^3 - b^3)}$$

$$\text{With } p_i \text{ only, } \sigma_{\theta\theta} = \frac{a^3 p_i \left(-\frac{b^3}{2} - r^3\right)}{r^3 (a^3 - b^3)}$$

Let us examine the limits of these solutions as $b \to a$. For this purpose, we set fixed values for Young's modulus E and Poisson's ratio ν. Since $r \to a$ when $b \to a$, we also set $r = a$ when evaluating the limits:

```
ur_pi_bba = sp.limit(ur_pi.subs({E:1, ν:1/3}), b, a).simplify()
display(Math(f" \\text{{With $p_i$ and $b→a$, $u_r$}}= \
                                                    {latex(ur_pi_bba)}"))
σrr_pi_bba = sp.limit(σrr_pi.subs({r:a}), b, a)
display(Math(f" \\text{{With $p_i$ and $b→a$, $σ_{{rr}}$}}=\
                                                   {latex(σrr_pi_bba)}"))
σθθ_pi_bba = sp.limit(σθθ_pi, b, a)
display(Math(f" \\text{{With $p_i$ and $b→a$, $σ_{{θθ}}$}}=\
                                                   {latex(σθθ_pi_bba)}"))
```

$$\text{With } p_i \text{ and } b \to a,\ u_r = \frac{\infty p_i}{|p_i|}$$

$$\text{With } p_i \text{ and } b \to a,\ \sigma_{rr} = -p_i$$

$$\text{With } p_i \text{ and } b \to a,\ \sigma_{\theta\theta} = \infty \operatorname{sign}(p_i)$$

The limit of the radial stress σ_{rr} is $-p_i$, which matches the boundary condition and is therefore expected. The limits for the radial displacement u_r and the hoop stress $\sigma_{\theta\theta}$ reveal the following observation:

> As $b \to a$, both the displacement u_r and the hoop stress $\sigma_{\theta\theta}$ tend to infinity. This indicates that the solution becomes unstable in this limit. A similar phenomenon is observed when inflating a balloon: as the material becomes very thin, it eventually bursts.

When $b \to \infty$, the following limits for these solutions are obtained:

```
ur_pi_boo = sp.limit(ur_pi, b, oo).simplify()
display(Math(f" \\text{{With $p_i$ and $b→∞$, $u_r$}}= \
                                          {latex(ur_pi_boo)}"))
σrr_pi_boo = sp.limit(σrr_pi, b, oo)
display(Math(f" \\text{{With $p_i$ and $b→∞$, $σ_{{rr}}$}}=\
                                          {latex(σrr_pi_boo)}"))
σθθ_pi_boo = sp.limit(σθθ_pi, b, oo)
display(Math(f" \\text{{With $p_i$ and $b→∞$, $σ_{{θθ}}$}}=\
                                          {latex(σθθ_pi_boo)}"))
```

With p_i and $b \to \infty$, $u_r = \dfrac{a^3 p_i (\nu + 1)}{2Er^2}$

With p_i and $b \to \infty$, $\sigma_{rr} = -\dfrac{a^3 p_i}{r^3}$

With p_i and $b \to \infty$, $\sigma_{\theta\theta} = \dfrac{a^3 p_i}{2r^3}$

It is found that:

1. When the inner surface is subjected to a pressure p_i and $b \to \infty$, the displacement decays rapidly to zero at a rate of $1/r^2$, with the maximum displacement occurring at $r = a$, given by $u_r^{\max} = \frac{ap_i(\nu+1)}{2E}$.
2. The radial stress σ_{rr} decays even faster, at a rate of $1/r^3$, reaching its maximum compressive value of $-p_i$ at $r = a$.
3. The hoop stress $\sigma_{\theta\theta}$ also decays at a rate of $1/r^3$, with the maximum tensile stress $\sigma_{\theta\theta} = p_i/2$ occurring at $r = a$, even when the outer radius is infinitely large. This may explain potential failure in materials with low tensile strength.

These results are consistent with the observations made for internally pressurized disks or cylinders discussed earlier.

4.7 Review question

Consider the same problem presented in Section 4.3.1. Derive the displacement solution using a similar approach as in Section 4.3, employing the following displacement function:

$$\xi(r, z) = c_0 \sqrt{r^2 + z^2} \tag{4.75}$$

together with the displacement potential function (see Chapter 5 of Ref. [1]):

$$\psi(r, z) = c_1 \ln(\sqrt{r^2 + z^2} + z) \tag{4.76}$$

Compare your derived solution with the one obtained in Section 4.3.

4.8 Remarks

This chapter discussed the displacement method for problems exhibiting axisymmetric features. The following remarks can be made:

1. Polar (for 2D), cylindrical (for 3D), and spherical (for 3D) coordinate systems are employed for different problems. Effective methodologies and procedures are presented in a clear, step-by-step manner.
2. Several fundamental problems have been solved analytically in closed form. It is shown that, in the displacement method, computing strains — and thus stresses — is straightforward and requires only differentiation, which can always be carried out.
3. To the author's best knowledge, this is the first time a displacement method has been proposed for spinning disk problems.
4. The main challenge lies in appropriately assuming displacement fields that satisfy all necessary conditions.

The following chapter discusses stress methods for stress concentration problems.

References

[1] G. R. Liu, *Solid Mechanics with Python: Cartesian Coordinates*, World Scientific, New Jersey, 2025.
[2] G. R. Liu, *Mechanics of Materials: Formulations and Solutions with Python*, World Scientific, New Jersey, 2024.
[3] G. R. Liu, *Numbers and Functions: Theory, Formulation, and Python Codes*, World Scientific, New Jersey, 2024.
[4] Z. L. Xu, *Elasticity*, People's Publisher, China, 1979.

Chapter 5

Stress Method for Stress Concentration on a Hole

Let us import the necessary external modules or dependencies for later use in this chapter:

```
# Place cursor in this cell, and press Ctrl+Enter to import dependencies.
import sys                          # for accessing the computer system
sys.path.append('../grbin/')          # Add in the path to your system

from commonImports import *        # Import dependencies from '../grbin/'
import grcodes as gr                  # Import the module of the author
importlib.reload(gr)                # When grcodes is modified, reload it

from continuum_mechanics import vector
from continuum_mechanics.solids import sym_grad, strain_stress
init_printing(use_unicode=True)        # For latex-like quality printing
np.set_printoptions(precision=4,suppress=True)    # Digits in print-outs
```

When a defect exists in a solid, stress tends to concentrate along the edge or surface of the defect. It is therefore particularly important to study stress concentrations in solids. One of the most significant studies conducted in the past concerns a plate with a circular hole, which led to a set of closed-form solutions for stresses. This chapter presents this study using the stress method, considering various types of loadings, including axial tension, biaxial loading, pure shearing, and general loading. Stress concentration factors for all these cases are obtained. It is found that the stress concentration factor can be as large as 6 for a general loading case.

This chapter is written with reference to Refs. [1–3].

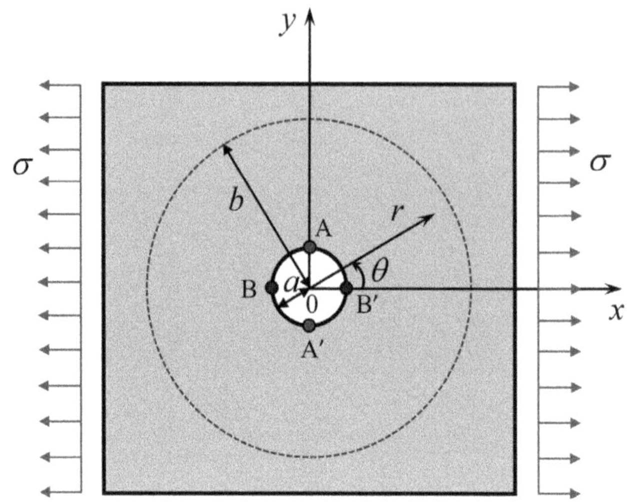

Figure 5.1. A plate with a small circular hole of radius a in the middle, subjected to a uniform stress σ at the left and right edges.

5.1 Plates with a hole subjected to uniaxial loading

Consider a practically important problem: a plate with a small circular hole in the middle. It is subjected to unidirectional tension with a stress level of σ in the x-direction at the ends of the plate, as shown in Fig. 5.1.

Without the hole, the plate is in a simple tensile stress state with only one nonzero component, $\sigma_{xx} = \sigma$. Naturally, the presence of the hole alters the stress distribution especially near the hole, and the stress state at point (r, θ) may have all four components. This problem is not axisymmetric but can be properly formulated and solved in the polar coordinate system. The procedure is as follows.

5.1.1 *Stress state in the polar coordinate system*

According to Saint-Venant's principle, the presence of the hole mainly affects the area around the hole. Assuming $b >> a$, the stress σ at radius b should be effectively the same as that in a plate without a hole. Thus, we know the stress state at any point on the circle of radius b, which can be expressed in Cartesian coordinates as

$$\begin{bmatrix} \sigma & 0 \\ 0 & 0 \end{bmatrix} \tag{5.1}$$

This stress state can be transformed into polar coordinates for further analysis. Let us perform this transformation using symbolic computation [4]:

```
from sympy import pi
a, b, r = sp.symbols('a, b, r', nonnegative=True)
θ, σ = symbols('θ, σ', real=True)

σ_Cart = Matrix([[σ, 0], [0, 0]]) # stress tensor, Cartesian coordinates

T = Matrix([[ sp.cos(θ), sp.sin(θ)],             # T: Transformation matrix
            [-sp.sin(θ), sp.cos(θ)]])              # Cartesian to Polar

σ_polar = T@σ_Cart@T.T
gr.printM(σ_polar, "Stress state on r=b in the polar coordinates:")
```

Stress state on r=b in the polar coordinates:

$$\begin{bmatrix} \sigma\cos^2(\theta) & -\sigma\sin(\theta)\cos(\theta) \\ -\sigma\sin(\theta)\cos(\theta) & \sigma\sin^2(\theta) \end{bmatrix}$$

```
# Import TR7 that reduces sine & cosine power, by increasing the angle:
from sympy.simplify.fu import TR7
σ_rθb = TR7(σ_polar[0,1]).expand().trigsimp()                       # at r = b
display(Math(f" \\text{{Shear stress at $r=b$: σ$_{{rθb}}$}}=\
                                            {latex(σ_rθb)}"))
from sympy.simplify.fu import TR7
σ_rrb = TR7(σ_polar[0,0]).expand()                                   # at r = b
display(Math(f" \\text{{Normal stress at $r=b$: σ$_{{rrb}}$}}=\
                                             {latex(σ_rrb)}"))
```

Shear stress at $r = b$: $\sigma_{r\theta b} = -\dfrac{\sigma\sin(2\theta)}{2}$

Normal stress at $r = b$: $\sigma_{rrb} = \dfrac{\sigma\cos(2\theta)}{2} + \dfrac{\sigma}{2}$

The results show that the stress possesses all components in polar coordinates. Note that the stress component $\sigma_{\theta\theta}$ does not present on the boundary at $r = b$ (it is only inside the boundary). Therefore, it needs not be considered there. Based on the characteristics of the stresses on the boundary at $r = b$, the stress solution in the disk $a < r < b$ can be expressed as a combination of solutions to two subproblems, as shown in Fig. 5.2:

1. **Subproblem-1** is a disk with inner pressure $p_i = 0$ and outer pressure $p_o = -\sigma/2$ (the constant part of σ_{rrb} obtained from the previous code cell). Since σ_{rrb} acts in the r-direction, this is an axisymmetric problem, and we have already derived its solution in Chapters 3 and 4.

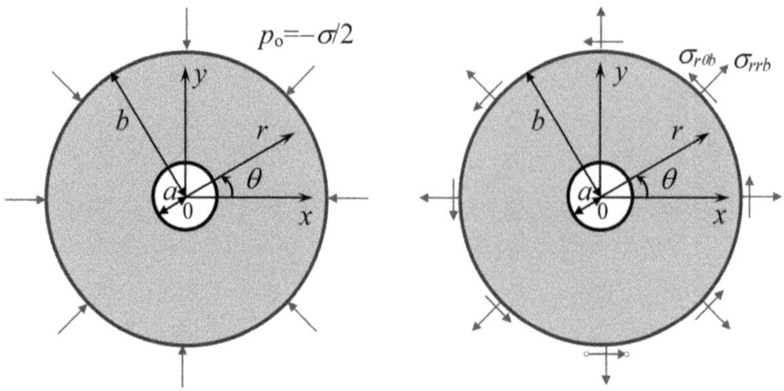

Figure 5.2. Two subproblems for the problem of a plate with a small circular hole.

2. **Subproblem-2** is a disk subjected to normal and shear stresses on the circular boundary $r = b$, given by

$$\begin{aligned} \sigma_{rrb} &= \frac{\sigma}{2}\cos 2\theta \quad \text{at} \quad r = b \\ \sigma_{r\theta b} &= -\frac{\sigma}{2}\sin 2\theta \quad \text{at} \quad r = b \end{aligned} \tag{5.2}$$

Let us now obtain the solution to the simpler Subproblem-1.

5.1.2 *Solution to Subproblem-1*

The solution for Subproblem-1, a disk with $p_o = -\sigma/2$ and $p_i = 0$, is given in Eq. (3.18) with $b \to \infty$. The corresponding expressions are generated using the following code snippet:

```
# final solution of stress concentration on a hole in plates:
p_o = -σ/2
σ_rrpo = b**2*p_o*(-a + r)*(a + r)/(r**2*(a - b)*(a + b))
σ_θθpo = b**2*p_o*(a**2 + r**2)/(r**2*(a - b)*(a + b))

σ_rr1 = σ_rrpo.limit(b,oo).expand()
display(Math(f"\\text{{ The solution of $σ_{{rr1}}$}}={latex(σ_rr1)}"))

σ_θθ1 = σ_θθpo.limit(b,oo).expand()
display(Math(f"\\text{{ The solution of $σ_{{θθ1}}$}}={latex(σ_θθ1)}"))
```

The solution of $\sigma_{rr1} = -\frac{a^2\sigma}{2r^2} + \frac{\sigma}{2}$

The solution of $\sigma_{\theta\theta 1} = \frac{a^2\sigma}{2r^2} + \frac{\sigma}{2}$

5.1.3 *Stress function for Subproblem-2*

Subproblem-2 is clearly not axisymmetric; however, we can obtain its solution using the following procedure. From Eq. (2.59), we know that σ_{rr} is related to the second derivative, and $\sigma_{r\theta}$ is related to the first derivative of the stress function ϕ with respect to θ. Therefore, it is reasonable to assume that ϕ should contain a term involving $\cos 2\theta$, so that the expressions in Eq. (5.2) can be produced. We thus write

$$\phi = f(r)\cos 2\theta \tag{5.3}$$

where $f(r)$ is an unknown function of r. Note also that Subproblem-2 possesses mirror symmetry with respect to $\theta = 0$ and $\theta = \pi/2$. The term $\cos 2\theta$ ensures this symmetry property.

Since there are no body forces in the plate, the assumed function ϕ must satisfy $\nabla^4\phi = 0$, which yields

$$\left(\frac{\partial^2}{\partial r^2} + \frac{1}{r}\frac{\partial}{\partial r} - \frac{4}{r^2}\right)\left(\frac{\partial^2\phi}{\partial r^2} + \frac{1}{r}\frac{\partial\phi}{\partial r} + \frac{4f}{r^2}\right)\cos 2\theta = 0 \tag{5.4}$$

which must be satisfied for all θ. This can happen only if

$$\left(\frac{\partial^2}{\partial r^2} + \frac{1}{r}\frac{\partial}{\partial r} - \frac{4}{r^2}\right)\left(\frac{\partial^2\phi}{\partial r^2} + \frac{1}{r}\frac{\partial\phi}{\partial r} + \frac{4f}{r^2}\right) = 0 \tag{5.5}$$

This is a fourth-order, one-dimensional differential equation (DE) for $f(r)$. Its general solution can be found using the following code:

```
r = sp.symbols('r', nonnegative=True)          # define symbolic variable
f = sp.Function('f')(r)
f2 = diff(diff(f,r),r)+(1/r)*diff(f, r)-4*f/r**2
f4 = diff(diff(f2,r),r)+ (1/r)*diff(f2,r)-4*f2/r**2

diff_eq = sp.Eq(f4, 0)                        # define differential equation
g_sln = sp.dsolve(diff_eq, f)                # solve the differential equation
display(Math(f" \\text{{ General solution for }}{latex(g_sln)}"))
```

General solution for $f(r) = C_1 + \frac{C_2}{r^2} + C_3 r^2 + C_4 r^4$

It has four monomial basis functions. We rewrite the general solution as follows for convenience in later use:

$$f = c_0 r^2 + c_1 r^4 + c_2 r^{-2} + c_3 \tag{5.6}$$

```
ϕ = sp.Function('ϕ')(r, θ)
c0, c1, c2, c3 = symbols('c0, c1, c2, c3')        # integration constants
f = c0*r**2 + c1*r**4 + c2*r**(-2) + c3
ϕ = f*sp.cos(2*θ)
display(Math(f" \\text{{ Stress function for the second problem, ϕ}}=\
                                                        {latex(ϕ)}"))
```

Stress function for the second problem, $\phi = \left(c_0 r^2 + c_1 r^4 + \frac{c_2}{r^2} + c_3\right) \cos(2\theta)$

```
## Check the biharmonic property of the stress function:

# The Laplacian of f(r):
df2 = (diff(diff(f,r),r)+diff(f,r)/r-4*f/r/r).expand()
display(Math(f" \\text{{ $∇^2ϕ$}}= {latex(df2)}"))

df4 = (diff(diff(df2,r),r)+diff(df2,r)/r-4*df2/r/r).expand()
display(Math(f" \\text{{ $∇^4ϕ$}}= {latex(df4)}"))
```

$$\nabla^2\phi = 12c_1 r^2 - \frac{4c_3}{r^2}$$

$$\nabla^4\phi = 0$$

This confirms that the assumed stress function satisfies the compatibility condition in Eq. (5.5).

5.1.4 *General solution to Subproblem-2*

We can now use the obtained stress function to compute the stresses, based on the relationship between the stresses and the stress function given in Eq. (2.59), assuming no body forces:

```
σ_rrg = (diff(diff(ϕ,θ),θ)/r/r+diff(ϕ,r)/r).expand().collect(cos(2*θ))
#display(Math(f" \\text{{ The general solution of σ$_{{rr}}$}}=\
#                                                  {latex(σ_rrg)}"))
σ_θθg = (diff(diff(ϕ,r),r)).expand().collect(cos(2*θ))
#display(Math(f" \\text{{ The general solution of σ$_{{θθ}}$}}=\
#                                                  {latex(σ_θθg)}"))
σ_rθg = (-diff(diff(ϕ,θ)/r,r)).expand().collect(sin(2*θ))
#display(Math(f" \\text{{ The general solution of σ$_{{rθ}}$}}=\
#                                                  {latex(σ_rθg)}"))
print('σ_rrg=', σ_rrg); print('σ_θθg=', σ_θθg); print('σ_rθg=', σ_rθg)
```

```
σ_rrg= (-2*c0 - 6*c2/r**4 - 4*c3/r**2)*cos(2*θ)
σ_θθg= (2*c0 + 12*c1*r**2 + 6*c2/r**4)*cos(2*θ)
σ_rθg= (2*c0 + 6*c1*r**2 - 6*c2/r**4 - 2*c3/r**2)*sin(2*θ)
```

We rewrite the general solution for these stresses in the following equation for future reference.

$$\begin{aligned}\sigma_{rr} &= -\left(2c_0 + \frac{6c_2}{r^4} + \frac{4c_3}{r^2}\right)\cos(2\theta)\\ \sigma_{\theta\theta} &= \left(2c_0 + 12c_1r^2 + \frac{6c_2}{r^4}\right)\cos(2\theta)\\ \sigma_{r\theta} &= \left(2c_0 + 6c_1r^2 - \frac{6c_2}{r^4} - \frac{2c_3}{r^2}\right)\sin(2\theta)\end{aligned} \tag{5.7}$$

The general solution contains a total of four constants. We need four boundary conditions to determine them.

5.1.5 *Boundary conditions for Subproblem-2*

Using the stress boundary condition, Eq. (5.2), we have the following four conditions:

BC-1: At $r = a$, $\sigma_{rr} = 0$

BC-2: At $r = a$, $\sigma_{r\theta} = 0$

BC-3: At $r = b$, $\sigma_{rr} - \frac{\sigma}{2}\cos 2\theta = 0$

BC-4: At $r = b$, $\sigma_{r\theta} + \frac{\sigma}{2}\sin 2\theta = 0$

Note that in each of these four equations, $\cos 2\theta$ or $\sin 2\theta$ can be factored out. Thus, we need only the coefficients of $\cos 2\theta$ or $\sin 2\theta$ to be zero. These coefficients can be located using `sp.arg`:

```
display(σ_rrg.args)
display(σ_θθg.args)
display(σ_rθg.args)
```

$$\left(-2c_0 - \frac{6c_2}{r^4} - \frac{4c_3}{r^2},\ \cos(2\theta)\right)$$

$$\left(2c_0 + 12c_1r^2 + \frac{6c_2}{r^4},\ \cos(2\theta)\right)$$

$$\left(2c_0 + 6c_1r^2 - \frac{6c_2}{r^4} - \frac{2c_3}{r^2},\ \sin(2\theta)\right)$$

We shall use only the first arguments to create these equations:

```
# Use BC-1:
eq1 = σ_rrg.args[0].subs(r, a)
display(Math(f" \\text{{ first equation as an example}}: {latex(eq1)}=0"))
# Use BC-2:
eq2 = σ_rθg.args[0].subs(r, a)
#BC-3:
eq3 = σ_rrg.args[0].subs(r, b ) - σ/2
# Use BC-4:
eq4 = σ_rθg.args[0].subs(r, b) + σ/2
```

Fist equation as an example : $-2c_0 - \frac{4c_3}{a^2} - \frac{6c_2}{a^4} = 0$

5.1.6 *Solution to Subproblem-2*

With these four equations derived from the boundary conditions, we are ready to solve them simultaneously using `sp.solve()`:

```
cs = [c0, c1, c2, c3]                # put four unknown constants in a list
sln_cs = sp.solve([eq1, eq2, eq3, eq4],cs)   # find the solution for cs

display(Math(f" \\text{{ Constant $c_{{0}}$}}=\
                                        {latex(sln_cs[c0].factor())}"))
display(Math(f" \\text{{ Constant $c_{{1}}$}}=\
                                        {latex(sln_cs[c1].factor())}"))
display(Math(f" \\text{{ Constant $c_{{2}}$}}=\
                                        {latex(sln_cs[c2].factor())}"))
display(Math(f" \\text{{ Constant $c_{{3}}$}}=\
                                        {latex(sln_cs[c3].factor())}"))
```

$$\text{Constant } c_0 = \frac{b^2\sigma\left(4a^4 + a^2b^2 + b^4\right)}{4\left(a-b\right)^3\left(a+b\right)^3}$$

$$\text{Constant } c_1 = -\frac{a^2b^2\sigma}{2\left(a-b\right)^3\left(a+b\right)^3}$$

$$\text{Constant } c_2 = \frac{a^4b^4\sigma\left(a^2 + b^2\right)}{4\left(a-b\right)^3\left(a+b\right)^3}$$

$$\text{Constant } c_3 = -\frac{a^2b^2\sigma\left(a^2 - ab + b^2\right)\left(a^2 + ab + b^2\right)}{2\left(a-b\right)^3\left(a+b\right)^3}$$

The constants obtained are functions of b. Since a is assumed to be very small compared to the overall dimensions of the plate, we set $b \to \infty$ to obtain simplified expressions for these constants. The corresponding code snippet is as follows:

```
cs_boo = dict((k, sp.limit(v, b, oo)) for k, v in sln_cs.items())
cs_boo
```

$$\left\{c_0 : -\frac{\sigma}{4},\ c_1 : 0,\ c_2 : -\frac{a^4\sigma}{4},\ c_3 : \frac{a^2\sigma}{2}\right\}$$

Now, we substitute these constants back into the expressions for the stresses, which gives the final stress solutions for Subproblem-2:

```
σ_rr2 = σ_rrg.subs(cs_boo)
display(Math(f"\\text{{Final solution of $σ_{{rr2}}$}}={latex(σ_rr2)}"))

σ_θθ2 = σ_θθg.subs(cs_boo)
display(Math(f"\\text{{Final solution of $σ_{{θθ2}}$}}={latex(σ_θθ2)}"))

σ_rθ2 = σ_rθg.subs(cs_boo)
display(Math(f"\\text{{Final solution of $σ_{{rθ2}}$}}={latex(σ_rθ2)}"))
```

$$\text{Final solution of } \sigma_{rr2} = \left(\frac{3a^4\sigma}{2r^4} - \frac{2a^2\sigma}{r^2} + \frac{\sigma}{2}\right)\cos(2\theta)$$

$$\text{Final solution of } \sigma_{\theta\theta 2} = \left(-\frac{3a^4\sigma}{2r^4} - \frac{\sigma}{2}\right)\cos(2\theta)$$

$$\text{Final solution of } \sigma_{r\theta 2} = \left(\frac{3a^4\sigma}{2r^4} - \frac{a^2\sigma}{r^2} - \frac{\sigma}{2}\right)\sin(2\theta)$$

```
# Redefine these formulas for later use without rerun the codes above:
σ_rr2_ = (3*a**4*σ/(2*r**4) - 2*a**2*σ/r**2 + σ/2)*cos(2*θ)
σ_θθ2_ = (-3*a**4*σ/(2*r**4) - σ/2)*cos(2*θ)
σ_rθ2_ = (3*a**4*σ/(2*r**4) - a**2*σ/r**2 - σ/2)*sin(2*θ)
```

5.1.7 *Final solution for plates subjected to uniaxial tension*

By adding up the solutions for both subproblems, we arrived at the final solution to the problem of stress in a plate with a circular hole:

$$\begin{aligned}
\sigma_{rr} &= \frac{\sigma}{2}\left(1 - \frac{a^2}{r^2}\right) + \frac{\sigma}{2}\left(1 - \frac{2a^2}{r^2} + \frac{3a^4}{r^4}\right)\cos(2\theta) \\
\sigma_{\theta\theta} &= \frac{\sigma}{2}\left(1 + \frac{a^2}{r^2}\right) - \frac{\sigma}{2}\left(1 + \frac{3a^4}{r^4}\right)\cos(2\theta) \\
\sigma_{r\theta} &= -\frac{\sigma}{2}\left(1 + \frac{2a^2}{r^2} - \frac{3a^4}{r^4}\right)\sin(2\theta)
\end{aligned} \tag{5.8}$$

Let us define the complete solutions using the formulas obtained, so that they can be used directly later without needing to rerun all the code above:

```
σ_rr = σ_rr2 + σ_rr1
σ_θθ = σ_θθ2 + σ_θθ1
σ_rθ = σ_rθ2 + 0

a, b, r = sp.symbols('a, b, r', nonnegative=True)
θ, σ = symbols('θ, σ', real=True)

r2 = r**2; r4 = r**4
σ_rr =-a**2*σ/(2*r2) + σ/2 + (3*a**4*σ/(2*r4)-2*a**2*σ/r2+σ/2)*cos(2*θ)
σ_θθ = a**2*σ/(2*r2) + σ/2 + (-3*a**4*σ/(2*r4) - σ/2)*cos(2*θ)
σ_rθ = (3*a**4*σ/(2*r4) - a**2*σ/r2 - σ/2)*sin(2*θ)
```

It is always good practice to verify the solution in special cases. When $r \to \infty$, the stress state should reduce to that of pure tension. We can also check whether the boundary conditions are satisfied by the final solution. First, let us compute the stresses in the far field in polar coordinates:

```
σ_rroo = sp.limit(σ_rr, r, oo)
display(Math(f" \\text{{ At $r→∞$, $σ_{{rr}}$}}={latex(σ_rroo)}"))

σ_θθoo = sp.limit(σ_θθ, r, oo)
display(Math(f" \\text{{ At $r→∞$, $σ_{{θθ}}$}}={latex(σ_θθoo)}"))

σ_rθoo = sp.limit(σ_rθ, r, oo)
display(Math(f" \\text{{ At $r→∞$, $σ_{{rθ}}$}}={latex(σ_rθoo)}"))
```

At $r \to \infty$, $\sigma_{rr} = \sigma \cos^2(\theta)$

At $r \to \infty$, $\sigma_{\theta\theta} = \sigma \sin^2(\theta)$

At $r \to \infty$, $\sigma_{r\theta} = -\dfrac{\sigma \sin(2\theta)}{2}$

It is clear that, at the far field, the normal stresses σ_{rr} and $\sigma_{\theta\theta}$ are bounded by σ, while the shear stress $\sigma_{r\theta}$ is bounded by $\pm\sigma/2$. This agrees with the specified condition that the normal stress $\sigma_{xx} = \sigma$ in the far field.

Next, we transform these stresses into Cartesian coordinates using the same transformation matrix, but applying the inverse transformation formula for a second-order tensor (see Chapter 3 of Ref. [4]):

```
σ_polar = Matrix([[σ_rroo, σ_rθoo],
                  [σ_rθoo, σ_θθoo]]) # stress tensor, polar coordinates

σ_Cart = sp.simplify((T.T@σ_polar@T))          # inverse transformation
display(Math(f" \\text{{ Stress state on $r=b$ in the Cartesian\
                                    coordinates}}={latex(σ_Cart)}"))
```

Stress state on $r = b$ in the Cartesian coordinates $= \begin{bmatrix} \sigma & 0 \\ 0 & 0 \end{bmatrix}$

We have now returned to the original problem setup: a plate with a hole subjected to pure tension, thus completing the entire study.

We can also verify whether the boundary conditions are satisfied by the final solution:

```
display(Math(f" \\text{{ At $r=a$, $σ_{{rr}}$}}=\
                                    {latex(σ_rr.subs({r:a}))}"))
display(Math(f" \\text{{ At $r=a$, $σ_{{rθ}}$}}=\
                                    {latex(σ_rθ.subs({r:a}))}"))
```

At $r = a$, $\sigma_{rr} = 0$

At $r = a$, $\sigma_{r\theta} = 0$

The boundary conditions at $r = a$ are confirmed to be satisfied.

5.1.8 *Stress concentration factor*

Using the obtained stress solution, we can now compute the stress concentration factor (SCF) around the hole. First, we derive the expression for the stress at $r = a$:

```
σ_rra = σ_rr.subs({r:a})
display(Math(f" \\text{{ Hoop stress at $r=a$, $σ_{{rra}}$}}=\
                                              {latex(σ_rra)}"))
σ_θθa = σ_θθ.subs({r:a})
display(Math(f" \\text{{ Hoop stress at $r=a$, $σ_{{θθa}}$}}=\
                                              {latex(σ_θθa)}"))
```

Hoop stress at $r = a$, $\sigma_{rra} = 0$

Hoop stress at $r = a$, $\sigma_{\theta\theta a} = -2\sigma\cos(2\theta) + \sigma$

It is observed that only the hoop stress, $\sigma_{\theta\theta}$, is of interest. The hoop stress depends on θ and therefore varies around the circle at $r = a$. Its maximum and minimum values are found as follows:

1. When $\theta = \pi/2$ or $3\pi/2$, corresponding to points A and A′ shown in Fig. 5.1, the maximum hoop stress is obtained: $\sigma_{\theta\theta}^{\max} = 3\sigma$. The **stress concentration factor** (SCF) is thus 3, which is a widely cited result in stress analysis.
2. When $\theta = 0$ or π, corresponding to points B and B′ shown in Fig. 5.1, the hoop stress is $\sigma_{\theta\theta} = -\sigma$, indicating compression.

```
display(Math(f" \\text{{ Maximum hoop stress at ($r=a,θ=\pi/2$)}}=\
                                    {latex(σ_θθ.subs({r:a, θ:pi/2}))}"))
display(Math(f" \\text{{ Minimum hoop stress at ($r=a,θ=0$)}}=\
                                       {latex(σ_θθ.subs({r:a, θ:0}))}"))
```

Maximum hoop stress at $(r = a, \theta = \pi/2) = 3\sigma$

Minimum hoop stress at $(r = a, \theta = 0) = -\sigma$

5.1.9 *SCF of 4 under a combined stress state*

Note that if the loading condition changes, higher stress concentrations may occur. For example, if a load of $-\sigma$ is also applied in the y-direction, as shown in Fig. 5.3, the SCF at points A and A′ becomes 4.

The code snippet to compute this is as follows:

```
σ_factor = σ_θθ.subs({r:a, θ:pi/2}) - σ_θθ.subs({r:a, θ:0})
display(Math(f" \\text{{ Hoop stress concentration factor at point A\
                              in combined loading}}={latex(σ_factor)}"))

σ_factor = σ_θθ.subs({r:a, θ:0}) - σ_θθ.subs({r:a, θ:pi/2})
display(Math(f" \\text{{ Hoop stress concentration factor at point B\
                              in combined loading}}={latex(σ_factor)}"))
```

Hoop stress concentration factor at point A in combined loading $= 4\sigma$

Hoop stress concentration factor at point B in combined loading $= -4\sigma$

In the case of equally combined loads, point A has a stress concentration factor of 4, while point B has a stress concentration factor of -4.

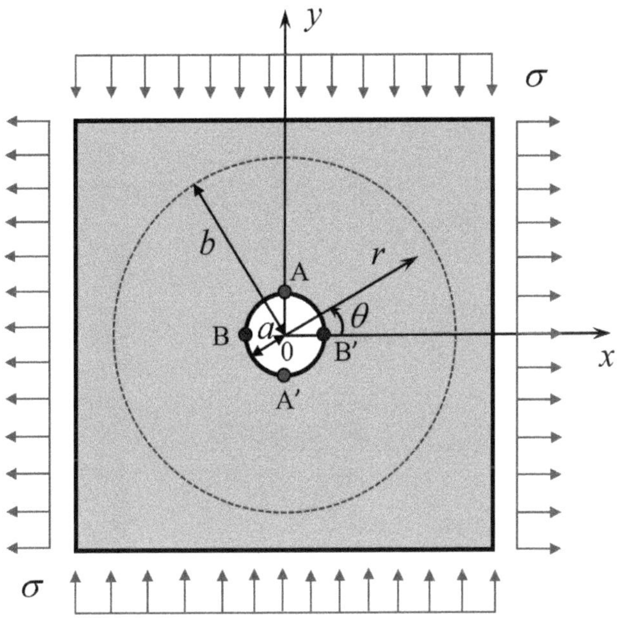

Figure 5.3. A plate with a small circular hole of radius a in the middle, subjected to uniform tensile stress σ on the left and right edges and uniform compressive stress σ on the top and bottom edges.

5.1.10 *Stress variation along r*

5.1.10.1 *Stress variation along r at $\theta = \pi/2$*

Along the vertical cross-section where $\theta = \pi/2$, we have:

```
σ_rr_along_r = σ_rr.subs({θ:pi/2})
display(Math(f" \\text{{ Stress $σ_{{rr}}$ (at $θ=π/2$ along $r$)}}=\
                                            {latex(σ_rr_along_r)}"))
σ_θθ_along_r = σ_θθ.subs({θ:pi/2})
display(Math(f" \\text{{ Stress $σ_{{θθ}}$ (at $θ=π/2$ along $r$)}}=\
                                            {latex(σ_θθ_along_r)}"))
σ_rθ_along_r = σ_rθ.subs({θ:pi/2})
display(Math(f" \\text{{ Stress $σ_{{rθ}}$ (at $θ=π/2$ along $r$)}}=\
                                            {latex(σ_rθ_along_r)}"))
```

$$\text{Stress } \sigma_{rr} \text{ (at } \theta = \pi/2 \text{ along } r) = -\frac{3a^4\sigma}{2r^4} + \frac{3a^2\sigma}{2r^2}$$

$$\text{Stress } \sigma_{\theta\theta} \text{ (at } \theta = \pi/2 \text{ along } r) = \frac{3a^4\sigma}{2r^4} + \frac{a^2\sigma}{2r^2} + \sigma$$

$$\text{Stress } \sigma_{r\theta} \text{ (at } \theta = \pi/2 \text{ along } r) = 0$$

The nonzero stresses decay with a leading order of r^2.

5.1.10.2 *Stress variation along r with $\theta = 0$*

Along the horizontal cross-section where $\theta = 0$, we have:

```
σ_rr_along_r = σ_rr.subs({θ:0})
display(Math(f" \\text{{ Stress $σ_{{rr}}$ (at $θ=0$ along $r$)}}=\
                                        {latex(σ_rr_along_r)}"))
σ_θθ_along_r = σ_θθ.subs({θ:0})
display(Math(f" \\text{{ Stress $σ_{{θθ}}$ (at $θ=0$ along $r$)}}=\
                                        {latex(σ_θθ_along_r)}"))
σ_rθ_along_r = σ_rθ.subs({θ:0})
display(Math(f" \\text{{ Stress $σ_{{rθ}}$ (at $θ=0$ along $r$)}}=\
                                        {latex(σ_rθ_along_r)}"))
```

$$\text{Stress } \sigma_{rr} \text{ (at } \theta = 0 \text{ along } r) = \frac{3a^4\sigma}{2r^4} - \frac{5a^2\sigma}{2r^2} + \sigma$$

$$\text{Stress } \sigma_{\theta\theta} \text{ (at } \theta = 0 \text{ along } r) = -\frac{3a^4\sigma}{2r^4} + \frac{a^2\sigma}{2r^2}$$

$$\text{Stress } \sigma_{r\theta} \text{ (at } \theta = 0 \text{ along } r) = 0$$

5.1.10.3 *Plotting the stress distribution*

Finally, we use the following code to plot the stress distribution. The SymPy expressions are first converted to Python functions, which are then plotted using `plt.plot()`:

```
# SymPy expressions to Python functions:
func1   = sp.lambdify(r, σ_rr.subs({a:1, σ:1, θ:pi/2}), 'numpy')
func2   = sp.lambdify(r, σ_θθ.subs({a:1, σ:1, θ:pi/2}), 'numpy')
func3   = sp.lambdify(r, σ_rr.subs({a:1, σ:1, θ:0}), 'numpy')
func4   = sp.lambdify(r, σ_θθ.subs({a:1, σ:1, θ:0}), 'numpy')
func5   = sp.lambdify(r, σ_rr.subs({a:1, σ:1, θ:pi/4}), 'numpy')
func6   = sp.lambdify(r, σ_θθ.subs({a:1, σ:1, θ:pi/4}), 'numpy')
func7   = sp.lambdify(r, σ_rθ.subs({a:1, σ:1, θ:pi/4}), 'numpy')
labels=[r'$σ_{rr}, θ=\pi/2$',r'$σ_{θθ}, θ=\pi/2$',
        r'$σ_{rr}, θ=0$',     r'$σ_{θθ}, θ=0$',
        r'$σ_{rr}, θ=\pi/4$',r'$σ_{θθ}, θ=\pi/4$', r'$σ_{rθ}, θ=\pi/4$']

r_data  = np.linspace(1, 5, 400)                 # r values for the plot
x_data = [r_data, r_data, r_data, r_data, r_data, r_data, r_data]
y_data = [func1(x_data[0]), func2(x_data[1]), func3(x_data[2]),
          func4(x_data[3]), func5(x_data[4]), func6(x_data[5]),
          func7(x_data[6])]

#gr.plot_fs(x_data, y_data, labels, '$r$', 'stresses', \
#                                  'imagesSMp/plateHoleSij_r',ioff=True)
                                   # Use ioff=False to display the plot
```

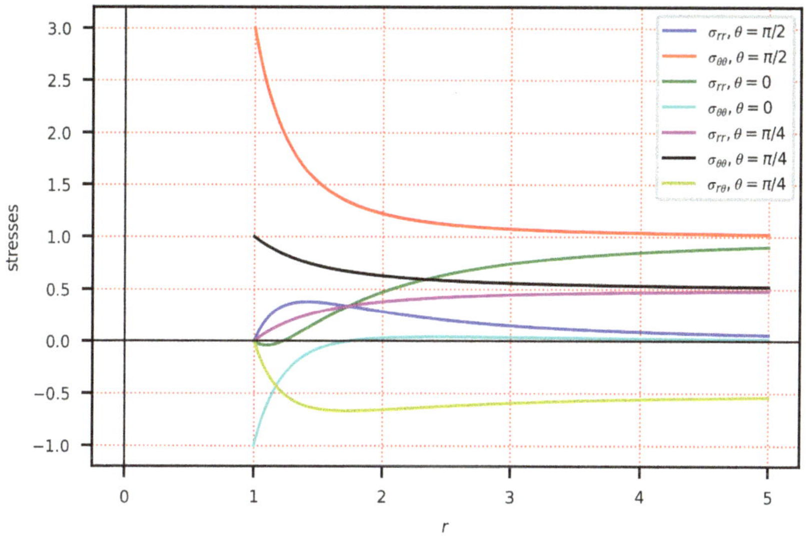

Figure 5.4. Distribution of the normal stresses σ_{rr} and $\sigma_{\theta\theta}$ along both vertical and horizontal sections.

The two normal stresses, σ_{rr} and $\sigma_{\theta\theta}$, along both the vertical and horizontal sections are plotted in Fig. 5.4. The stress concentration around the hole is clearly visible, along with its variation with respect to the radial direction, r.

It is also observed that at $r = a$, the normal stress $\sigma_{rr} = 0$, and in the far field ($r \to \infty$), the normal stresses σ_{rr} and $\sigma_{\theta\theta}$ are bounded by σ, consistent with the specified condition, $\sigma_{xx} = \sigma$, at the far field. The shear stress $\sigma_{r\theta}$ is bounded by $\pm\sigma/2$ at the far field. Therefore, over the entire plate, the critical stress component is the hoop stress $\sigma_{\theta\theta}$ along the hole edge.

These nonzero stresses decay quadratically with increasing r. If the plate width is larger than four times the hole diameter, the error in the maximum hoop stress, $\sigma_{\theta\theta}^{\max}$, is less than 6%.

5.1.10.4 *Contour plot of stresses*

To plot the stress contours, we first express the stresses in Cartesian coordinates using the relations (see Chapter 2)

$$r = \sqrt{x^2 + y^2}, \quad \theta = \arctan\left(\frac{y}{x}\right) \tag{5.9}$$

The expressions are then converted to NumPy functions, and **gr.contour_plot()** is used to generate the contour plots. The corresponding code snippet is as follows:

```
# Contour of σ_θθ:
x, y = sp.symbols('x, y', real=True)
a_val = 1
dic_rt = {σ: 1, a:a_val, r:sp.sqrt(x**2+y**2), θ:sp.atan(y/x)}
σ_θθxy = σ_θθ.subs(dic_rt)
display(Math(f" \\text{{$σ_{{θθ}}$ written in $(x, y)$}} \
                                                ={latex(σ_θθxy)}"))
```

$$\sigma_{\theta\theta} \text{ written in } (x, y) = \left(-\frac{1}{2} - \frac{3}{2\left(x^2+y^2\right)^2}\right)\cos\left(2\operatorname{atan}\left(\frac{y}{x}\right)\right) + \frac{1}{2} + \frac{1}{2\left(x^2+y^2\right)}$$

```
# Create a mesh grid for a rectangular domain in x-y plane:
dX = 3.0*a_val;     dY = 2.0*a_val                   # plot 2dX × 2dY area
x_val = np.linspace(-dX, dX, 800)
y_val = np.linspace(-dY, dY, 800)
X, Y = np.meshgrid(x_val, y_val)

# define a circular hole to be masked in plot:
mask = (X/a_val)**2 + (Y/a_val)**2 <= 1

# Hoop stress:
np_σθθxy = lambdify((x,y), σ_θθxy, 'numpy')            # convert to numpy
Z = np_σθθxy(X, Y)                      # compute the stress over the mesh grid
Z[mask] = np.nan                          # mask the hole, print nothing there

gr.contour_plot(X,Y, Z, x_lbl='x-axis',y_lbl='y-axis',
                title='Stress $σ_{{θθ}}$',lvls=80, ioff=True);
plt.savefig('imagesSMp/plateHoleStt_rt.png', dpi=500)     # save to file
#plt.show()
```

```
Minimum value in the domain: -0.987528
Maximum value in the domain: 2.98499
```

A contour plot of the hoop stress, $\sigma_{\theta\theta}$, within a plate containing a circular hole is shown in Fig. 5.5. The stress distribution across the plate and the concentration around the hole are clearly illustrated.

Note that in practical structures, holes or defects are not necessarily circular. In general, the stress concentration around a circular hole is less severe due to its constant radius. For other hole shapes, the stress concentration factor can be higher. If the defect is a sharp crack, the stress concentration factor can theoretically become infinite. Therefore, converting a

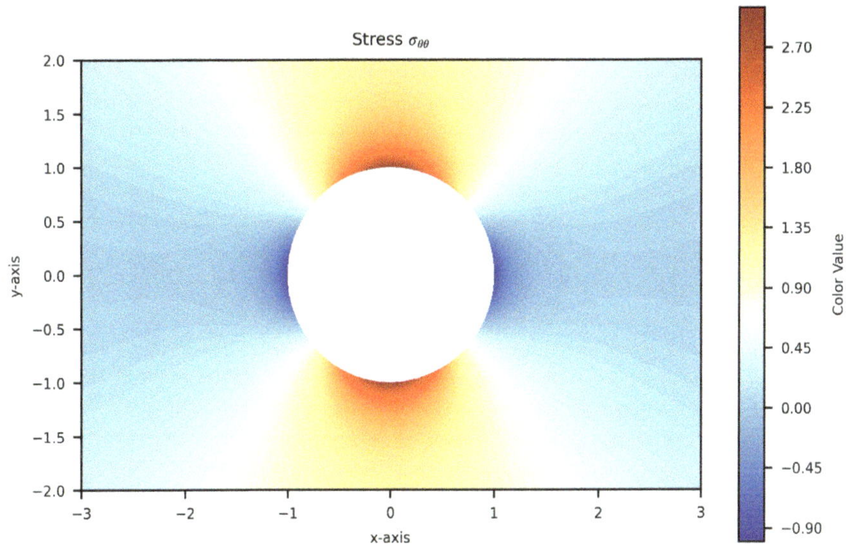

Figure 5.5. Contour of the hoop stress $\sigma_{\theta\theta}$ in a plate with a circular hole.

defect to a circular hole by drilling may serve as a quick (though temporary) remedy. If the material removed is minimal, the stress concentration can be moderately reduced.

```
# Contour of σ_rr:
σ_rrxy = σ_rr.subs(dic_rt)
display(Math(f" \\text{{ $σ_{{rr}}$ }} \
                                                    ={latex(σ_rrxy)}"))
np_σrrxy = lambdify((x,y), σ_rrxy, 'numpy')          # convert to numpy
Z = np_σrrxy(X, Y)                       # compute the stress over the mesh grid
title = 'Displacement u for $σ$ applied in the $y$-direction'
Z[mask] = np.nan                            # mask the hole, print nothing there

gr.contour_plot(X,Y,Z,x_lbl='x-axis',y_lbl='y-axis',
                title='Stress $σ_{{rr}}$',lvls=80, ioff=True);
plt.savefig('imagesSMp/plateHoleSrr_rt.png', dpi=500)     # save to file
```

$$\sigma_{rr} = \left(\frac{1}{2} - \frac{2}{x^2+y^2} + \frac{3}{2\left(x^2+y^2\right)^2}\right)\cos\left(2\,\mathrm{atan}\left(\frac{y}{x}\right)\right) + \frac{1}{2} - \frac{1}{2\left(x^2+y^2\right)}$$

```
Minimum value in the domain: -0.0416374
Maximum value in the domain: 0.74074
```

A contour plot of the radial normal stress, σ_{rr}, within a plate containing a circular hole is shown in Fig. 5.6. The stress distribution across the plate and the zero value of σ_{rr} along the hole are clearly visible. Along the left and right edges, the stress approaches the far-field value, σ:

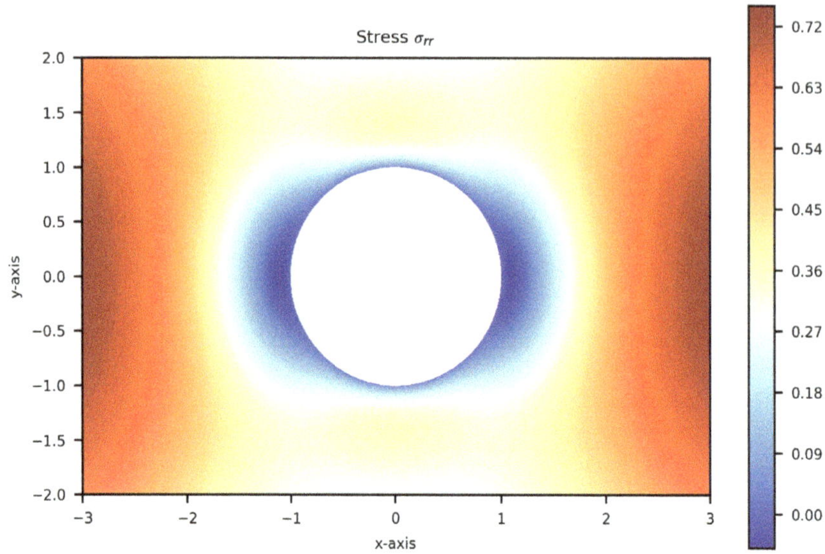

Figure 5.6. Contour of the radial normal stress σ_{rr} in a plate with a circular hole.

```
# Contour of σ_rθ:
σ_rθxy = σ_rθ.subs(dic_rt)
display(Math(f" \\text{{ $σ_{{rθ}}$ }}={latex(σ_rθxy)}"))
np_σrθxy = lambdify((x,y), σ_rθxy, 'numpy')          # convert to numpy
Z = np_σrθxy(X, Y)                    # compute the stress over the mesh grid
title = 'Displacement u for $σ$ applied in the $y$-direction'
Z[mask] = np.nan                         # mask the hole, print nothing there

gr.contour_plot(X,Y,Z,x_lbl='x-axis',y_lbl='y-axis',
                title='Stress $σ_{{rθ}}$',lvls=80, ioff=True);
plt.savefig('imagesSMp/plateHoleSrθ_rt.png', dpi=500)     # save to file
```

$$\sigma_{r\theta} = \left(-\frac{1}{2} - \frac{1}{x^2+y^2} + \frac{3}{2\left(x^2+y^2\right)^2}\right)\sin\left(2\,\mathrm{atan}\left(\frac{y}{x}\right)\right)$$

```
Minimum value in the domain: -0.666663
Maximum value in the domain: 0.666663
```

A contour plot of the shear stress, $\sigma_{r\theta}$, within a plate containing a circular hole is shown in Fig. 5.7. The stress distribution across the plate and the zero value of $\sigma_{r\theta}$ along the hole are clearly illustrated. It is also observed that the shear stress reaches its maximum and minimum along lines at angles of $\theta = \pm 45°$ to the axes.

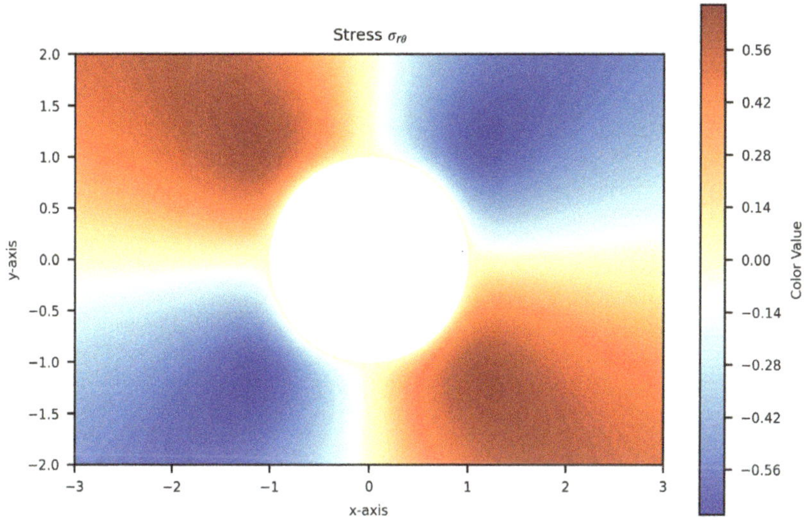

Figure 5.7. Contour of the shear stress $\sigma_{r\theta}$ in a plate with a circular hole.

5.2 Plates subjected to biaxial loading

Consider another practical problem: a plate with a small circular hole at its center. It is subjected to biaxial tension, with stress levels of σ_{xx} in the x-direction and σ_{yy} in the y-direction at the plate edges, as shown in Fig. 5.8.

Without the hole, the plate is in a principal stress state with two nonzero components, σ_{xx} and σ_{yy}, in Cartesian coordinates:

$$\begin{bmatrix} \sigma_{xx} & 0 \\ 0 & \sigma_{yy} \end{bmatrix} \tag{5.10}$$

Naturally, the presence of the hole alters the stress distribution near it, and the stress state at a point (r, θ) can generally have all four stress components. This problem can be solved similar to the uniaxial tension case. However, a simpler approach uses superposition by employing the solution for uniaxial tension.

5.2.1 *Solution by superposition*

The solution to the biaxial problem can be naturally expressed as the sum of two contributions, one from σ_{xx} and the other from σ_{yy}:

1. The contribution from σ_{xx} is the same as that given in Eq. (5.8), simply replacing σ with σ_{xx}.

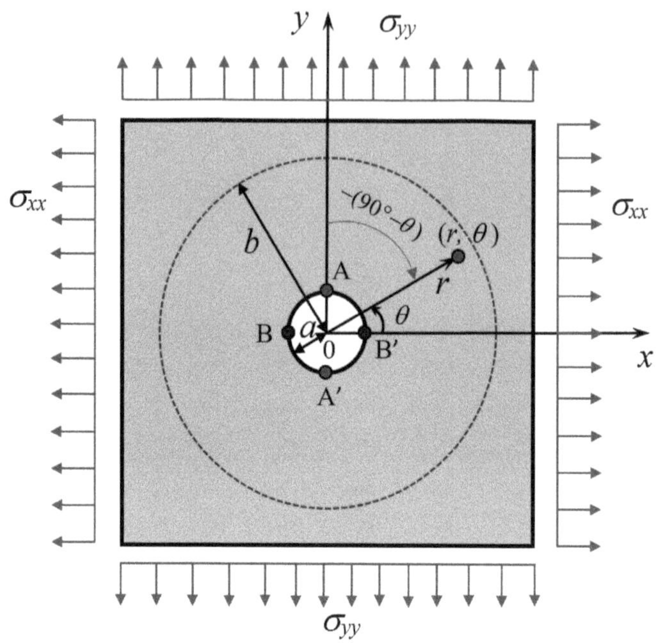

Figure 5.8. A plate with a small circular hole of radius a at its center, subjected to biaxial tension at its edges.

2. The contribution from σ_{yy} is obtained using Eq. (5.8) by replacing σ with σ_{yy} and θ with $-(90° - \theta)$, as illustrated in Fig. 5.8.

The sum of these two contributions gives the final stress solution. For more readable formulas, we introduce the average stress, $\sigma_{\text{ave}} = (\sigma_{xx}+\sigma_{yy})/2$, and the difference stress, $\sigma_{\text{dif}} = (\sigma_{xx}-\sigma_{yy})/2$, at the far field. The corresponding code snippet to compute these quantities is as follows:

```
σxx, σyy = symbols('σ_xx, σ_yy', real=True)     # Stresses at far field
σave, σdif = symbols('σ_ave, σ_dif', real=True)

eqn1 = σave - (σxx+σyy)/2            # σave: average stress at far field
eqn2 = σdif - (σxx-σyy)/2          # σdif difference stress at far field
dic_σ = sp.solve([eqn1, eqn2], [σxx, σyy])
dic_σ
```

$\{\sigma_{xx}: \sigma_{\text{ave}} + \sigma_{\text{dif}},\ \sigma_{yy}: \sigma_{\text{ave}} - \sigma_{\text{dif}}\}$

```
# The contribution from σxx, needs a simple substitution:
σ_polar_σxx = Matrix([[σ_rr, σ_rθ],
                      [σ_rθ, σ_θθ]]).subs(σ, σxx)

# The contribution from σyy, needs two substitutions:
θy = pi/2
σ_polar_σyy = Matrix([[σ_rr, σ_rθ],
                      [σ_rθ, σ_θθ]]).subs(σ, σyy).subs(θ, θ-θy)

display(Math(f" \\text{{Contribution from $σ_{{yy}}$:}} "))
display(Math(f" \\text{{ }}\\small {latex(σ_polar_σyy)}"))

# The final combined solution:
σ_polar_biaxial = sp.simplify((σ_polar_σxx + σ_polar_σyy)\
                                        .subs(dic_σ)).expand()
# simplify the expressions:
σ_polar_biaxial = σ_polar_biaxial.applyfunc(\
                          lambda expr: sp.collect(expr,cos(2*θ)))
σ_polar_biaxial = σ_polar_biaxial.applyfunc\
                         (lambda expr: sp.collect(expr,sin(2*θ)))

display(Math(f" \\text{{Solution for plate under biaxial tension:}} "))
display(Math(f" \\text{{$σ_{{rr}}$}}= {latex(σ_polar_biaxial[0,0])}"))
display(Math(f" \\text{{$σ_{{θθ}}$}}= {latex(σ_polar_biaxial[1,1])}"))
display(Math(f" \\text{{$σ_{{rθ}}$}}= {latex(σ_polar_biaxial[0,1])}"))
```

Contribution from σ_{yy}:

$$\begin{bmatrix} -\frac{a^2\sigma_{yy}}{2r^2} + \frac{\sigma_{yy}}{2} - \left(\frac{3a^4\sigma_{yy}}{2r^4} - \frac{2a^2\sigma_{yy}}{r^2} + \frac{\sigma_{yy}}{2}\right)\cos(2\theta) & -\left(\frac{3a^4\sigma_{yy}}{2r^4} - \frac{a^2\sigma_{yy}}{r^2} - \frac{\sigma_{yy}}{2}\right)\sin(2\theta) \\ -\left(\frac{3a^4\sigma_{yy}}{2r^4} - \frac{a^2\sigma_{yy}}{r^2} - \frac{\sigma_{yy}}{2}\right)\sin(2\theta) & \frac{a^2\sigma_{yy}}{2r^2} + \frac{\sigma_{yy}}{2} - \left(-\frac{3a^4\sigma_{yy}}{2r^4} - \frac{\sigma_{yy}}{2}\right)\cos(2\theta) \end{bmatrix}$$

Solution for plate under biaxial tension:

$$\sigma_{rr} = -\frac{a^2\sigma_{\text{ave}}}{r^2} + \sigma_{\text{ave}} + \left(\frac{3a^4\sigma_{\text{dif}}}{r^4} - \frac{4a^2\sigma_{\text{dif}}}{r^2} + \sigma_{\text{dif}}\right)\cos(2\theta)$$

$$\sigma_{\theta\theta} = \frac{a^2\sigma_{\text{ave}}}{r^2} + \sigma_{\text{ave}} + \left(-\frac{3a^4\sigma_{\text{dif}}}{r^4} - \sigma_{\text{dif}}\right)\cos(2\theta)$$

$$\sigma_{r\theta} = \left(\frac{3a^4\sigma_{\text{dif}}}{r^4} - \frac{2a^2\sigma_{\text{dif}}}{r^2} - \sigma_{\text{dif}}\right)\sin(2\theta)$$

This is the final solution for a plate with a small circular hole subjected to biaxial tension.

5.2.2 *Stress concentration factor under biaxial loading*

5.2.2.1 *SCF of 2 under biaxial tension*

The stress concentration factor along the hole edge in a plate subjected to biaxial tension of equal magnitude can be derived by setting $r = a$, $\sigma_{\text{ave}} = \sigma_{xx} = \sigma_{yy}$, and $\sigma_{\text{dif}} = 0$:

```
display(Math(f"\\text{{Solution around the hole for biaxial load:}}"))
display(Math(f" {latex(σ_polar_biaxial.subs({r:a,σave:σxx,σdif:0}))}"))
```

Solution around the hole for biaxial load:

$$\begin{bmatrix} 0 & 0 \\ 0 & 2\sigma_{xx} \end{bmatrix}$$

It is again observed that the hoop stress is twice the nominal stress at any point along the hole edge. The stress concentration factor (SCF) is therefore 2. This indicates that when a plate with a hole is subjected to hydrostatic pressure on its outer edges, the hoop stress is twice the nominal pressure everywhere along the hole edge.

5.2.2.2 *SCF of 4 under biaxial tension-compression*

When the biaxial load consists of tension in the x-direction and compression in the y-direction of equal magnitude, the stresses along the hole edge can be obtained by setting $r = a$, $\sigma_{\text{ave}} = 0$, and $\sigma_{\text{dif}} = \sigma_{xx} = -\sigma_{yy}$. This corresponds to the case studied earlier, with the stress state illustrated in Fig. 5.3. This solution is computed as follows:

```
display(Math(f"\\text{{Solution around the hole for biaxial load:}}"))
display(Math(f" {latex(σ_polar_biaxial.subs({r:a,σave:0,σdif:σxx}))}"))
```

Solution around the hole for biaxial load:

$$\begin{bmatrix} 0 & 0 \\ 0 & -4\sigma_{xx}\cos(2\theta) \end{bmatrix}$$

We find:

1. When $\theta = \pi/2$ or $3\pi/2$, corresponding to points A and A′ in Fig. 5.3, the maximum hoop stress is $\sigma_{\theta\theta}^{\max} = 4\sigma_{xx}$. The **stress concentration factor (SCF)** is therefore 4.

2. When $\theta = 0$ or π, corresponding to points B and B′ in Fig. 5.3, the minimum hoop stress is $\sigma_{\theta\theta} = -4\sigma_{xx}$.

The result agrees with previous findings, but now provides a precise and simple formula to compute the hoop stress along the hole edge.

5.3 Plates subjected to pure shear

Consider another fundamental problem: a plate with a small circular hole at its center, subjected to uniform shear stress τ along all edges, as shown in Fig. 5.9.

Without the hole, the plate is in a simple pure shear stress state with only one nonzero component, $\sigma_{xy} = \tau$. Naturally, the presence of the hole affects the stress distribution near it, and the stress state at a point (r, θ) can involve all four stress components. This problem can be properly set up and solved in the polar coordinate system, following essentially the same procedure as for the plate under uniaxial tension.

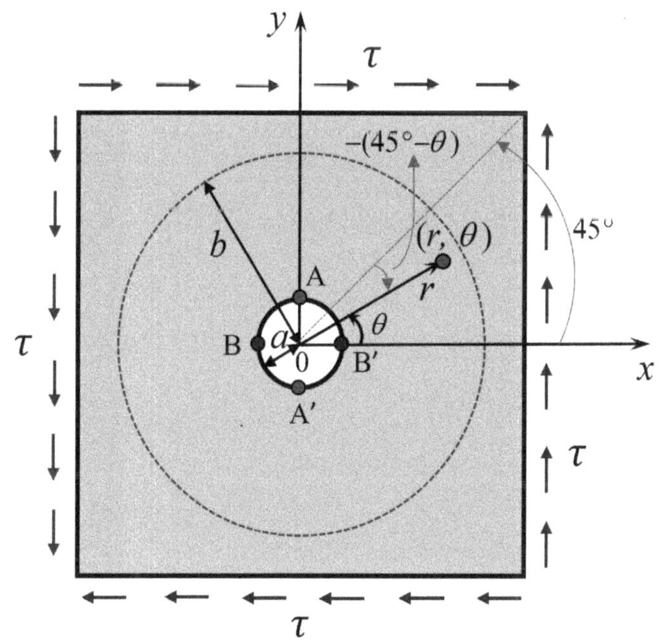

Figure 5.9. A plate with a small circular hole of radius a at its center, subjected to uniform shear stress τ along all edges.

5.3.1 *Stress state in the polar coordinate system*

Assuming $b \gg a$, the shear stress τ at radius b should be effectively the same as in the plate without a hole. Therefore, the stress state at any point on the circle of radius b can be expressed in Cartesian coordinates as

$$\begin{bmatrix} 0 & \tau \\ \tau & 0 \end{bmatrix} \tag{5.11}$$

This stress state can then be transformed into polar coordinates for further analysis [4]:

```
from sympy import pi
a, b, r = sp.symbols('a, b, r', nonnegative=True)
θ, τ = symbols('θ, τ', real=True)

σ_Cart = Matrix([[0, τ], [τ, 0]])  # stress tensor, Cartesian coordinate

T = Matrix([[ sp.cos(θ), sp.sin(θ)],           # T: Transformation matrix
            [-sp.sin(θ), sp.cos(θ)]])               # Cartesian to Polar

σ_polar = T@σ_Cart@T.T
gr.printM(σ_polar, "Stress state on r=b in the polar coordinates:")
```

Stress state on r=b in the polar coordinates:

$$\begin{bmatrix} 2\tau\sin(\theta)\cos(\theta) & -\tau\sin^2(\theta)+\tau\cos^2(\theta) \\ -\tau\sin^2(\theta)+\tau\cos^2(\theta) & -2\tau\sin(\theta)\cos(\theta) \end{bmatrix}$$

```
# Import TR7 that reduces sine & cosine power, using double angle:
from sympy.simplify.fu import TR7
σ_rθb = TR7(σ_polar[0,1]).collect(τ).trigsimp()                # at r = b
display(Math(f" \\text{{Shear stress at $r=b$: σ$_{{rθb}}$}}=\
                                            {latex(σ_rθb)}"))
from sympy.simplify.fu import TR7
σ_rrb = TR7(σ_polar[0,0]).trigsimp()                           # at r = b
display(Math(f" \\text{{Normal stress at $r=b$: σ$_{{rrb}}$}}=\
                                            {latex(σ_rrb)}"))
```

Shear stress at $r = b$: $\sigma_{r\theta b} = \tau\cos(2\theta)$

Normal stress at $r = b$: $\sigma_{rrb} = \tau\sin(2\theta)$

The differences between pure shear and uniaxial tension are as follows:

1. There is no constant term for σ_{rr}.
2. The terms $\sin 2\theta$ and $\cos 2\theta$ are swapped.

Due to item 1, we only need to consider the second subproblem, shown in the left figure of Fig. 5.2. The stress boundary condition on the circular boundary $r = b$ becomes

$$\begin{aligned} \sigma_{rrb} &= \tau \sin 2\theta \quad \text{at} \quad r = b \\ \sigma_{r\theta b} &= \tau \cos 2\theta \quad \text{at} \quad r = b \end{aligned} \tag{5.12}$$

Due to the sine–cosine swapping, the stress function ϕ should include a term in $\sin 2\theta$, expressed as

$$\phi = f(r) \sin 2\theta \tag{5.13}$$

The rest of the procedure leading to the final solutions is very similar to the uniaxial tension problem and requires only minor modifications to the code. We stop here, leaving it as a good exercise for the reader to complete the process.

Alternatively, we can use the superposition technique to obtain the solution for the pure shear problem. This provides a useful exercise in stress state analysis, as discussed in detail in Ref. [4].

Note that the pure shear problem in Fig. 5.2 can be converted to a biaxial loading problem, with a tensile stress σ_{XX} along the X-direction and a compressive stress σ_{YY} along the Y-direction, as shown in Fig. 5.10.

This equivalence can be confirmed using coordinate transformation, as performed earlier, along with the substitution of τ by σ_{xy}:

```
σxy = symbols('σ_xy', real=True)                    # Stresses at far field
σ_rb = σ_polar.subs(θ,pi/4).subs(τ, σxy)
display(Math(f" \\text{{Pure shear state at $r=b$ (far field)}} = \
                                                    {latex(σ_rb)}"))
```

$$\text{Pure shear state at } r = b \text{ (far field)} = \begin{bmatrix} \sigma_{xy} & 0 \\ 0 & -\sigma_{xy} \end{bmatrix}$$

5.3.2 *Solution by superposition*

We can therefore use the formulas for biaxial tension obtained in the previous section, with the following modifications:

1. Replace σ_{xx} with σ_{xy}, and σ_{yy} with $-\sigma_{xy}$.
2. Replace θ with $-(45° - \theta)$, as shown in Fig. 5.10.

For these substitutions, the formulas expressed in terms of average and difference stresses must first be converted to formulas in terms of σ_{xx} and σ_{yy}.

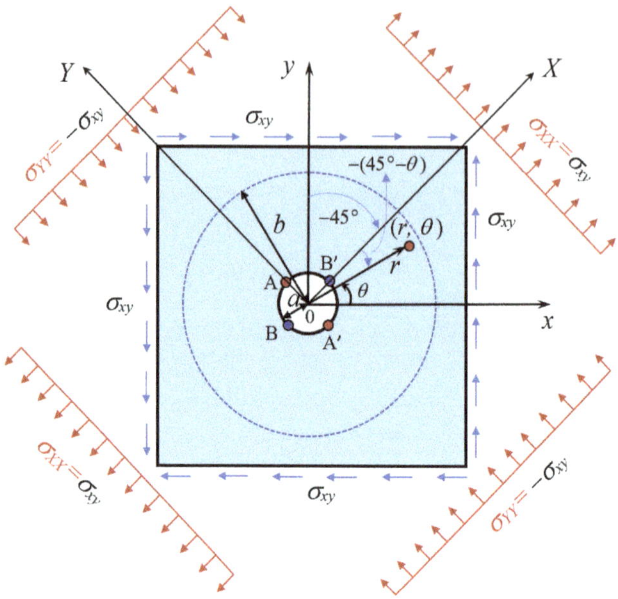

Figure 5.10. A plate with a small circular hole of radius a at its center, subjected to uniform stress at all edges, equivalent to equal tension and compression in the far field.

After obtaining the solution for the pure shear stress, the formulas can be converted back to average and difference stresses, providing concise expressions. The code snippet for these computations is as follows:

```
# Using the biaxial formula:
θy = pi/4
dic_σI = {σave:(σxx+σyy)/2 , σdif: (σxx-σyy)/2}      # Inverse of dic_σ
σ_polar_pureShear = σ_polar_biaxial.subs(dic_σI)\
                                        .subs({σxx:σxy, σyy:-σxy, θ:θ-θy})

# Convert back in terms of average and difference stresses:
σ_polar_pureShear = σ_polar_pureShear.subs(dic_σ).simplify().expand()

# simplify the expressions:
σ_polar_pureShear = σ_polar_pureShear.applyfunc(\
                                 lambda expr: sp.collect(expr,cos(2*θ)))
σ_polar_pureShear = σ_polar_pureShear.applyfunc\
                                (lambda expr: sp.collect(expr,sin(2*θ)))
σ_polar_pureShear = σ_polar_pureShear.applyfunc\
                                (lambda expr: sp.collect(expr,σxy))

display(Math(f" \\text{{Solution for pure shear at far field:}} "))
display(Math(f" \\text{{ }} {latex(σ_polar_pureShear)}"))
```

Solution for pure shear at far field:

$$\begin{bmatrix}\sigma_{xy}\left(\frac{3a^4}{r^4}-\frac{4a^2}{r^2}+1\right)\sin(2\theta) & \sigma_{xy}\left(-\frac{3a^4}{r^4}+\frac{2a^2}{r^2}+1\right)\cos(2\theta)\\ \sigma_{xy}\left(-\frac{3a^4}{r^4}+\frac{2a^2}{r^2}+1\right)\cos(2\theta) & \sigma_{xy}\left(-\frac{3a^4}{r^4}-1\right)\sin(2\theta)\end{bmatrix}$$

5.3.3 *Final solution for plates subjected to pure shear*

We arrived at the final solution for plates subjected to pure shear stress σ_{xy}:

$$\begin{aligned}\sigma_{rr} &= \sigma_{xy}\left(\frac{3a^4}{r^4}-\frac{4a^2}{r^2}+1\right)\sin(2\theta)\\ \sigma_{\theta\theta} &= -\sigma_{xy}\left(\frac{3a^4}{r^4}+1\right)\sin(2\theta)\\ \sigma_{r\theta} &= \sigma_{xy}\left(-\frac{3a^4}{r^4}+\frac{2a^2}{r^2}+1\right)\cos(2\theta)\end{aligned} \tag{5.14}$$

Let us define the solution formulas in Python for later use, so that it is not necessary to run all of the previous code snippets:

```
# Final solution for plates subjected to pure shear:
σrr_by_σxy = σxy*(3*a**4/r**4 - 4*a**2/r**2 + 1)*sin(2*θ)
σθθ_by_σxy =-σxy*(3*a**4/r**4 + 1)*sin(2*θ)
σrθ_by_σxy = σxy*(-3*a**4/r**4 + 2*a**2/r**2 + 1)*cos(2*θ)
```

5.3.4 *Stress concentration factor of 4, pure shearing*

The stress concentration factor on the hole edge in a plate subjected to pure shearing can be found by setting $r = a$:

```
display(Math(f" \\text{{Solution for pure shear at far field:}} "))
display(Math(f" \\small {latex(σ_polar_pureShear.subs(r,a))}"))
```

Solution for pure shear at far field:

$$\begin{bmatrix}0 & 0\\ 0 & -4\sigma_{xy}\sin(2\theta)\end{bmatrix}$$

It is again observed that only the hoop stress is of concern. It varies with θ around the circle at $r = a$. We find:

1. When $\theta = \pi/4$ or $5\pi/4$, corresponding to points A and A′ in Fig. 5.10, the minimum hoop stress is $\sigma_{\theta\theta}^{\min} = -4\sigma_{xy}$. The stress concentration factor (SCF) is -4.

2. When $\theta = 3\pi/4$ or $7\pi/4$, corresponding to points B and B′ in Fig. 5.10, the maximum hoop stress is $\sigma_{\theta\theta}^{\max} = 4\sigma_{xy}$. The SCF is 4.

5.4 Plates subjected to general loading

Consider a general problem: a plate with a small circular hole at its center, subjected to general stresses (tractions) σ_{xx}, σ_{yy}, and σ_{xy} on all edges, as shown in Fig. 5.11.

The same procedure used for a plate under uniaxial tension can be applied to find the solution. The stress function ϕ should include both $\sin 2\theta$ and $\cos 2\theta$ terms, expressed as

$$\phi = f(r)\sin 2\theta + g(r)\cos 2\theta \tag{5.15}$$

5.4.1 *Solution via superposition*

Since the solutions for biaxial and pure shear stresses have already been obtained, the simplest approach is to use superposition of these solutions. The code snippet for this procedure is as follows:

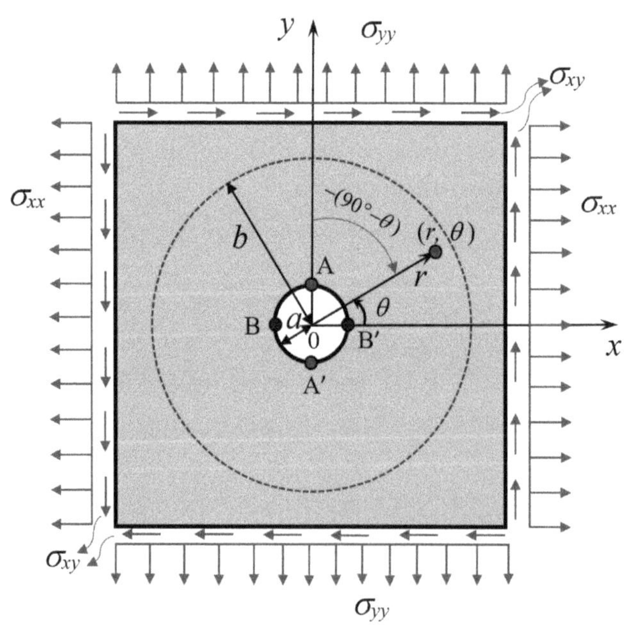

Figure 5.11. A plate with a small circular hole of radius a at its center, subjected to general stresses (tractions) on all edges.

```
# Superimpose solutions for biaxial and pure shear:
σ_polar_general = σ_polar_biaxial + σ_polar_pureShear
σ_polar_general = σ_polar_general.applyfunc(\
                                lambda expr: sp.collect(expr,cos(2*θ)))
σ_polar_general = σ_polar_general.applyfunc\
                               (lambda expr: sp.collect(expr,sin(2*θ)))
σ_polar_general = σ_polar_general.applyfunc\
                               (lambda expr: sp.collect(expr,σave))
σ_polar_general = σ_polar_general.applyfunc\
                               (lambda expr: sp.collect(expr,σdif))

display(Math(f"\\text{{Solution for plates under general stresses:}} "))
display(Math(f" \\text{{$σ_{{rr}}$}}= \
                               \\small {latex(σ_polar_general[0,0])}"))
display(Math(f" \\text{{$σ_{{θθ}}$}}= {latex(σ_polar_general[1,1])}"))
display(Math(f" \\text{{$σ_{{rθ}}$}}= {latex(σ_polar_general[0,1])}"))
```

Solution for plates under general stresses:

$$\sigma_{rr} = \sigma_{\text{ave}}\left(-\frac{a^2}{r^2}+1\right)+\sigma_{\text{dif}}\left(\frac{3a^4}{r^4}-\frac{4a^2}{r^2}+1\right)\cos(2\theta)$$
$$+\sigma_{xy}\left(\frac{3a^4}{r^4}-\frac{4a^2}{r^2}+1\right)\sin(2\theta)$$

$$\sigma_{\theta\theta} = \sigma_{\text{ave}}\left(\frac{a^2}{r^2}+1\right)+\sigma_{\text{dif}}\left(-\frac{3a^4}{r^4}-1\right)\cos(2\theta)+\sigma_{xy}\left(-\frac{3a^4}{r^4}-1\right)\sin(2\theta)$$

$$\sigma_{r\theta} = \sigma_{\text{dif}}\left(\frac{3a^4}{r^4}-\frac{2a^2}{r^2}-1\right)\sin(2\theta)+\sigma_{xy}\left(-\frac{3a^4}{r^4}+\frac{2a^2}{r^2}+1\right)\cos(2\theta)$$

We define the solutions in Python using the formulas obtained for future use, without the need to run all the codes above:

```
# Final solution for plates subjected to general stress loading:
a, b, r = sp.symbols('a, b, r', nonnegative=True)
θ = symbols('θ', real=True)

σxx, σyy, σxy = symbols('σ_xx,σ_yy,σ_xy',real=True) # far field stresses

σave_ = (σxx+σyy)/2                   # σave: average stress at far field
σdif_ = (σxx-σyy)/2                 # σdif difference stress at far field

σrr_general = σave_*(-a**2/r**2 + 1)+σdif_*(3*a**4/r**4-4*a**2/r**2+1)\
                *cos(2*θ) + σxy*(3*a**4/r**4 - 4*a**2/r**2 + 1)*sin(2*θ)
σθθ_general = σave_*(a**2/r**2 + 1) + σdif_*(-3*a**4/r**4-1)*cos(2*θ)\
                                        + σxy*(-3*a**4/r**4-1)*sin(2*θ)
σrθ_general = σdif_*(3*a**4/r**4 - 2*a**2/r**2 - 1)*sin(2*θ) \
            + σxy*(-3*a**4/r**4 + 2*a**2/r**2 + 1)*cos(2*θ)

σ_polar_general_ = sp.Matrix([[σrr_general, σrθ_general],
                               [σrθ_general, σθθ_general]])
```

This is the final solution for plates with a small circular hole subjected to general loading.

5.4.2 *Stress concentration factor under general loading*

5.4.2.1 *SCF of 6: Biaxial tension with shearing*

The stress concentration factor on the hole edge for a plate subjected to biaxial tension combined with shear of equal amplitude can be obtained by setting $r = a$, $\sigma_{\text{ave}} = \sigma_{xx} = \sigma_{yy} = \sigma$, $\sigma_{xy} = \sigma$, and $\sigma_{\text{dif}} = 0$, where σ is the nominal stress. The code snippet for this computation is as follows:

```
dicσ = {r:a, σdif:0, σave:σ, σxx:σ, σyy:σ, σxy:σ}
σ_polar_general_σ= σ_polar_general_.subs(dicσ)
display(Math(f"\\text{{Solution on hole, biaxial tension+shearing:}}"))
display(Math(f" {latex(σ_polar_general_σ)}"))
```

Solution on hole, biaxial tension+shearing:

$$\begin{bmatrix} 0 & 0 \\ 0 & -4\sigma\sin(2\theta) + 2\sigma \end{bmatrix}$$

It is again observed that only the hoop stress is of concern. It varies with θ around the circle at $r = a$. Specifically:

1. When $\theta = \pi/4$ or $5\pi/4$, corresponding to points B and B$'$ in Fig. 5.10, the minimum hoop stress is $\sigma_{\theta\theta}^{\min} = -2\sigma$. The stress concentration factor (SCF) is -2.
2. When $\theta = 3\pi/4$ or $7\pi/4$, corresponding to points A and A$'$ in Fig. 5.10, the maximum hoop stress is $\sigma_{\theta\theta}^{\max} = 6\sigma$. The SCF is 6, which is the sum of the SCF of 4 from pure shear and the SCF of 2 from hydrostatic pressure on the outer edges. This represents the largest SCF observed so far for a general stress state with a common nominal stress σ:

```
σθθ_min = σ_polar_general_σ[1,1].subs(θ, pi/4)
display(Math(f"\\text{{$ σ_{{θθ}}$ on the hole = }} {latex(σθθ_min)}"))
σθθ_max = σ_polar_general_σ[1,1].subs(θ, 3*pi/4)
display(Math(f"\\text{{$ σ_{{θθ}}$ on the hole = }} {latex(σθθ_max)}"))
```

$\sigma_{\theta\theta}$ on the hole $= -2\sigma$
$\sigma_{\theta\theta}$ on the hole $= 6\sigma$

```
# Create a mesh grid for a rectangular domain in x-y plane:
x, y = sp.symbols('x, y', real=True)
dX = 3.0*a_val;     dY = 2.0*a_val                    # plot 2dX × 2dY area
x_val = np.linspace(-dX, dX, 800)
y_val = np.linspace(-dY, dY, 800)
X, Y = np.meshgrid(x_val, y_val)

# define a circular hole to be masked in plot:
mask = (X/a_val)**2 + (Y/a_val)**2 <= 1

# Hoop stress:
dicσ = {σdif:0, σave:1, σxx:1, σyy:1, σxy:1}
dic_rt = {a:1, r:sp.sqrt(x**2+y**2), θ:sp.atan(y/x)}

σ_θθxy_general_σ = σ_polar_general_[1, 1].subs(dicσ).subs(dic_rt)
np_σθθxy = lambdify((x,y), σ_θθxy_general_σ,'numpy') # convert to numpy
Z = np_σθθxy(X, Y)                   # compute the stress over the mesh grid
Z[mask] = np.nan                       # mask the hole, print nothing there

gr.contour_plot(X,Y, Z, x_lbl='x-axis',y_lbl='y-axis',
                title='Stress $σ_{{θθ}}$',lvls=80, ioff=True);
plt.savefig('imagesSMp/plateHoleSttg_rt.png', dpi=500)    # save to file
#plt.show()
```

```
Minimum value in the domain: -1.98997
Maximum value in the domain: 5.98615
```

A contour plot of the hoop stress $\sigma_{\theta\theta}$ in a plate with a circular hole subjected to general loading is shown in Fig. 5.12. It is observed that the hoop stress reaches its minimum on the hole edge at $\theta = 45°$ relative to the x-axis, and the maximum occurs at $\theta = -45°$ relative to the x-axis, resulting in a stress concentration factor of 6.

5.4.2.2 *SCF of* $4\sqrt{2}$*: Tension-compression-shearing*

When the biaxial load consists of tension in the x-direction and compression in the y-direction of equal amplitude, the stresses along the hole edge can be

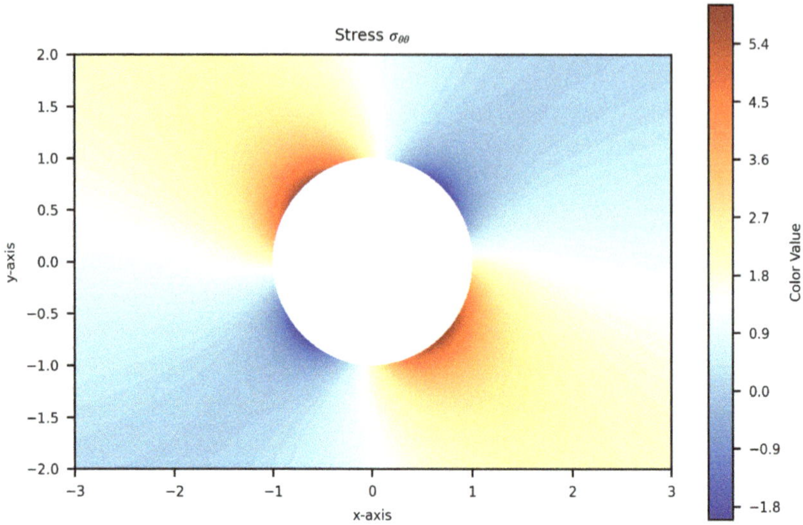

Figure 5.12. Contour of the hoop stress $\sigma_{\theta\theta}$ in a plate with a circular hole subjected to general loading.

determined by setting $r = a$, $\sigma_{\text{ave}} = 0$, $\sigma_{\text{dif}} = \sigma_{xx} = -\sigma_{yy}$, and $\sigma_{xy} = \sigma_{xx}$. This solution is obtained using:

```
σ = sp.symbols('σ')
dicσ = {r:a, σave:0, σxy:σ, σdif:-σ}                    # σxx=-σ, σyy=σ, σxy=σ
σθθ_max = σ_polar_general.subs(dicσ)[1,1]
display(Math(f"\\text{{Solution on the hole \
        $(σ_{{xx}}=-σ, σ_{{yy}}=σ, σ_{{xy}}=σ)$}} = {latex(σθθ_max)}"))
σθθ_max_location = sp.solve(σθθ_max.diff(θ), θ)[0]    # stationary point
display(Math(f"\\text{{Maximum hoop stress location, θ =}}\
                                                {latex(σθθ_max_location)}"))
display(Math(f"\\text{{Maximum hoop stress =}}\
                                    {latex(σθθ_max.subs(θ,σθθ_max_location+pi))}"))
```

Solution on the hole $(\sigma_{xx} = -\sigma, \sigma_{yy} = \sigma, \sigma_{xy} = \sigma) = -4\sigma \sin(2\theta) + 4\sigma \cos(2\theta)$

Maximum hoop stress location, $\theta = -\dfrac{\pi}{8}$

Maximum hoop stress $= 4\sqrt{2}\sigma$

When the far-field loading is $\sigma_{xx} = -\sigma$, $\sigma_{yy} = \sigma$, and $\sigma_{xy} = \sigma$, at $\theta = \pi/8$ or $9\pi/8$, the maximum hoop stress is $\sigma_{\theta\theta}^{\max} = 4\sqrt{2}\,\sigma$. The stress concentration factor is therefore $4\sqrt{2} \approx 5.567$.

If the sign of σ is reversed, we have:

```
dicσ = {r:a, σave:0, σxy:-σ, σdif:σ}
σθθ_max = σ_polar_general.subs(dicσ)[1,1]
display(Math(f"\\text{{Solution on the hole (σxx=σ, σyy=-σ, σxy=-σ)=}}\
                                              {latex(σθθ_max)}"))
σθθ_max_location = sp.solve(σθθ_max.diff(θ), θ)[0]   # stationary point
display(Math(f"\\text{{Maximum hoop stress location, θ =}}\
                                         {latex(σθθ_max_location)}"))
display(Math(f"\\text{{Maximum hoop stress =}}\
                          {latex(σθθ_max.subs(θ,σθθ_max_location+pi))}"))
```

Solution on the hole $(\sigma xx = \sigma, \sigma yy = -\sigma, \sigma xy = -\sigma) = 4\sigma \sin(2\theta)$

$-4\sigma \cos(2\theta)$

Maximum hoop stress location, $\theta = -\frac{\pi}{8}$

Maximum hoop stress $= -4\sqrt{2}\sigma$

When the far-field loading is $\sigma_{xx} = \sigma$, $\sigma_{yy} = -\sigma$, and $\sigma_{xy} = -\sigma$, at $\theta = -\pi/8$ or $7\pi/8$, the minimum hoop stress is $\sigma_{\theta\theta}^{\min} = -4\sqrt{2}, \sigma$. The stress concentration factor is therefore $-4\sqrt{2} \approx \quad 5.567$.

5.5 Review questions

1. Consider a plate with a small hole subjected to biaxial stress at the far field, as shown in Fig. 5.8 and described in Section 5.2. Using the solution procedure described in that section, find the stress solution.
2. Consider a plate with a small hole subjected to pure shear stress at the far field, as shown in Fig. 5.9 and described in Section 5.3. Using the solution procedure described in Section 5.1, find the stress solution.

5.6 Remarks

This chapter presented a study of stress concentration around a hole in a plate using the stress method. We conclude with the following remarks:

1. Various types of loading were considered, including axial tension, biaxial loading, pure shearing, and general loading.
2. Stress concentration factors for all these cases were determined. It was found, for the first time, that the stress concentration factor can reach as high as 6 under a general loading condition.

3. Superposition proves to be an effective approach for obtaining solutions to many problems using existing results.

We have found negative SCFs in several cases, even when the applied loading is tensile. These negative SCFs can often be more **damaging** because the material is subjected to compression in those regions. This can lead to failures in the form of local **buckling**, particularly when the plate is thin.

The following chapter develops a displacement method to study the same stress concentration problem, enabling the determination of a complete set of solutions, including displacements, stresses, and strains. More complex problems, such as those involving curved beams, will also be examined to provide a comprehensive set of solutions.

References

[1] S. Timoshenko and J. N. Goodier, *Theory of Elasticity*, McGraw-Hill, New York, 1970.
[2] Z. L. Xu, *Elasticity*, People's Publisher, China, 1979.
[3] A. P. Boresi, K. Chong and J. D. Lee, *Elasticity in Engineering Mechanics*, John Wiley & Sons, New York, 2010.
[4] G. R. Liu, *Mechanics of Materials: Formulations and Solutions with Python*, World Scientific, New Jersey, 2024.

Chapter 6

Displacement Methods for Non-axisymmetric Problems

Let us import the necessary external modules or dependencies for later use in this chapter:

```
# Place cursor in this cell, and press Ctrl+Enter to import dependencies.
import sys                              # for accessing the computer system
sys.path.append('../grbin/')              # Add in the path to your system

from commonImports import *       # Import dependencies from '../grbin/'
import grcodes as gr                     # Import the module of the author
importlib.reload(gr)                    # When grcodes is modified, reload it

from continuum_mechanics import vector
from continuum_mechanics.solids import sym_grad, strain_stress
init_printing(use_unicode=True)          # For latex-like quality printing
np.set_printoptions(precision=4,suppress=True)    # Digits in print-outs
```

In the previous chapter, a stress method was used to study non-axisymmetric problems, particularly the stress concentration around a hole in a plate subjected to various types of loading. Stress solutions were obtained, but displacement solutions were not. This chapter proposes a novel displacement method for such problems, which naturally provides the displacement solutions in addition to stresses and strains.

The strategy for the displacement method in polar coordinates is similar to that in Cartesian coordinates, as discussed in Chapter 3 of Ref. [1]. However, the procedure is more intricate due to the use of radial and circumferential coordinates. Analysis of symmetry requires careful consideration. The following section presents the procedure step by step.

6.1 Stress concentration on a hole in a plate

The setup of this problem is the same as in the previous chapter. A plate with a hole subjected to uniaxial stress σ in the far field is shown in Fig. 6.1.

The stress state in Cartesian coordinates is

$$\begin{bmatrix} \sigma_{xx} & \sigma_{xy} \\ \sigma_{xy} & \sigma_{yy} \end{bmatrix} = \begin{bmatrix} \sigma & 0 \\ 0 & 0 \end{bmatrix} \tag{6.1}$$

The same stress state can be expressed in polar coordinates via coordinate transformation [2], as found in Section 5.1.1:

$$\begin{bmatrix} \sigma_{rr} & \sigma_{r\theta} \\ \sigma_{r\theta} & \sigma_{\theta\theta} \end{bmatrix} = \begin{bmatrix} \sigma\cos^2(\theta) & -\sigma\sin(\theta)\cos(\theta) \\ -\sigma\sin(\theta)\cos(\theta) & \sigma\sin^2(\theta) \end{bmatrix} \tag{6.2}$$

The problem was decomposed into two subproblems, as detailed in Section 5.1.1. **Subproblem-1** is a disk with zero inner pressure and an outer pressure of $-\frac{\sigma}{2}$, which is axially symmetric and can be addressed using the displacement method presented in Chapter 4. This chapter focuses on **Subproblem-2**, a disk subjected to normal and shear stresses on the

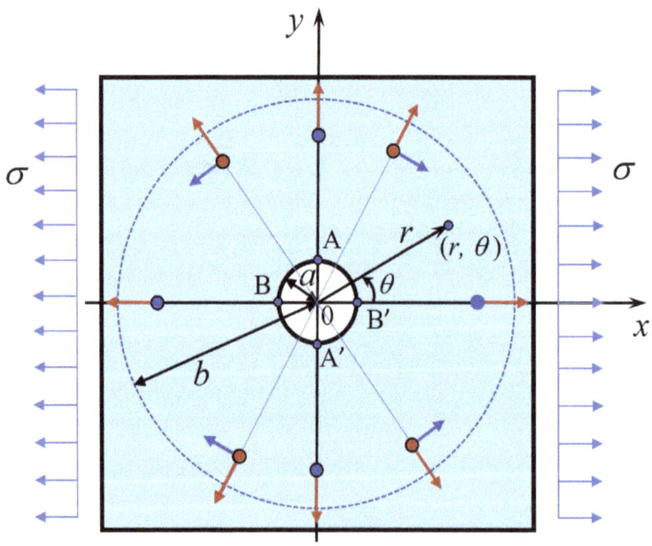

Figure 6.1. A plate with a small circular hole of radius a at the center. It is subjected to a uniform stress σ at the left and right edges. The resulting displacement components at points in the four quadrants are shown. Red arrows indicate the displacement components in the radial direction, u_r, and blue arrows indicate the displacement components in the circumferential direction, u_θ.

circular boundary $r = b$, given by

$$
\begin{aligned}
\sigma_{rrb} &= \frac{\sigma}{2}\cos 2\theta \quad \text{at} \quad r = b \\
\sigma_{r\theta b} &= -\frac{\sigma}{2}\sin 2\theta \quad \text{at} \quad r = b
\end{aligned}
\tag{6.3}
$$

Both stresses depend on θ, making the problem asymmetric. This chapter proposes a displacement method to obtain the complete set of solutions, including displacements, stresses, and strains. Python code will be developed step by step to achieve the full solution set.

6.2 Strategy for solution

The procedure and concept of this displacement method are largely similar to those discussed in Chapter 3 for mixed axially symmetric problems. It mainly consists of three major steps:

1. Assume expressions for the displacements with unknown constants.
2. Compute the expressions for the stresses using the assumed displacements.
3. Determine the unknown constants using the available conditions, including equilibrium equations, stress and displacement boundary conditions, symmetry conditions, etc.

The main difference here is the treatment of the additional tangential coordinate θ, making the problem fully two-dimensional (2D).

In any displacement method, the key is how to properly assume the displacements. Figure 6.1 shows the displacement components (red and blue arrows) in the plate with a hole under axial tension.

6.2.1 *Symmetry with respect to θ*

1. The tangential displacement u_θ is directly related to the circumferential coordinate θ, and we use u_θ to examine symmetry in θ.
2. As shown in Fig. 6.1, u_θ (blue arrows) is antisymmetric with respect to θ: the direction of u_θ is opposite to the θ-direction in the first quadrant, but aligns with the θ-direction in the fourth quadrant. Therefore, u_θ should carry a $\sin 2\theta$ term.
3. Considering the boundary conditions in Eq. (6.3), it is reasonable to assume that u_r (orthogonal to u_θ) carries $\cos 2\theta$, an even function of θ.
4. The problem has two symmetric planes, $x = 0$ and $y = 0$, so the trigonometric argument should be 2θ, giving a period of π rather than 2π.

6.2.2 *Symmetry with respect to r*

1. The radial displacement u_r depends directly on the radial coordinate r, and we examine symmetry in r using u_r.
2. As shown in Fig. 6.1, u_r (red arrows) is antisymmetric with respect to r: u_r in the first quadrant is positive (radially outward), while u_r in the third quadrant is negative (radially inward). Thus, the sign of u_r follows that of r.
3. If monomials of r are used as basis functions, only odd-order terms should be included.
4. Moreover, since r can extend to infinity, the monomial basis functions should have negative powers and be at most first order in r.

6.3 Construction of displacements

Based on the analysis above, both displacement components can be assumed as

$$\begin{aligned} u_r(r,\theta) &= \left(\frac{a_0}{r^7} + \frac{a_1}{r^5} + \frac{a_2}{r^3} + \frac{a_3}{r} + a_4 r\right)\cos(2\theta) \\ u_\theta(r,\theta) &= \left(\frac{b_0}{r^7} + \frac{b_1}{r^5} + \frac{b_2}{r^3} + \frac{b_3}{r} + b_4 r\right)\sin(2\theta) \end{aligned} \tag{6.4}$$

where a_i and b_i are arbitrary unknown constants. Additional or fewer negative-order terms can be included as needed. In the following example, terms up to r^{-7} are used. Readers are encouraged to experiment by reducing the expansion to r^{-3} or extending it to r^{-9} as a useful exercise.

This problem also exhibits mirror symmetry with respect to $\theta = 0$ and $\theta = \pi/2$. The term $\cos 2\theta$ ensures this symmetry for the radial displacement component u_r, while the term $\sin 2\theta$ ensures the antisymmetry for the tangential component u_θ. Therefore, the displacement assumptions in Eq. (6.4) satisfy the primary requirements of this problem, with the arbitrary constants available to enforce all detailed boundary and equilibrium conditions.

As a sanity check, we know from the stress results obtained in the previous chapter that all the stresses σ_{rr}, $\sigma_{\theta\theta}$, and $\sigma_{r\theta}$ are even functions of r. Therefore, the displacements should be odd functions, as shown in Eq. (6.4), since stresses are essentially obtained by differentiation of the displacements.

The code snippet to generate the expressions for the displacements is as follows:

```
# Define variables for displacement method for asymmetric problems:
E, ν = symbols('E, ν', positive=True)          # given material constants
r = sp.symbols('r', positive=True)  # symbolic variable for radial axis
θ = sp.symbols('θ', real=True)                          # for tangential axis

ur_rbs_list =[1/r**7, 1/r**5, 1/r**3, 1/r, r]               # basis functions
ur_rbs = sp.Matrix(ur_rbs_list)                                # form a vector
_, order = ur_rbs[0].as_coeff_exponent(r)  # order of the leading basis
order = -order                                        # convert it to positive
print(f"The highest inverse order of the monomial = {order}")

uθ_rbs_list = ur_rbs_list                              # using the same bases
uθ_rbs = sp.Matrix(uθ_rbs_list)

# unknown constants for displacements:
ai = sp.symbols(f'a:{len(ur_rbs)}')                       # unknown constants
display(Math(f" \\text{{Unknown constants for $u$}}={latex(ai)}"))
ai = Matrix([ai[i] for i in range(len(ur_rbs))]) #form coeffs in vector

bi = sp.symbols(f'b:{len(uθ_rbs)}')                       # unknown constants
display(Math(f" \\text{{Unknown constants for $v$}}={latex(bi)}"))
bi = Matrix([bi[i] for i in range(len(uθ_rbs))])             # form a vector

cs = ai.col_join(bi)                      # form a single vector of constants

urg = (ai.T@Matrix(ur_rbs))[0]*cos(2*θ) #Form the general displacements
uθg = (bi.T@Matrix(uθ_rbs))[0]*sin(2*θ)
display(Math(f"\\text{{Assumed displacements, $u_r(r,θ)$}}=\
                                                             {latex(urg)}"))
display(Math(f"\\hspace{{4.1cm}} \\text{{ $u_θ(r,θ)$}}= {latex(uθg)}"))
```

```
The highest inverse order of the monomial = 7
```

Unknown constants for $u = (a_0,\ a_1,\ a_2,\ a_3,\ a_4)$

Unknown constants for $v = (b_0,\ b_1,\ b_2,\ b_3,\ b_4)$

Assumed displacements, $u_r(r,\theta) = \left(\frac{a_0}{r^7} + \frac{a_1}{r^5} + \frac{a_2}{r^3} + \frac{a_3}{r} + a_4 r\right)\cos(2\theta)$

$$u_\theta(r,\theta) = \left(\frac{b_0}{r^7} + \frac{b_1}{r^5} + \frac{b_2}{r^3} + \frac{b_3}{r} + b_4 r\right)\sin(2\theta)$$

The constructed displacements contain a total of 10 unknown constants. These constants can now be determined using the equilibrium equations and the stress boundary conditions, all of which require explicit expressions for the stresses.

6.3.1 *Expressions for stresses*

We next compute the strain components from the assumed displacements and then determine the corresponding stresses:

```
# Compute strains using the displacements:
εrr = urg.diff(r)
εθθ = urg/r + uθg.diff(θ)/r
εrθ = (urg.diff(θ)/r + uθg.diff(r) - uθg/r)/2
εvg = sp.Matrix([εrr, εθθ, εrθ])                # define strains as a vector

# Compute stress vector using the strain vector:
σvg = sp.Matrix(gr.strain2stress2D(E, ν, εvg, Ptype= 'stress'))
σvg = σvg.expand()

display(Math(f" \\text{{General solution for stresses:}} "))
display(Math(f" \\text{{$σ_{{v}}$=}} \small {latex(sp.simplify(σvg))}"))
display(Math(f" \\text{{General solution for strains:}} "))
display(Math(f" \\text{{$ε_{{v}}$=}} {latex(sp.simplify(εvg))}"))
```

General solution for stresses:

$$\sigma_v = \begin{bmatrix} \frac{E(-a_0\nu+7a_0-a_1r^2\nu+5a_1r^2-a_2r^4\nu+3a_2r^4-a_3r^6\nu+a_3r^6-a_4r^8\nu-a_4r^8 - 2b_0\nu-2b_1r^2\nu-2b_2r^4\nu-2b_3r^6\nu-2b_4r^8\nu)\cos(2\theta)}{r^8(\nu^2-1)} \\ \frac{E(7a_0\nu-a_0+5a_1r^2\nu-a_1r^2+3a_2r^4\nu-a_2r^4+a_3r^6\nu-a_3r^6 - a_4r^8\nu-a_4r^8-2b_0-2b_1r^2-2b_2r^4-2b_3r^6-2b_4r^8)\cos(2\theta)}{r^8(\nu^2-1)} \\ \frac{E\left(-a_0-a_1r^2-a_2r^4-a_3r^6-a_4r^8-4b_0-3b_1r^2-2b_2r^4-b_3r^6\right)\sin(2\theta)}{r^8(\nu+1)} \end{bmatrix}$$

General solution for strains:

$$\varepsilon_v = \begin{bmatrix} \frac{\left(-7a_0-5a_1r^2-3a_2r^4-a_3r^6+a_4r^8\right)\cos(2\theta)}{r^8} \\ \frac{\left(a_0+a_1r^2+a_2r^4+a_3r^6+a_4r^8+2b_0+2b_1r^2+2b_2r^4+2b_3r^6+2b_4r^8\right)\cos(2\theta)}{r^8} \\ \frac{\left(-a_0-a_1r^2-a_2r^4-a_3r^6-a_4r^8-4b_0-3b_1r^2-2b_2r^4-b_3r^6\right)\sin(2\theta)}{r^8} \end{bmatrix}$$

We now need to solve a system of expressions for the unknown constants. It is practical to write a Python function to construct and solve such systems, which can then be reused multiple times:

```
def constantUpdate(eqns, r, order, cs, prnt=False):
    '''Compute and update constants `cs` for a given 1D sympy `eqns`.
    return: sln_eqn - solution of cs in dictionary.
    '''
    exprs = []
    for eqn in eqns:
        for i in range(order):
            exprs.append(eqn.coeff(r,i))

    exprs = gr.remove_zero_exprs(exprs)
    if prnt: display(exprs)
    sln_eqns = sp.solve(exprs, cs)       # solve for unknown constants
    return sln_eqns
```

Let us define the equilibrium equations in Python and then invoke `constantUpdate()` to determine some of the constants:

```
# Define equilibrium equation that needs to be satisfied:
br = 0; bθ = 0                                   # Consider zero body forces
equi1=(σvg[0].diff(r)+σvg[2].diff(θ)/r+(σvg[0]-σvg[1])/r+br)\
                                                .simplify().factor()
equi2=(σvg[2].diff(r)+σvg[1].diff(θ)/r+2*σvg[2]/r+bθ)\
                                                .simplify().factor()
equi1 = sp.numer(equi1).expand().coeff(cos(2*θ))
equi2 = sp.numer(equi2).expand().coeff(sin(2*θ))
#display(equi1, equi2)          # uncomment this to view these equations

sln_cs = {}                 # initialize solution of unknown constants, cs
sln_equi = gr.constantUpdate([equi1, equi2], r, order+1, cs, prnt=False)

# Update unknown constants:
sln_cs = {**sln_equi}
sln_cs = {key: value.subs(sln_cs) for key, value in sln_cs.items()}
print(f" Current len of sln_cs={len(sln_cs)}; Total cs={len(cs)}")
display(sln_cs)

σv = sp.simplify(σvg.subs(sln_cs))                      # update stresses
display(Math(f" \\text{{ Current stresses $σ_{{v}}$=}} \
                                          {latex(sp.simplify(σv))}"))
```

```
Total number of expressions = 16
Number of nonzero expressions = 8
Current len of sln_cs=6; Total cs=10
```

$$\left\{a_0 : 0,\ a_1 : 0,\ a_2 : b_2,\ a_3 : \frac{2b_3}{\nu - 1},\ b_0 : 0,\ b_1 : 0\right\}$$

$$\text{Current stresses } \sigma_v = \begin{bmatrix} \frac{E\left(-a_4r^4\nu - a_4r^4 - 3b_2\nu + 3b_2 - 2b_3r^2\nu - 2b_3r^2 - 2b_4r^4\nu\right)\cos(2\theta)}{r^4(\nu^2-1)} \\ \frac{E\left(-a_4r^4\nu - a_4r^4 + 3b_2\nu - 3b_2 - 2b_4r^4\right)\cos(2\theta)}{r^4(\nu^2-1)} \\ -\frac{E\left(2b_3r^2 + (\nu-1)\left(a_4r^4 + 3b_2 + b_3r^2\right)\right)\sin(2\theta)}{r^4(\nu-1)(\nu+1)} \end{bmatrix}$$

We have successfully found several unknown constants, which can be used to update the displacements. We can also use them to update the stresses and/or strains if needed:

```
# update the displacement:
ur = urg.subs(sln_cs)#.collect(q)
display(Math(f"\\text{{$u_r$ satisfying equilibrium = }}{latex(ur)}"))

uθ = uθg.subs(sln_cs)#.collect(q)
display(Math(f"\\text{{$u_θ$ satisfying equilibrium = }}{latex(uθ)}"))

# Check how many constants are found and how many unknowns remain:
list_keys = list(sln_cs.keys())
cs_Eql = list(set(list(cs))-set(list_keys))
print(f"Constants yet to be found: {cs_Eql}")
```

$$u_r \text{ satisfying equilibrium} = \left(a_4r + \frac{b_2}{r^3} + \frac{2b_3}{r(\nu-1)}\right)\cos(2\theta)$$

$$u_\theta \text{ satisfying equilibrium} = \left(\frac{b_2}{r^3} + \frac{b_3}{r} + b_4r\right)\sin(2\theta)$$

```
Constants yet to be found: [b2, b4, b3, a4]
```

We observe that the constants for the highest- and lowest-order terms, as well as the constant term, are automatically set to zero when the equilibrium equation is imposed.

6.3.2 *General displacements*

We have obtained an important equation for the general displacement in this type of asymmetric problem, which satisfies the equilibrium equation:

$$\begin{aligned} u_r &= \left(a_4r + \frac{b_2}{r^3} + \frac{2b_3}{r(\nu-1)}\right)\cos(2\theta) \\ u_\theta &= \left(\frac{b_2}{r^3} + \frac{b_3}{r} + b_4r\right)\sin(2\theta) \end{aligned} \tag{6.5}$$

These displacements contain four unknown constants, which indicates that for similar problems, the assumed displacement requires only four unknown constants. The involved monomial bases are $\frac{1}{r}$, r, and r^3. In the future, Eq. (6.5) should be used directly for similar problems instead of Eq. (6.4).

The remaining constants must be determined using the boundary conditions of a specific problem. Since our problem has two boundaries, $r = a$ and $r = b$, each with two stress boundary conditions, we obtain a total of four conditions to determine these four constants. Therefore, a complete solution can always be found for this problem.

The code snippet is given as follows:

```
a, b = sp.symbols('a, b', positive=True)        # variable for boundary
σ = sp.symbols('σ', real=True)                 # given stress at far field

sln_cs = {**sln_equi}
σv = sp.simplify(σvg.subs(sln_cs))                       # update stresses

# Define stress boundary conditions (SBCs) at r=b:
SBCrr_ra = (σv[0].subs(r,a).coeff(cos(2*θ)) ).simplify()   # SBC on r=a
SBCrr_ra = sp.numer(SBCrr_ra).expand()
SBCrθ_ra = (σv[2].subs(r,a).coeff(sin(2*θ)) ).simplify()
SBCrθ_ra = sp.numer(SBCrθ_ra).expand()

SBCrr_rb = (σv[0].subs(r,b).coeff(cos(2*θ))-σ/2).simplify()# SBC on r=b
SBCrr_rb = sp.numer(SBCrr_rb).expand()
SBCrθ_rb = (σv[2].subs(r,b).coeff(sin(2*θ))+σ/2).simplify()
SBCrθ_rb = sp.numer(SBCrθ_rb).expand()

SBCs = [SBCrr_ra, SBCrθ_ra, SBCrr_rb, SBCrθ_rb]
sln_SBCs = gr.constantUpdate(SBCs, r, order, cs)
sln_SBCs_ = gr.split_dict_n(sln_SBCs, 2)
print(f"Constants found: ")
display(sln_SBCs_[0])
display(sln_SBCs_[1])

sln_cs = {**sln_equi, **sln_SBCs}
sln_cs = {key: value.subs(sln_cs) for key, value in sln_cs.items()}
print(f"Current len of sln_cs = {len(sln_cs)}; Total cs = {len(cs)}")
#display(sln_cs)                                                # lengthy

# Check how many constants are found and how many unknowns remain:
list_keys = list(sln_cs.keys())
cs_Eql = list(set(list(cs))-set(list_keys))
print(f"Constants yet to be found: {cs_Eql}")
```

```
 Total number of expressions = 28
 Number of nonzero expressions = 4
Constants found:
```

$$\left\{a_4: \frac{2a^2b^2\nu\sigma + 2a^2b^2\sigma - b^4\nu\sigma - b^4\sigma}{2Ea^4 - 2Eb^4},\ b_2: \frac{a^4b^4\nu\sigma + a^4b^4\sigma}{2Ea^4 - 2Eb^4}\right\}$$

$$\left\{b_3: \frac{-a^2b^2\nu\sigma + a^2b^2\sigma}{Ea^2 - Eb^2},\right.$$

$$\left. b_4: \frac{a^2b^2\nu^2\sigma - 2a^2b^2\nu\sigma - 3a^2b^2\sigma + b^4\nu^2\sigma + b^4\nu\sigma}{2Ea^4\nu - 2Eb^4\nu}\right\}$$

```
Current len of sln_cs = 10; Total cs = 10
Constants yet to be found: []
```

The constants found are related to b. Since we assume a is very small compared to the dimensions of the plate, we set b to ∞ to simplify the solutions for these constants. The snippet is as follows:

```
cs_boo = dict((k, sp.limit(v, b, oo)) for k, v in sln_cs.items() )
cs_boo_ = gr.split_dict_n(cs_boo, 2)
display(Math(f" \\text{{ Constants found without $b$:}} "))
display(Math(f" \\text{{ }} \\small {latex(cs_boo_[0])}"))
display(Math(f" \\text{{ }} \\small {latex(cs_boo_[1])}"))
```

Constants found without b:

$$\left\{a_0 : 0,\ a_1 : 0,\ a_2 : -\frac{a^4\nu\sigma + a^4\sigma}{2E},\ a_3 : -\frac{2\left(-a^2\nu\sigma + a^2\sigma\right)}{E\left(\nu - 1\right)},\ b_0 : 0\right\}$$

$$\left\{a_4 : -\frac{-\nu\sigma - \sigma}{2E},\ b_1 : 0,\ b_2 : -\frac{a^4\nu\sigma + a^4\sigma}{2E},\right.$$

$$\left. b_3 : -\frac{-a^2\nu\sigma + a^2\sigma}{E},\ b_4 : -\frac{\nu^2\sigma + \nu\sigma}{2E\nu}\right\}$$

6.3.3 *Final solution for stresses*

By substituting the constants found into the general expression for stresses, we obtain the final solution for a plate with a hole subjected to uniaxial loading. We can also display the formulas for stresses at both boundaries $r = a$ and $r = b(\infty)$:

```
# Compute the final solution for the stresses:
σv = sp.simplify(σvg.subs(sln_cs)) # Update the stress solutions
σv_final = σv.applyfunc(lambda expr: sp.limit(expr, b, oo))

# Simplify the solutions:
σv_final = σv_final.applyfunc(lambda expr: \
        sp.collect(expr.expand(),cos(2*θ)).collect(σ).collect(sin(2*θ)))
σv_final_b=σv_final.subs(r,b).applyfunc(lambda expr:sp.limit(expr,b,oo))
display((σv_final, σv_final.subs(r,a), σv_final_b))    # any r, r=a, r=b
```

$$\left(\begin{bmatrix}\sigma\left(\frac{3a^4}{2r^4}-\frac{2a^2}{r^2}+\frac{1}{2}\right)\cos(2\theta)\\ \sigma\left(-\frac{3a^4}{2r^4}-\frac{1}{2}\right)\cos(2\theta)\\ \sigma\left(\frac{3a^4}{2r^4}-\frac{a^2}{r^2}-\frac{1}{2}\right)\sin(2\theta)\end{bmatrix}, \begin{bmatrix}0\\ -2\sigma\cos(2\theta)\\ 0\end{bmatrix}, \begin{bmatrix}\frac{\sigma\cos(2\theta)}{2}\\ -\frac{\sigma\cos(2\theta)}{2}\\ -\frac{\sigma\sin(2\theta)}{2}\end{bmatrix}\right)$$

The final stress solution is given in a tuple of three vectors. The first is the formula for these three stress components, the second is the stresses on the inner circle $r = a$, and the last is the stresses in the far field. It is clear that all these conditions for the stresses at these boundaries are satisfied.

6.3.4 *Validation of the solution*

It is good practice to double check whether the equilibrium conditions are satisfied by the final solution:

```
# check whether the equilibrium conditions:
br = 0; bθ = 0
equi1=(σv_final[0].diff(r)+σv_final[2].diff(θ)/r\
                    +(σv_final[0]-σv_final[1])/r+br).simplify().factor()
equi2=(σv_final[2].diff(r)+σv_final[1].diff(θ)/r\
                                 +2*σv_final[2]/r+bθ).simplify().factor()
display(Math(f" \\text{{ Are the equilibrium conditions satisfied? }}\
                              \\; {latex((abs(equi1)+abs(equi2))==0)}"))
```

Are the equilibrium conditions satisfied? True

We can also compare the stress solution obtained via our displacement method with that obtained in the previous chapter using the stress method:

```
# Stresses obtained in the previous chapter using the stress method:
σ_rr2_strssMthd = (3*a**4*σ/(2*r**4) - 2*a**2*σ/r**2 + σ/2)*cos(2*θ)
σ_θθ2_strssMthd = (-3*a**4*σ/(2*r**4) - σ/2)*cos(2*θ)
σ_rθ2_strssMthd = (3*a**4*σ/(2*r**4) - a**2*σ/r**2 - σ/2)*sin(2*θ)
σ_rr2_strssMthd.simplify().expand()

σv_final[0].simplify().expand()      # obtained via displacement method

display(Math(f"\\text{{Agree with $σ_{{rr}}$ from the stress method?}}\
         \\;\\; {latex((σv_final[0]-σ_rr2_strssMthd).simplify()==0)}"))
display(Math(f"\\text{{Agree with $σ_{{rr}}$ from the stress method?}}\
         \\;\\; {latex((σv_final[1]-σ_θθ2_strssMthd).simplify()==0)}"))
display(Math(f"\\text{{Agree with $σ_{{rr}}$ from the stress method?}}\
         \\;\\; {latex((σv_final[2]-σ_rθ2_strssMthd).simplify()==0)}"))
```

Agree with σ_{rr} from the stress method? True
Agree with σ_{rr} from the stress method? True
Agree with σ_{rr} from the stress method? True

6.3.5 *Final solution for strains*

By substituting the constants found into the general expression for strains, we obtain the final solution for a plate with a hole subjected to uniaxial load. We also display the strains at both boundaries $r = a$ and $r = b(\infty)$:

```
# Compute the final solution for the strains:
εv = sp.simplify(εvg.subs(sln_cs))          # Update the strain solutions
εv_final = εv.applyfunc(lambda expr: sp.limit(expr, b, oo))

# Simplify the solutions:
εv_final = εv_final.applyfunc(lambda expr: \
     sp.collect(expr.expand(),cos(2*θ)).collect(σ/E).collect(sin(2*θ)))
εv_final_b=εv_final.subs(r,b).applyfunc(lambda expr:sp.limit(expr,b,oo))
display((εv_final, εv_final.subs(r,a), εv_final_b))    # any r, r=a, r=b
```

$$\left(\begin{bmatrix}\frac{\sigma\left(\frac{3a^4\nu}{2r^4}+\frac{3a^4}{2r^4}-\frac{2a^2}{r^2}+\frac{\nu}{2}+\frac{1}{2}\right)\cos(2\theta)}{E}\\ \frac{\sigma\left(-\frac{3a^4\nu}{2r^4}-\frac{3a^4}{2r^4}+\frac{2a^2\nu}{r^2}-\frac{\nu}{2}-\frac{1}{2}\right)\cos(2\theta)}{E}\\ \frac{\sigma\left(\frac{3a^4\nu}{2r^4}+\frac{3a^4}{2r^4}-\frac{a^2\nu}{r^2}-\frac{a^2}{r^2}-\frac{\nu}{2}-\frac{1}{2}\right)\sin(2\theta)}{E}\end{bmatrix}, \begin{bmatrix}\frac{2\nu\sigma\cos(2\theta)}{E}\\ -\frac{2\sigma\cos(2\theta)}{E}\\ 0\end{bmatrix}, \begin{bmatrix}\frac{\sigma\left(\frac{\nu}{2}+\frac{1}{2}\right)\cos(2\theta)}{E}\\ \frac{\sigma\left(-\frac{\nu}{2}-\frac{1}{2}\right)\cos(2\theta)}{E}\\ \frac{\sigma\left(-\frac{\nu}{2}-\frac{1}{2}\right)\sin(2\theta)}{E}\end{bmatrix}\right)$$

6.3.6 Final solution for displacements

By substituting the constants found into the general expression for displacements, we obtain the final solution for a plate with a hole subjected to uniaxial load. We also display the displacements at both boundaries $r = a$ and $r = b(\infty)$:

```
# Final solution for the displacements:
ur_final = sp.limit(urg.subs(sln_cs), b, oo).collect(a).collect(r)\
                                          .collect(σ).collect(ν+1)
display(Math(f"\\text{{Final solution, $u_r$ = }}{latex(ur_final)}"))

uθ_final = sp.limit(uθg.subs(sln_cs), b, oo).collect(a).collect(r)\
                                          .collect(σ).collect(ν+1)
display(Math(f"\\text{{Final solution, $u_θ$ = }}{latex(uθ_final)}"))
#print(ur_final); print(uθ_final)

ur_final_  = σ*(4*a**2*r**2 + (r**4-a**4)*(ν + 1))*cos(2*θ)/(2*E*r**3)
uθ_final_ =-σ*(a**2*r**2*2*(1-ν)+(a**4+r**4)*(1+ν))*sin(2*θ)/(2*E*r**3)

# Displacement at r = a:
ur_final_a = ur_final_.subs(r,a).simplify()
uθ_final_a = uθ_final_.subs(r,a).simplify()
display(Math(f"\\text{{ At $r=a$, $u_r$ = }}{latex(ur_final_a)}"))
display(Math(f"\\text{{ At $r=a$, $u_θ$ = }}{latex(uθ_final_a)}"))

# Displacement at r = b (∞):
ur_final_b = sp.limit(urg.subs(sln_cs).subs(r,b),b,oo)
uθ_final_b = sp.limit(uθg.subs(sln_cs).subs(r,b),b,oo)
display(Math(f"\\text{{ At $r=b$, $u_r$ = }}{latex(ur_final_b)}"))
display(Math(f"\\text{{ At $r=b$, $u_θ$ = }}{latex(uθ_final_b)}"))
```

$$\text{Final solution, } u_r = \frac{\sigma\left(4a^2r^2 + \left(-a^4 + r^4\right)(\nu+1)\right)\cos(2\theta)}{2Er^3}$$

$$\text{Final solution, } u_\theta = \frac{\sigma\left(a^2r^2(2\nu-2) + \left(-a^4 - r^4\right)(\nu+1)\right)\sin(2\theta)}{2Er^3}$$

$$\text{At } r = a,\ u_r = \frac{2a\sigma\cos(2\theta)}{E}$$

$$\text{At } r = a,\ u_\theta = -\frac{2a\sigma\sin(2\theta)}{E}$$

$$\text{At } r = b,\ u_r = \infty\,\text{sign}\left((\nu\sigma + \sigma)\cos(2\theta)\right)$$

$$\text{At } r = b,\ u_\theta = \infty\,\text{sign}\left((-\nu\sigma - \sigma)\sin(2\theta)\right)$$

We found that along the inner circle $r = a$:

1. When $\theta = 0$ or π, corresponding to points B and B′ shown in Fig. 6.1, u_r reaches its maximum, directed horizontally outward. When $\theta = \pi/2$

or $3\pi/2$, corresponding to points A and A′ in Fig. 6.1, u_r reaches its minimum, directed vertically downward.

2. When $\theta = 0$ or π (points B and B′), u_θ is zero, indicating no vertical movement at those points. Similarly, when $\theta = \pi/2$ or $3\pi/2$ (points A and A′), u_θ is zero, indicating no horizontal movement. At $\theta = \pi/4$ or $3\pi/4$, which are 45° from the x-axis, u_θ attains its maximum or minimum values of $\pm\frac{2a\sigma}{E}$.

At $r = b$, representing infinity, the displacement becomes unbounded, even though the stresses remain finite. This is consistent with physical reasoning: a solid subjected to uniform stress at infinity must exhibit infinite displacements in the far field.

We now rewrite the formula for the final displacement solutions for future reference:

$$
\begin{aligned}
u_r &= \frac{\sigma}{2Er^3}\left(4a^2r^2 + \left(r^4 - a^4\right)(\nu+1)\right)\cos(2\theta) \\
u_\theta &= -\frac{\sigma}{2Er^3}\left(2a^2r^2(1-\nu) + \left(a^4 + r^4\right)(\nu+1)\right)\sin(2\theta)
\end{aligned}
\tag{6.6}
$$

We note that u_r is always positive (for positive $\cos 2\theta$), and u_θ is always negative (for positive $\sin 2\theta$), as schematically drawn in Fig. 6.1 at the beginning of the discussion.

6.3.7 *Plot for displacement distribution*

The stress solutions are all plotted in Chapter 5. We can now plot the distribution of the displacements.

```
# Convert SymPy expressions to Python functions:
dic_data = {E:1, ν:0.3, a:1, σ:1}
func1 = sp.lambdify(r, ur_final_.subs(dic_data).subs({θ:   0}),'numpy')
func2 = sp.lambdify(r, ur_final_.subs(dic_data).subs({θ:pi/2}),'numpy')
func3 = sp.lambdify(r, uθ_final_.subs(dic_data).subs({θ:pi/4}),'numpy')
labels=[r'$u_{r}, θ=0$',r'$u_{r}, θ=\pi/2$', r'$u_{θ}, θ=\pi/4$' ]

r_data  = np.linspace(1, 5, 400)                # r values for the plot
x_data = [r_data, r_data, r_data ]
y_data = [func1(x_data[0]), func2(x_data[1]), func3(x_data[2]) ]
#plt.ion()
#gr.plot_fs(x_data, y_data, labels, '$r$', 'stresses', \
#                               'imagesSMp/plateHoleUrUt',ioff=True)
                               # Use ioff=False to display the plot
```

The two displacement components u_r and u_θ are plotted in Fig. 6.2. The absolute values of these displacements increase roughly linearly with respect to the radial direction r.

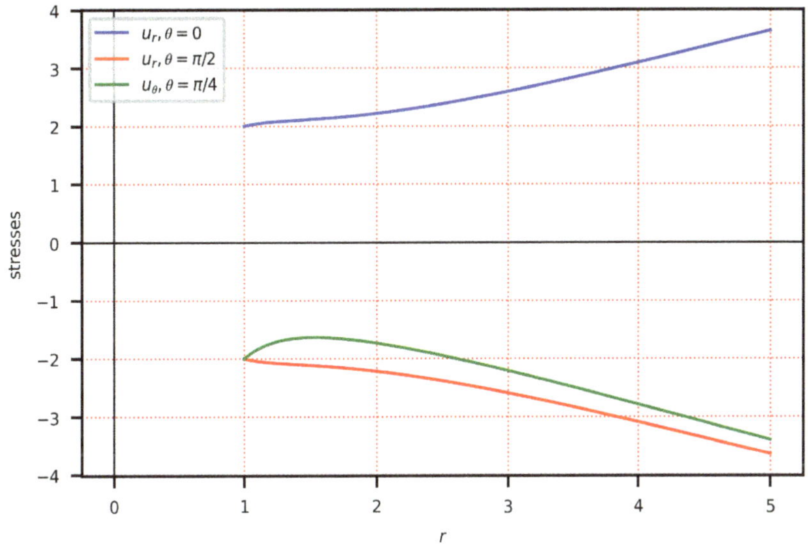

Figure 6.2. Distribution of the two displacement components u_r and u_θ in the vertical, horizontal, and 45° directions.

6.3.8 *Contour plot of displacements*

To plot the contour of the displacements, we first need to express the displacements in Cartesian coordinates using the relations

$$r = \sqrt{x^2 + y^2} \quad \text{and} \quad \theta = \arctan\left(\frac{-x}{y}\right) \tag{6.7}$$

Note that here $\theta = 0$ is along the y-axis. We then convert the displacement expressions to NumPy functions and use `gr.contour_plot()` to plot the contours. The snippet is as follows:

```
# Contour of displacement component u_r:
x, y = sp.symbols('x, y', real=True)
a_val = 1
dic_rt = {E:1,ν:0.3,σ:1, a:a_val, r:sp.sqrt(x**2+y**2), θ:sp.atan(y/x)}
ur_xy = ur_final.subs(dic_rt)
display(Math(f" \\text{{Radial displacement $u_{{r}}(x, y)$}} \
                                                    ={latex(ur_xy)}"))
```

Radial displacement $u_r(x, y)$

$$= \frac{\left(4x^2 + 4y^2 + 1.3\left(x^2 + y^2\right)^2 - 1.3\right)\cos\left(2\,\mathrm{atan}\left(\frac{y}{x}\right)\right)}{2\left(x^2 + y^2\right)^{\frac{3}{2}}}$$

```
# Create a mesh grid for a rectangular domain in x-y plane:
dX = 3.0*a_val;    dY = 2.0*a_val                      # plot 2dX × 2dY area
x_val = np.linspace(-dX, dX, 800)
y_val = np.linspace(-dY, dY, 800)
X, Y = np.meshgrid(x_val, y_val)

# define a circular hole to be masked in plot:
mask = (X/a_val)**2 + (Y/a_val)**2 <= 1

np_ur_xy = lambdify((x,y), ur_xy, 'numpy')            # convert to numpy
Z = np_ur_xy(X, Y)          # compute the displacement over the mesh grid
Z[mask] = np.nan                        # mask the hole, print nothing there

cntur = gr.contour_plot(X,Y, Z, x_lbl='x-axis',y_lbl='y-axis',
                title='Displacement $u_{{r}}$',lvls=80, ioff=True);
#plt.savefig('imagesSMp/plateHoleCntur_ur.png', dpi=500)  # save to file
#plt.show()
```

A contour plot of the radial displacement u_r within the plate with a circular hole is shown in Fig. 6.3. The mirror symmetry of the u_r distribution across the plate is clearly visible.

The periodicity of u_r, with a period of π in the tangential direction, is also evident, while it varies roughly linearly in the radial direction. The following snippet plots the tangential displacement component u_θ:

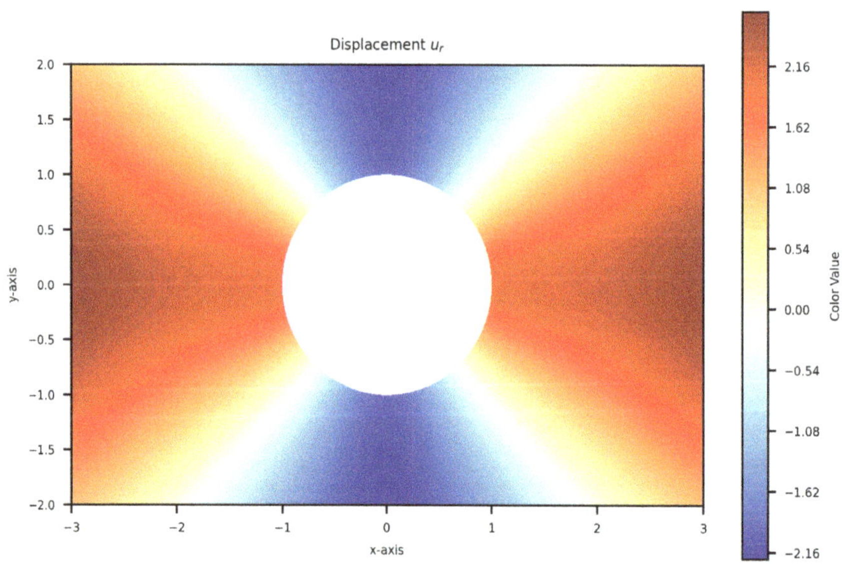

Figure 6.3. Contour of the radial displacement u_r in a plate with a circular hole.

```
# Create a mesh grid for a rectangular domain in x-y plane:
uθ_xy = uθ_final.subs(dic_rt)
np_uθ_xy = lambdify((x,y), uθ_xy, 'numpy')            # convert to numpy
Z = np_uθ_xy(X, Y)         # compute the displacement over the mesh grid
Z[mask] = np.nan                          # mask the hole, print nothing there

cntur = gr.contour_plot(X,Y, Z, x_lbl='x-axis',y_lbl='y-axis',
                title='Displacement $u_{{θ}}$',lvls=80, ioff=True);
#plt.savefig('imagesSMp/plateHoleCntur_ut.png', dpi=500)  # save to file
#plt.show()
```

```
Minimum value in the domain: -2.35534
Maximum value in the domain: 2.35534
```

A contour plot of the tangential displacement component u_θ within the plate with a circular hole is shown in Fig. 6.4.

The antisymmetry of the distribution of u_θ across the plate is clearly visible. The periodicity of π along the tangential direction is also evident, while the displacement varies approximately linearly in the radial direction.

Note that the effect of Poisson's ratio on the plate with a hole is noteworthy. The displacement is positive along the x-direction (the loading direction) and negative along the y direction (transverse to the loading direction).

Compared to the stress method discussed in Chapter 5, the displacement method is much more straightforward to implement and can produce the

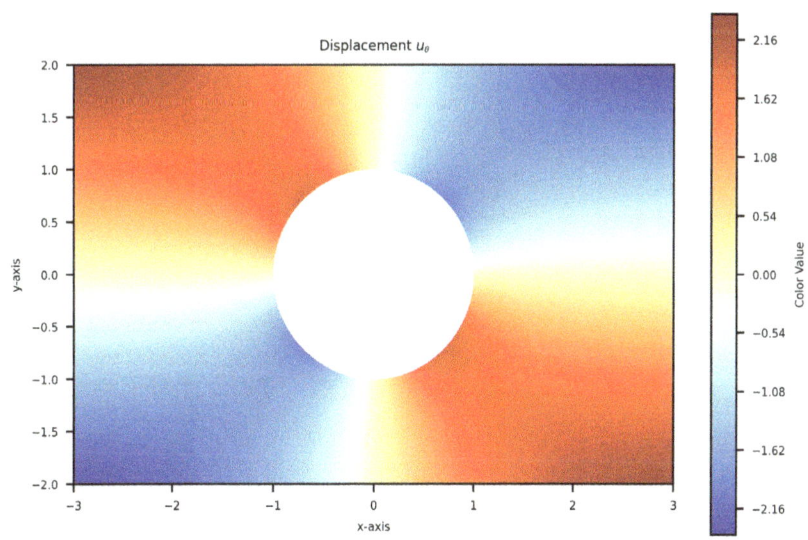

Figure 6.4. Contour of the tangential displacement component u_θ in a plate with a circular hole.

complete set of solutions: displacements, strains, and stresses. The major difficulty lies in finding a proper set of basis functions in the polar coordinate system. The following section presents a displacement method for a more challenging problem.

6.4 Curved beams under shearing at two ends

Consider the same problem studied in Section 3.4: a thin, semicircular curved beam of uniform thickness with an inner radius a and an outer radius b. It is subjected to a pair of horizontal shear forces P acting on the cross-sections at its two ends, as shown in Fig. 6.5.

All the settings are exactly the same, but here we propose a displacement method to find the complete set of solutions, including displacements, stresses, and strains. The strategy, concept, and procedure are similar to those discussed in the previous section.

6.4.1 *Construction of displacements*

As mentioned earlier, in any displacement method, the key is to properly assume the displacements. We shall make use of the symmetry properties of the problem:

Symmetry with respect to θ: It is clear that the displacement component in the radial direction, u_r, is symmetric with respect to the axis $\theta = 0$, and hence it must be an even function of θ. On the other hand, the tangential component, u_θ, must be antisymmetric and therefore an odd function of θ.

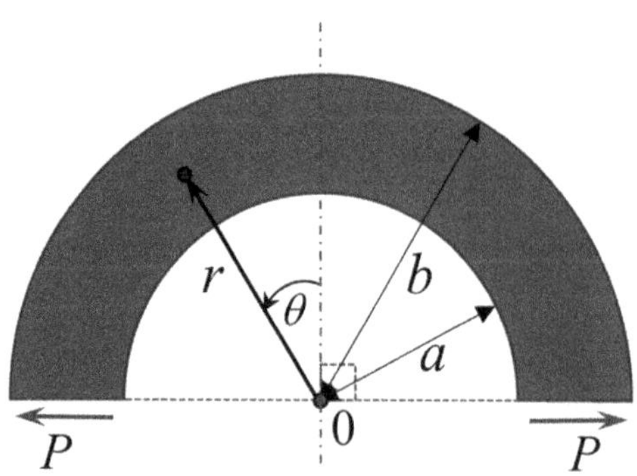

Figure 6.5. A thin, semicircular curved beam of uniform thickness loaded by a pair of horizontal shear forces acting on the cross-sections at its two ends.

Thus, the displacements can be assumed as

$$u_r(r,\theta) = \frac{a_0}{r^2}\cos(\theta) + a_1\cos(\theta) + a_2\theta\sin(\theta) + a_3 + a_4\cos(\theta)\log(r) + a_5 r\cos(\theta) + a_6 r\log(r)\cos(\theta) + a_7 r^2\cos(\theta)$$

$$[3pt]u_\theta(r,\theta) = \frac{b_0}{r^2}\sin(\theta) + b_1\sin(\theta) + b_2\theta\cos(\theta) + b_3 r\theta + b_4\sin(\theta)\log(r) + b_5 r\sin(\theta) + b_6 r\log(r)\sin(\theta) + b_7 r^2\sin(\theta) \tag{6.8}$$

where a_i and b_i are unknown constants, making a total of 16, corresponding to 16 basis functions, all of which are elementary functions. The choice of these bases largely follows the considerations below, similar to those used in the displacement method in the Cartesian coordinate system [1]:

1. **Lowest-order bases**: Use the lowest-order possible bases, assuming that the displacements lie in a space spanned by these basis functions. If this assumption is not valid, one must increase the order and include more bases.
2. **Symmetry conditions**: The chosen bases should satisfy the symmetry requirements. This drastically reduces the number of possible bases while maintaining the desired orders. For this problem, u_r must be an even function of θ, and u_θ must be an odd function of θ. The choices in Eq. (6.8) satisfy these conditions. Note that the lowest constant basis, 1, is included only in u_r, not in u_θ, consistent with the symmetry rules.
3. **Sine and cosine functions**: The selection of sine and cosine functions is based on the sinusoidal nature of the problem. The choice of a single angle θ (rather than the double angle 2θ used in the previous section) is due to the presence of only one plane of mirror symmetry in this problem. The inclusion of logarithmic basis functions is motivated by the fact that the stresses in this problem (see Section 3.4) contain a term $1/r$, whose antiderivative is $\log r$. This arises because stresses are related to the derivatives of displacements.
4. **Radial coordinate** r: Terms with both negative and positive powers of r are included.

The displacements assumed in Eq. (6.8) satisfy the major requirements for this problem and are equipped with arbitrary constants that can be adjusted to meet all detailed boundary conditions.

It is important to note that whether this assumption works depends on the specific problem. If the true displacements do not lie in the space spanned by the assumed basis functions, the displacement method will fail. However, if they do lie in this space, a valid solution can be obtained. Once the displacements are found, the strains and stresses follow naturally, since they involve only differentiation, which can always be performed for any linear combination of elementary basis functions [1,3].

Since the assumption of displacements is essentially based on educated guessing, experience and a deep understanding of the problem are essential.

The code snippet for generating the expressions for the displacements is as follows:

```
# Define variables for displacement method for curved beam problems:
E, ν = symbols('E, ν', positive=True)          # given material constants
r = sp.symbols('r', positive=True)  # symbolic variable for radial axis
θ = sp.symbols('θ', real=True)                        # for tangential axis

ur_rbs_list = [cos(θ)/r**2, cos(θ), θ*sin(θ), 1, cos(θ)*log(r),\
                                r*cos(θ), r*log(r)*cos(θ), r**2*cos(θ)]
ur_rbs = sp.Matrix(ur_rbs_list)

_, order_n = ur_rbs[0].as_coeff_exponent(r)
_, order_p = ur_rbs[-1].as_coeff_exponent(r)
order = order_p - order_n + 1               # compute the order difference
print(f" The highest inverse order of the monomial = {order}")

uθ_rbs_list = [sin(θ)/r**2, sin(θ), θ*cos(θ), θ*r, sin(θ)*log(r),\
                                r*sin(θ), r*log(r)*sin(θ), r**2*sin(θ)]
uθ_rbs = sp.Matrix(uθ_rbs_list)

# unknown constants for displacements:
ai = sp.symbols(f'a:{len(ur_rbs)}')                     # unknown constants
display(Math(f" \\text{{Unknown constants for $u$}}={latex(ai)}"))
ai = Matrix([ai[i] for i in range(len(ur_rbs))]) #form coeffs in vector

bi = sp.symbols(f'b:{len(uθ_rbs)}')                     # unknown constants
display(Math(f" \\text{{Unknown constants for $v$}}={latex(bi)}"))
bi = Matrix([bi[i] for i in range(len(uθ_rbs))])           # form a vector

cs = ai.col_join(bi)                 # form a single vector of constants

urg = (ai.T@Matrix(ur_rbs))[0]
uθg = (bi.T@Matrix(uθ_rbs))[0]
display(Math(f"\\text{{Assumed displacements}}:"))
display(Math(f" \\text{{ $u_θ$}}=\\scriptsize {latex(urg)}"))
display(Math(f" \\text{{ $u_θ$}}=\\scriptsize {latex(uθg)}"))
```

```
The highest inverse order of the monomial = 5
```

Unknown constants for $u = (a_0,\ a_1,\ a_2,\ a_3,\ a_4,\ a_5,\ a_6,\ a_7)$

Unknown constants for $v = (b_0,\ b_1,\ b_2,\ b_3,\ b_4,\ b_5,\ b_6,\ b_7)$

Assumed displacements:

$$u_\theta = \frac{a_0 \cos(\theta)}{r^2} + a_1 \cos(\theta) + a_2\theta \sin(\theta) + a_3 + a_4 \log(r) \cos(\theta) + a_5 r \cos(\theta) + a_6 r \log(r) \cos(\theta) + a_7 r^2 \cos(\theta)$$

$$u_\theta = \frac{b_0 \sin(\theta)}{r^2} + b_1 \sin(\theta) + b_2\theta \cos(\theta) + b_3 r\theta + b_4 \log(r) \sin(\theta) + b_5 r \sin(\theta) + b_6 r \log(r) \sin(\theta) + b_7 r^2 \sin(\theta)$$

The constructed displacements contain a total of 16 unknown constants. We now determine these constants using equilibrium and stress boundary conditions, all of which require stresses.

6.4.2 *Enforcing differential equilibrium equations*

Next, we compute the expressions for strains using the displacements and then compute the stresses:

```
# Compute strains using the displacements:
εrr = urg.diff(r)
εθθ = urg/r + uθg.diff(θ)/r
γrθ = urg.diff(θ)/r + uθg.diff(r) - uθg/r
εvg = sp.Matrix([εrr, εθθ, γrθ/2])           # define strains as a vector

# Compute stress vector using the strain vector:
σvg = sp.Matrix(gr.strain2stress2D(E, ν, εvg, Ptype= 'stress')).expand()

# To take a look, but lengthy:
#display(Math(f" \\text{{General solution for stresses:}} "))
#display(Math(f" \\text{{$σ_{{v}}$=}} {latex(sp.simplify(σvg))}"))
#display(Math(f" \\text{{$ε_{{v}}$=}} {latex(sp.simplify(εvg))}"))
#executed in 33.8s, finished 20:12:23 2024-12-05
```

These expressions for stresses and strains are very lengthy. Readers may uncomment these lines to display them. The stresses are then used to create equilibrium equations, as shown in the following snippet:

```
# Define equilibrium equation that needs to be satisfied:
br = 0; bθ = 0                                    # Consider zero body forces
equi1=(σvg[0].diff(r)+σvg[2].diff(θ)/r+(σvg[0]-σvg[1])/r+br)\
                                                     .simplify().factor()
equi2=(σvg[2].diff(r)+σvg[1].diff(θ)/r+2*σvg[2]/r+bθ)\
                                                     .simplify().factor()
#display(equi1, equi2)
```

Typical terms in the equilibrium equations are

$$-\frac{E(a_0\nu\cos(\theta)+\cdots+3a_1r^2\cos(\theta) + a_2r^2\theta\nu\sin(\theta)+\cdots+a_4r^2\nu\log(r)\cos(\theta)+\cdots)}{2r^4(\nu-1)(\nu+1)} \tag{6.9}$$

The expressions for the equilibrium equations are half a page long, so we display only a few terms. To ensure that these equilibrium equations are satisfied, their numerators must be zero. Each numerator is a linear combination of the combined bases, as shown above. We group these terms in the following steps for each of the two equilibrium equations:

Step 1: Group all terms that contain $\sin(\theta)$ and $\cos(\theta)$ (including their zeroth-order terms). The code snippet is as follows:

```
exprs = []
exprs.append(sp.numer(equi1).expand().coeff(sin(θ)))
exprs.append(sp.numer(equi2).expand().coeff(cos(θ)))
exprs.append(sp.numer(equi1).expand().coeff(cos(θ)))
exprs.append(sp.numer(equi2).expand().coeff(sin(θ)))
exprs.append(sp.numer(equi1).expand().coeff(cos(θ),0).coeff(sin(θ),0))
exprs.append(sp.numer(equi2).expand().coeff(cos(θ),0).coeff(sin(θ),0))
# display(exprs)                                                # Lengthy
```

Typical terms in a grouped expression are

$$[-Ea_2r^2\theta\nu+3Ea_2r^2\theta+\cdots-Ea_4r^2\nu\log(r) + 3Ea_4r^2\log(r)-Ea_5r^3\nu+\cdots] \tag{6.10}$$

Step 2: For each grouped expression, further group the terms that contain bases of θ, $\log(r)$, and their zeroth-order terms. These grouped terms are concatenated, forming a set of expressions that are purely polynomials in r. Some expressions in this set may be zero and are removed. The remaining expressions are then solved for the unknown constants.

All these steps are implemented using **gr.constantUpdate()**, which we developed earlier. The code snippet is as follows:

```
exprsθ   = [e.coeff(θ) for e in exprs]
exprsLog = [e.coeff(log(r)) for e in exprs]
exprs00  = [e.coeff(θ,0).coeff(log(r),0) for e in exprs]
exprs000 = [e.coeff(r,0) for e in exprs]
all_exprs= exprsθ + exprsLog + exprs00  + exprs000    # concatenate all
#display(all_exprs) # lengthy

sln_cs = {}                # initialize solution of unknown constants, cs
sln_equi = gr.constantUpdate(all_exprs, r, order, cs, prnt=False)

# Update unknown constants:
sln_cs = {**sln_equi}
sln_cs = {key: value.subs(sln_cs) for key, value in sln_cs.items()}
print(f" Current len of sln_cs={len(sln_cs)}; Total cs={len(cs)}")

# split sln_cs into 2 dics just for printing:
cs_tmp = gr.split_dict_n(sln_cs, 2)
display(Math(f"\\text{{Constants found: }}"))
display(Math(f"\\text{{ }} \\small {latex(cs_tmp[0])}"))
display(Math(f"\\text{{ }} \\small {latex(cs_tmp[1])}"))

σv = sp.simplify(σvg.subs(sln_cs))                    # update stresses
#display(Math(f"\\text{{ Current stresses: }}"))
#display(Math(f"\\text{{$σ_{{v}}$=}} \\small {latex(sp.simplify(σv))}"))
```

```
Total number of expressions = 120
Number of nonzero expressions = 18
Current len of sln_cs=11; Total cs=16
```

Constants found:

$$\left\{a_0 : b_0,\ a_1 : -b_1 + \frac{b_2\,(\nu+1)}{\nu-3} + \frac{b_4\,(-\nu-1)}{\nu-3},\ a_2 : b_2,\ a_3 : 0,\ a_4 : -b_4\right\}$$

$$\left\{a_5 : 0,\ a_6 : 0,\ a_7 : \frac{b_7\,(1-3\nu)}{\nu+5},\ b_3 : 0,\ b_5 : 0,\ b_6 : 0\right\}$$

We have successfully determined several unknown constants and used them to update the stress expressions. This process, though complex, can be fully automated in code, eliminating the need for manual calculations.

The resulting general stress expressions satisfy the equilibrium equations. These expressions are significantly simplified, as many unknown constants were found to be zero. The following snippet verifies that these stresses satisfy the equilibrium equations:

```
# Define equilibrium equation that needs to be satisfied:
equi1_=(σv[0].diff(r)+σv[2].diff(θ)/r+(σv[0]-σv[1])/r + br)\
                                                   .simplify().factor()
equi2_=(σv[2].diff(r)+σv[1].diff(θ)/r+2*σv[2]/r + bθ)\
                                                   .simplify().factor()
display((equi1_, equi2_)) # must be (0, 0)
```

(0, 0)

6.4.3 *General expressions for displacements*

We can also update the displacements (or the strains if needed):

```
# Compute the general expression for the displacements:
urg_ = urg.subs(sln_cs)
#display(Math(f"\\text{{$u_r$ satisfying equilibrium = }}{latex(urg)}"))
uθg_ = uθg.subs(sln_cs)

# Check how many constants are found and how many unknowns remain:
list_keys = list(sln_cs.keys())
cs_Eql = list(set(list(cs))-set(list_keys))
print(f"Constants yet to be found: {cs_Eql}")
print(f"Current displacement uθg: \n{uθg_}")
#print(f"Current displacement urg: \n{urg_}") # lengthy
```

```
Constants yet to be found: [b0, b7, b4, b1, b2]
Current displacement uθg:
b0*sin(θ)/r**2 + b1*sin(θ) + b2*θ*cos(θ) + b4*log(r)*sin(θ) + b7*r**2*sin
(θ)
```

We obtained the general displacement expressions that satisfy the equilibrium equations. These are rearranged as follows for future reference:

$$
\begin{aligned}
u_r &= \frac{b_0 \cos\theta}{r^2} + b_2\theta\sin\theta - b_4 \log r \cos\theta \\
&\quad + \frac{b_7 r^2 (1-3\nu)\cos\theta}{\nu+5} + \left(-b_1 + \frac{b_2(\nu+1)}{\nu-3} - \frac{b_4(\nu+1)}{\nu-3}\right)\cos\theta \qquad (6.11) \\
u_\theta &= \frac{b_0 \sin\theta}{r^2} + b_1\sin\theta + b_2\theta\cos\theta + b_4\log r\sin\theta + b_7 r^2 \sin\theta
\end{aligned}
$$

These displacement expressions still contain five unknown constants, which must be determined using the boundary conditions for stresses, forces, and displacements specific to the problem.

6.4.4 General expressions for stresses

To facilitate further analysis without rerunning the previous computations, we explicitly redefine the general displacements. The following snippet also computes the corresponding stresses based on these displacements:

```
E, ν = symbols('E, ν', positive=True)          # given material constants
r = sp.symbols('r', positive=True)  # symbolic variable for radial axis
θ, p = sp.symbols('θ, p', real=True)                  # for tangential axis

b0, b1, b2, b4, b7 = sp.symbols("b0, b1, b2, b4, b7")
cs = [b0, b1, b2, b4, b7]                                # unknown constants

urg_= b0*cos(θ)/r**2 + b2*θ*sin(θ) - b4*log(r)*cos(θ) +b7*r**2*(1-3*ν)\
      *cos(θ)/(ν+5) + (-b1 + b2*(ν + 1)/(ν-3) + b4*(-ν-1)/(ν-3))*cos(θ)

uθg_= b0*sin(θ)/r**2 + b1*sin(θ) + b2*θ*cos(θ) + b4*log(r)*sin(θ) \
                                                    + b7*r**2*sin(θ)
R = sp.Matrix([r, θ])                                  # position vector
u_g = [urg_, uθg_]        # general displacements with unknown constants

εg = gr.displacement2StrainPolar(u_g, R)                 # general strains

# Compute the general stresses with with unknown constants and check:
σrrg, σθθg, σrθg = gr.strain2stress2D(E, ν, εg, Ptype= 'stress')
display(Math(f"\\text{{General stresses in vector form $σ_{{v}}$:}}"))
display(Math(f" {latex(sp.simplify(sp.Matrix([σrrg, σθθg, σrθg])))}"))
# Check:
#display(Math(f" \\text{{Agrees with $σ_{{rr}}$ obtained earlier? }}\
#                          {latex((σv[0]-σrrg).simplify()==0)}"))
#display(Math(f" \\text{{Agrees with $σ_{{θθ}}$ obtained earlier? }}\
#                          {latex((σv[1]-σθθg).simplify()==0)}"))
#display(Math(f" \\text{{Agrees with $σ_{{rθ}}$ obtained earlier? }}\
#                          {latex((σv[2]-σrθg).simplify()==0)}"))
```

General stresses in vector form σ_v:

$$\begin{bmatrix} \dfrac{E(-2b_0\nu^2-4b_0\nu+30b_0-2b_2r^2\nu^2-10b_2r^2\nu+b_4r^2\nu^2+8b_4r^2\nu+15b_4r^2+2b_7r^4\nu^2-4b_7r^4\nu-6b_7r^4)\cos(\theta)}{r^3(\nu^3+3\nu^2-13\nu-15)} \\ \dfrac{E(2b_0\nu^2+4b_0\nu-30b_0-2b_2r^2\nu-10b_2r^2+b_4r^2\nu^2+4b_4r^2\nu-5b_4r^2+6b_7r^4\nu^2-12b_7r^4\nu-18b_7r^4)\cos(\theta)}{r^3(\nu^3+3\nu^2-13\nu-15)} \\ \dfrac{E(-2b_0\nu^2-4b_0\nu+30b_0-2b_2r^2\nu-10b_2r^2+b_4r^2\nu^2+4b_4r^2\nu-5b_4r^2+2b_7r^4\nu^2-4b_7r^4\nu-6b_7r^4)\sin(\theta)}{r^3(\nu^3+3\nu^2-13\nu-15)} \end{bmatrix}$$

6.4.5 *Enforcing stress boundary conditions*

The inner and outer curved surfaces of the curved beam must satisfy the following stress boundary conditions:

$$\begin{aligned} \sigma_{rr} = \sigma_{r\theta} = 0 \quad \text{at} \quad r = a \\ \sigma_{rr} = \sigma_{r\theta} = 0 \quad \text{at} \quad r = b \end{aligned} \tag{6.12}$$

The following snippet applies the stress boundary conditions (SBCs) to determine some of the unknown constants, starting with σ_{rr}:

```
# Solve SBCs for some of the unknown constants:
sln_cs = {}
a, b = symbols('a, b', positive=True)                          # radii
sln_SBC = sp.solve([σrrg.subs(r,a), σrrg.subs(r,b),
                    σrθg.subs(r,a), σrθg.subs(r,b)], cs)

# Update unknown constants:
sln_cs = {**sln_SBC}
sln_cs = {key: value.subs(sln_cs).simplify() for key, value in \
                                                     sln_cs.items()}
display(sln_cs)
```

$$\left\{b_0 : -\frac{a^2b^2b_7(\nu+1)}{\nu+5},\ b_2 : -\frac{4b_7(a^2+b^2)}{\nu+5},\ b_4 : \frac{2b_7(-a^2\nu+a^2-b^2\nu+b^2)}{\nu+5}\right\}$$

We find that the constants b_0, b_2, and b_4 are expressed in terms of b_7. Hence, one constant, b_7, remains to be determined. We will now use the integral equilibrium conditions or the force boundary conditions (FBCs).

6.4.6 *Enforcing integral equilibrium conditions*

At the cross-section of an end of the beam ($\theta = \pm\pi/2$), as shown in Fig. 6.5, the following integral equilibrium conditions hold:

$$\begin{aligned} \sum M_z = 0 : \int_a^b \sigma_{\theta\theta}\left(r - \frac{a+b}{2}\right) dr = 0 \\ \sum F_r = 0 : \int_a^b \sigma_{r\theta} dr - P = 0 \end{aligned} \tag{6.13}$$

Note that the moment is measured with respect to the centroid axis of the beam cross-section. The following snippet enforces these conditions. First, for $\sum M_z$:

```
# Compute the stresses with known constants and compute moment:
Mz = sp.integrate(σθθg.subs(θ, pi/2).subs(sln_cs)*(r-(a+b)/2), (r,a,b))
display(Math(f" \\text{{ Σ$M_z$}}=\\small {latex(Mz.simplify())}"))
```

$$\sum M_z = 0$$

It is found that the first equation in Eq. (6.13) is satisfied regardless of b_7. This is because $\sigma_{\theta\theta}$ has a factor of $\cos\theta$, which is zero at $\theta = \pm\pi/2$. Thus, we need to enforce only the second equation in Eq. (6.13):

```
# Compute the sum of the force F_r on a cross-section via integral:
P = symbols('P', real=True)
equI=sp.integrate(σrθg.subs(sln_cs).subs(θ,pi/2),(r,a,b)).simplify() - P
#display(equI)
sln_equI = sp.solve([equI], cs)                    # solve for the constant

# simplify the expression for the constant:
sln_equI = {key: value.subs(sln_equI).simplify() for key, \
                                          value in sln_equI.items()}
display(Math(f"\\text{{ Constant found:}}=\\small {latex(sln_equI)}"))

Cab = sp.symbols('C_ab')    # defined for further simplifying expressions
b7_numer = (sp.numer(sln_equI[b7])/P).expand().simplify()
b7_denom = sp.denom(sln_equI[b7]).expand().simplify()/(2*E)

Cab_ = (b**2/a**2-1) + (1+b**2/a**2)*(log(a)-log(b))      # simplification
display(Math(f"\\text{{Simplified denominator, $C_{{ab}}$}}={latex(Cab_)}"))

dffr = sp.expand_log((Cab_*a**2-b7_denom)).simplify()              # check
display(Math(f"\\text{{ Is simplification correct:}}={latex(dffr==0)}"))
```

$$\text{Constant found:} = \left\{ b_7 : \frac{P(\nu+5)}{2E\left(-a^2+b^2+\log\left(a^{a^2+b^2}b^{-a^2-b^2}\right)\right)} \right\}$$

$$\text{Simplified denominator, } C_{ab} = \left(1+\frac{b^2}{a^2}\right)(\log(a)-\log(b)) - 1 + \frac{b^2}{a^2}$$

$$\text{Is simplification correct:} = \text{True}$$

In the cell above, we applied simplifications due to the length of the expressions. The constants can now be expressed in concise forms as follows:

```
sln_equI[b7] = P*b7_numer/(2*E*Cab*a**2)

sln_cs = {**sln_SBC, **sln_equI}
print(f"Current len of sln_cs={len(sln_cs)}; Total cs={len(cs)}")
sln_cs = {key: value.subs(sln_cs).simplify() for key, value in \
                                                    sln_cs.items()}
#sln_cs = {key: value.subs(sln_equI).simplify() for key, \
#                                           value in sln_cs.items()}
#sln_cs.update(sln_equI)
display(Math(f"\\text{{Constants found:}} "))
print(f"  b0={sln_cs[b0]}, b2={sln_cs[b2]}")
print(f"  b4={sln_cs[b4]}")
print(f"  b7={sln_cs[b7]}")
```

```
Current len of sln_cs=4; Total cs=5
```

Constants found:

```
b0=-P*b**2*(ν + 1)/(2*C_ab*E), b2=-2*P*(a**2 + b**2)/(C_ab*E*a**2)
b4=P*(-a**2*ν + a**2 - b**2*ν + b**2)/(C_ab*E*a**2)
b7=P*(ν + 5)/(2*C_ab*E*a**2)
```

We have found all these constants. The formulas for these constants are rewritten as follows for future reference:

$$
\begin{aligned}
b_0 &= -\frac{Pb^2(\nu+1)}{2C_{ab}E}; \quad b_2 = -\frac{2P\left(a^2+b^2\right)}{C_{ab}Ea^2} \\
b_4 &= \frac{P\left(-a^2\nu + a^2 - b^2\nu + b^2\right)}{C_{ab}Ea^2}; \quad b_7 = \frac{P(\nu+5)}{2C_{ab}Ea^2}
\end{aligned}
\tag{6.14}
$$

6.4.7 *Final formulas for stresses*

Using the constants found, we compute the final formulas for the stresses:

```
σrr = σrrg.subs(sln_cs).simplify()
σθθ = σθθg.subs(sln_cs).simplify()
σrθ = σrθg.subs(sln_cs).simplify()

Cab_ = (b**2/a**2-1) + (1+b**2/a**2)*(log(a)-log(b))
σrr_ = P*( a**2*b**2-a**2*r**2-b**2*r**2+r**4)*cos(θ)/(Cab_*a**2*r**3)
σθθ_ = P*(-a**2*b**2-a**2*r**2-b**2*r**2+3*r**4)*cos(θ)/(Cab_*a**2*r**3)
σrθ_ = P*( a**2*b**2-a**2*r**2-b**2*r**2+r**4)*sin(θ)/(Cab_*a**2*r**3)

display(Math(f" \\text{{ Final solution for stresses:}}"))
display(Math(f" \\text{{ $σ_{{rr}}$}}=\\small {latex(σrr_)}"))
display(Math(f" \\text{{ $σ_{{θθ}}$}}=\\small {latex(σθθ_)}"))
display(Math(f" \\text{{ $σ_{{rθ}}$}}=\\small {latex(σrθ_)}"))
```

Final solution for stresses:

$$\sigma_{rr} = \frac{P\left(a^2b^2 - a^2r^2 - b^2r^2 + r^4\right)\cos(\theta)}{a^2r^3\left(\left(1+\frac{b^2}{a^2}\right)(\log(a) - \log(b)) - 1 + \frac{b^2}{a^2}\right)}$$

$$\sigma_{\theta\theta} = \frac{P\left(-a^2b^2 - a^2r^2 - b^2r^2 + 3r^4\right)\cos(\theta)}{a^2r^3\left(\left(1+\frac{b^2}{a^2}\right)(\log(a) - \log(b)) - 1 + \frac{b^2}{a^2}\right)}$$

$$\sigma_{r\theta} = \frac{P\left(a^2b^2 - a^2r^2 - b^2r^2 + r^4\right)\sin(\theta)}{a^2r^3\left(\left(1+\frac{b^2}{a^2}\right)(\log(a) - \log(b)) - 1 + \frac{b^2}{a^2}\right)}$$

6.4.8 *Neutral axis of semicircular beams*

Following the same concept as in standard straight beam theory, the neutral axis along the cross-section of a semicircular beam can be defined as the line where $\sigma_{\theta\theta} = 0$. From the formula for $\sigma_{\theta\theta}$, we can determine the value of r at which $\sigma_{\theta\theta} = 0$, independent of θ. This implies that the neutral axis along the cross-section of a semicircular beam is a circle. The formula for computing the radius of this neutral circle is given by

$$r_{\text{neutral}} = \sqrt{\frac{a^2}{6} + \frac{b^2}{6} + \frac{\sqrt{a^4 + 14a^2b^2 + b^4}}{6}} \tag{6.15}$$

Let $\alpha = \frac{b}{a}>1$. We have

$$r_{\text{neutral}} = a\sqrt{\frac{\alpha^2}{6} + \frac{\sqrt{\alpha^4 + 14\alpha^2 + 1}}{6} + \frac{1}{6}} \tag{6.16}$$

When b/a is close to 1, indicating that the beam height is small (referred to as a thin beam), Eq. (6.16) can be approximated in the following series form:

$$r_{\text{neutral}} = a\left[\frac{(1+\alpha)}{2} + \frac{(\alpha-1)^4}{128} - \frac{3(\alpha-1)^5}{256} + \frac{13(\alpha-1)^6}{1024} - \frac{25(\alpha-1)^7}{2048} + O\left((\alpha-1)^8\right)\right] \tag{6.17}$$

If b/a is very close to 1, we can retain only the linear term:

$$r_{\text{neutral}} \approx \frac{a}{2}(1+\alpha) = \frac{a+b}{2} \tag{6.18}$$

In this case, the neutral circle lies in the middle. Since the sign of the series in Eq. (6.16) oscillates, the cut-off error in this linear approximation is

$$e_{r_{\text{neutral}}} < \frac{(\alpha - 1)^4}{128} \tag{6.19}$$

For a thick beam with $\alpha = \frac{b}{a} = 1.5$, we find that $r_{\text{neutral}} \approx 1.25a$ (in the middle), with the error being less than 0.05%. This shows that even for a relatively thick beam, the neutral circle is still approximately in the middle.

All these formulas are derived using the following snippets:

```
roots = sp.solve(σθθ_, r)
display(Math(f"\\text{{Roots of $σ_{{θθ}}=0$ are}}:\
                                        \\scriptsize {latex(roots)}"))
display(Math(f"\\text{{The first root is complex, and discarded.}}"))
display(Math(f"\\text{{The real root}}=\\scriptsize {latex(roots[1])}"))

α = symbols('α')                                        # ratio of b/a
constraint = α > 1
rootα = a*roots[1].subs(b,α*a).subs(a,1)
display(Math(f"\\text{{Formula: $r_{{neutral}}$}}=\\scriptsize \
                                                    {latex(rootα)}"))
sp.series(rootα, α, 1, 7)
display(Math(f"\\text{{Series}}=\
                 \\scriptsize {latex(sp.series(rootα, α, 1, 7))}"))
r_1d5 = rootα.subs(α,1.5).evalf()
display(Math(f"\\text{{When $α=1.5, r_{{neutral}}$}}=\\small \
                                                    {latex(r_1d5)}"))
error = ((α-1)**4/128).subs(α,1.5).evalf()
display(Math(f"\\text{{ Error when using linear term}}=\\small \
                                                    {latex(error)}"))
```

$$\text{Roots of } \sigma_{\theta\theta} = 0 \text{ are: } \left[\frac{\sqrt{6}\sqrt{a^2 + b^2 - \sqrt{a^4 + 14a^2b^2 + b^4}}}{6}, \frac{\sqrt{6}\sqrt{a^2 + b^2 + \sqrt{a^4 + 14a^2b^2 + b^4}}}{6} \right]$$

The first root is complex, and discarded.

$$\text{The real root} = \frac{\sqrt{6}\sqrt{a^2 + b^2 + \sqrt{a^4 + 14a^2b^2 + b^4}}}{6}$$

$$\text{Formula: } r_{\text{neutral}} = \frac{\sqrt{6}a\sqrt{\alpha^2 + \sqrt{\alpha^4 + 14\alpha^2 + 1} + 1}}{6}$$

$$\text{Series} = a + \frac{a(\alpha-1)}{2} + \frac{a(\alpha-1)^4}{128} - \frac{3a(\alpha-1)^5}{256} + \frac{13a(\alpha-1)^6}{1024} + O((\alpha-1)^7; \alpha \to 1)$$

When $\alpha = 1.5, r_{\text{neutral}} = 1.25025499639511a$

Error when using linear term $= 0.00048828125$

6.4.9 *Internal forces in semicircular beams*

With the obtained stress components, we can now compute the internal forces on the cross-section of the beam through integration. The expressions for these forces are as follows:

$$\begin{aligned} M &= \int_a^b \sigma_{\theta\theta}\left(r - \frac{a+b}{2}\right) dr \\ N &= \int_a^b \sigma_{\theta\theta} dr \\ V &= \int_a^b \sigma_{r\theta} dr \end{aligned} \tag{6.20}$$

Here, M, N, and V represent the internal moment, normal force, and shear force, respectively, on the beam cross-section at a point along the beam's curved axis. Note that the moment is measured with respect to the centroid axis of the beam cross-section. The following snippet computes all of these internal forces:

```
## Compute internal forces on the cross-section:
#M = sp.integrate(σθθ_*(r-(a+b)/2), (r, a, b)).simplify()
#N = sp.integrate(σθθ_, (r, a, b)).simplify()
#V = sp.integrate(σrθ_, (r, a, b)).simplify()
#print('M = ', M); print('N = ', N); print('V = ', V)
## executed in 5m 19s, finished 06:15:35 2024-12-06
```

This first integral takes about five minutes on the author's laptop. The reader may uncomment the code above and run it. The results are given as follows:

```
M = P*(-a - b)*cos(θ)/2
N = P*cos(θ)
V = P*sin(θ)
#display(Math(f"\\text{{ Moment $M$}}=\\small {latex(M)}"))
#display(Math(f"\\text{{ Axial force $N$}}=\\small {latex(N)}"))
#display(Math(f"\\text{{ Shear force $V$}}=\\small {latex(V)}"))
```

We rewrite the formulas for these internal forces as follows. These are functions of only θ because r has been integrated over:

$$\begin{aligned} M &= -\frac{P}{2}(a+b)\cos\theta \\ N &= P\cos\theta \\ V &= P\sin\theta \end{aligned} \tag{6.21}$$

It is clear that a positive shear force P results in a negative internal moment M and a positive normal force N along the semicircular beam axis. Both are symmetric. The internal shear force V is positive in $0 \leq \theta \leq \pi/2$ and negative in $-\pi/2 \leq \theta \leq 0$, and it is antisymmetric.

6.4.10 *Final formulas for strains*

After obtaining the stresses, we compute the strains using the constitutive equation, assuming the beam is very thin (in the z-direction that perpendicular to the paper) and the problem is under plane stress conditions:

```
E, ν = symbols('E, ν', positive=True)          # given material constants

εrr  = ((σrr_  - ν*σθθ_)/E).simplify()
εθθ  = ((σθθ_  - ν*σrr_)/E).simplify()
γrθ  = σrθ_/E*(2*(1+ν))
display(Math(f" \\text{{ Final solution for strains}}:"))
display(Math(f" \\text{{$ε_{{rr}}$=}}\\small {latex(εrr)}"))
display(Math(f" \\text{{$ε_{{θθ}}$=}}\\small {latex(εθθ)}"))
display(Math(f" \\text{{$γ_{{rθ}}$=}}\\small {latex(γrθ)}"))
```

Final solution for strains:

$$\varepsilon_{rr} = \frac{P\left(a^2b^2 - a^2r^2 - b^2r^2 + r^4 + \nu\left(a^2b^2 + a^2r^2 + b^2r^2 - 3r^4\right)\right)\cos(\theta)}{Er^3\left(-a^2 + b^2 + \log\left(\left(\frac{a}{b}\right)^{a^2+b^2}\right)\right)}$$

$$\varepsilon_{\theta\theta} = \frac{P\left(-a^2b^2 - a^2r^2 - b^2r^2 + 3r^4 - \nu\left(a^2b^2 - a^2r^2 - b^2r^2 + r^4\right)\right)\cos(\theta)}{Er^3\left(-a^2 + b^2 + \log\left(\left(\frac{a}{b}\right)^{a^2+b^2}\right)\right)}$$

$$\gamma_{r\theta} = \frac{P\left(2\nu+2\right)\left(a^2b^2 - a^2r^2 - b^2r^2 + r^4\right)\sin\left(\theta\right)}{Ea^2r^3\left(\left(1+\frac{b^2}{a^2}\right)\left(\log\left(a\right)-\log\left(b\right)\right)-1+\frac{b^2}{a^2}\right)}$$

The following snippet confirms that all stresses and strains satisfy the necessary conditions and are therefore correct:

```
# Verfication of equilibrium constants (zero body force):
br = 0; bθ = 0
equi1_=(σrr_.diff(r)+σrθ_.diff(θ)/r+(σrr_-σθθ_)/r+br).simplify()
equi2_=(σrθ_.diff(r)+σθθ_.diff(θ)/r+2*σrθ_/r+bθ).simplify()
equis= [equi1_, equi2_]
display(Math(f" \\text{{Are the equilibrium equations satisfied? }}\
                {latex(abs(equis[0]+abs(equis[0])).simplify()==0)}"))

# The Laplacian of the stress trace:
σ_sum = (σrr_+σθθ_)
cmptbl = σ_sum.diff(r,2)+σ_sum.diff(r)/r + σ_sum.diff(θ,2)/r/r
display(Math(f" \\text{{Is the compatibility equation satisfied? }}\
                                     {latex(cmptbl.simplify()==0)}"))
```

Are the equilibrium equations satisfied? True
Is the compatibility equation satisfied? True
It is seen that all necessary conditions for the stresses are satisfied.

6.4.11 *Determining displacements using DBCs*

These displacements contain only one unknown constant, b_1, which needs to be determined using the displacement boundary conditions (DBCs) for a given problem. For our problem, assume the DBCs are, at point A ($r = a$, $\theta = 0$, see Fig. 3.5),

$$u_r = 0; \quad u_\theta = 0; \quad \frac{\partial u_r}{\partial \theta} = 0 \tag{6.22}$$

These give us three conditions. Note that, due to symmetry, the second and third conditions are already satisfied. Hence, only the first condition is effective for solving b_1. Nevertheless, there is no harm in including all three conditions in the solution process. The snippet is as follows. We first update the displacements, rewrite them with the constant b_1 explicitly, and confirm that the rewriting is correct:

```
# Update the displacements:
sln_cs = {**sln_SBC, **sln_equI}
urg = urg_.subs(sln_cs).simplify().collect(cos(θ))
uθg = uθg_.subs(sln_cs).simplify().collect(sin(θ))

aa = a**2; bb = b**2; rr = r**2;
abb= (aa*ν-aa+bb*ν-bb)
Cab_ = (b**2/a**2-1) + (1+b**2/a**2)*(log(a)-log(b))

# Rewrite the displacements with constant b1 explicitly:
urg_b1=(P*rr*(ν-3)*(-4*θ*(aa+bb)*sin(θ)+(-rr*(3*ν-1)+2*abb*log(r))\
    *cos(θ))+(-P*aa*bb*(ν-3)*(ν+1)-2*rr*(Cab_*E*aa*b1*(ν-3)+2*P*(aa+bb)\
    *(ν+1)-P*(ν+1)*abb))*cos(θ))/(2*Cab_*E*aa*rr*(ν-3))

uθg_b1=(P*rr*(-4*θ*(aa+bb)*cos(θ)+(rr*(ν+5)-2*abb*log(r))*sin(θ))\
      +(2*Cab_*E*aa*b1*rr-P*aa*bb*(ν+1))*sin(θ))/(2*Cab_*E*aa*rr)

display(((urg_b1-urg.subs(Cab,Cab_)).simplify()\
        ,(uθg_b1-uθg.subs(Cab,Cab_)).simplify()))  # check the rewriting

#display(urg_b1); display(uθg_b1)                               # Lengthy
```

(0, 0)

```
# Using DBCs to determine constants b1:

expr1 = uθg.subs({r:a, θ:0})
expr2 = urg.subs({r:a, θ:0})
expr3 = urg.diff(θ).subs({r:a, θ:0})

#display(expr1, expr2, expr3)
sln_DBCs = sp.solve([expr1, expr2, expr3], cs)
display(Math(f" \\text{{Constant found: }} "))
display(Math(f" \\text{{ }}\\scriptsize {latex(sln_DBCs)}"))

sln_cs = {**sln_SBC, **sln_equI, **sln_DBCs}
sln_cs = {key: value.subs(sln_DBCs).simplify() for key, \
                                              value in sln_cs.items()}
```

Constant found:

$$\left\{b_1 : \frac{P\nu\log(a)}{C_{ab}E} - \frac{P\nu}{2C_{ab}E} - \frac{P\log(a)}{C_{ab}E} + \frac{3P}{2C_{ab}E} + \frac{Pb^2\nu\log(a)}{C_{ab}Ea^2} + \frac{Pb^2\nu}{2C_{ab}Ea^2} - \frac{Pb^2\log(a)}{C_{ab}Ea^2} + \frac{Pb^2}{2C_{ab}Ea^2}\right\}$$

We have successfully found the last constant. Let's now update the displacements and strains and check the strain–displacement relations.

```
# Check strain–displacement relations:
εu_expr1 = (εrr - urg_.diff(r)).subs(sln_cs).subs(Cab,Cab_).simplify()
εu_expr2 = (r*εθθ - urg_- uθg.diff(θ)).subs(sln_cs).subs(Cab,Cab_)\
                                                         .simplify()
εu_expr3 = (r*γrθ - urg_.diff(θ) - r*uθg_.diff(r) + uθg_).subs(sln_cs)\
                                          .subs(Cab,Cab_).simplify()
(εu_expr1, εu_expr2, εu_expr3)                  # should be (0, 0, 0)
```

(0, 0, 0)

6.4.12 *Final formulas for displacements*

We can obtain the final solution for these displacements, which also satisfy the strain–displacement relations.

Let us compare these with those obtained in Chapter 3. The snippet is as follows:

```
ur = sp.expand_log(urg_.subs(sln_cs).simplify()).simplify()\
                              .collect(sin(θ)).collect(cos(θ)).collect(r)
uθ = sp.expand_log(uθg_.subs(sln_cs).simplify()).simplify()\
                              .collect(sin(θ)).collect(cos(θ)).collect(r)
#display(Math(f" \\text{{ Final $u_θ$}}=\\small {latex(uθ)}")) #Lengthy

# Displacements obtained in Chapter 3:
a2 = a**2; b2 = b**2; r2 = r**2
c0 =-P*b**2/(2*Cab); c2 = P/(2*Cab*a**2); c3 =-P*(a**2+b**2)/(Cab*a**2)
C2 = (a2*(3*a2*c2*ν - a2*c2 + c3*ν*log(a) - c3*log(a))-c0*(ν+1))/(E*a2)

ur_ = C2*cos(θ) + (-2*P*θ*(a2 + b2)*sin(θ))/(Cab*E*a2) -P*(a2*b2*(ν+1)
    + r2*(r2*(3*ν-1) - 2*(a2+b2)*(ν-1)*log(r)))*cos(θ)/(2*Cab*E*a2*r2)
uθ_ =-C2*sin(θ) + P*(-4*(a2+b2)*θ*cos(θ) + (4*a2+4*b2 + r2*(3*ν-1) \
    - 2*(a2+b2)*(ν-1)*log(r))*sin(θ))/(2*Cab*E*a2) + (P*a2*b2*(ν+1)/2 \
    - P*(a2*b2 - 3*r**4 + r2*(a2+b2) + ν*(a2*b2 + r**4 \
    - r2*(a2+b2))))*sin(θ)/(Cab*E*a2*r2)

# Compute the differences for checking:
(ur_-ur).simplify(), (uθ_-uθ).simplify()            # should be (0, 0)
```

(0, 0)

We obtained exactly the same solution as in Chapter 3.

6.4.13 *Plot of displacement distribution*

We can now plot the distribution of the displacements:

```
# Convert SymPy expressions to numpy functions:
dic_data = {E:1, ν:0.3, a:1, b:2, P:1}
Cab_val = ((b**2/a**2-1)+(1+b**2/a**2)*(log(a)-log(b))).subs(dic_data)
ur_sp = ur.subs(Cab, Cab_val).subs(dic_data)
uθ_sp = uθ.subs(Cab, Cab_val).subs(dic_data)

func1 = sp.lambdify(r, ur_sp.subs({θ:   0.}),'numpy')
func2 = sp.lambdify(r, ur_sp.subs({θ:np.pi/4.}),'numpy')
func3 = sp.lambdify(r, ur_sp.subs({θ:np.pi/2.}),'numpy')
func4 = sp.lambdify(r, uθ_sp.subs({θ:np.pi/4.}),'numpy')
func5 = sp.lambdify(r, uθ_sp.subs({θ:np.pi/2.}),'numpy')
labels=[r'$u_{r}, θ=0$',r'$u_{r}, θ=\pi/4$', r'$u_{r}, θ=\pi/2$',
        r'$u_{θ}, θ=\pi/4$', r'$u_{θ}, θ=\pi/2$' ]

r_data  = np.linspace(1, 2, 400)                    # r values for the plot
x_data = [r_data,  r_data, r_data,  r_data, r_data  ]
y_data = [func1(x_data[0]), func2(x_data[1]), func3(x_data[0]),
          func4(x_data[1]), func5(x_data[2]) ]
# plt.ion()
# gr.plot_fs(x_data, y_data, labels, '$r$', 'displacements', \
#                                   'imagesSMp/CurveBeamU',ioff=False)
#                                 # Use ioff=False to display the plot
# plt.show()
```

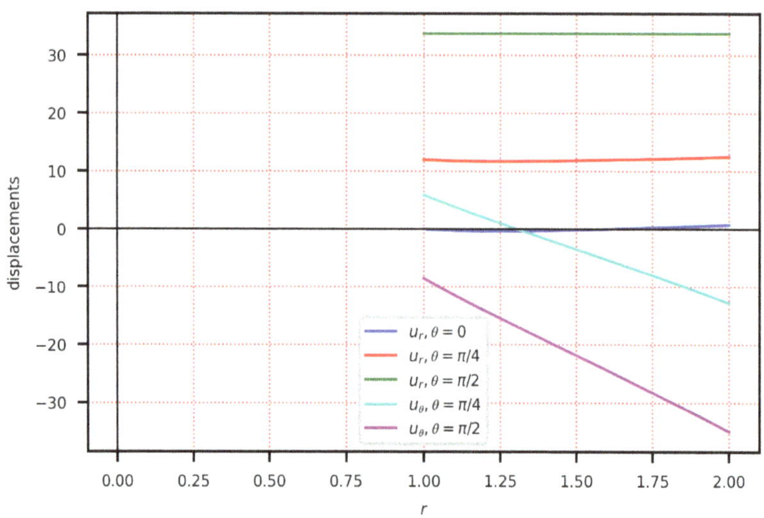

Figure 6.6. Distribution of the two displacement components u_r and u_θ along the vertical, horizontal, and 45° radial directions.

The two displacement components, u_r and u_θ, are plotted in Fig. 6.6. It is found that the component u_r does not change much with r, but varies significantly with θ. The largest value occurs at the beam ends, $\theta = \pm\pi/2$. The component u_θ varies with both r and θ. It changes with r almost linearly, which is quite similar to the Euler–Bernoulli beam theory [2]. The maximum of u_θ is also occurring at the beam ends, $\theta = \pm\pi/2$, on the upper surface.

6.4.14 *Plot of stress distribution*

We can now plot the distribution of the stresses:

```
# Convert SymPy expressions to Numpy functions:
dic_data = {E:1, ν:0.3, a:1, b:2, P:1}
Cab_val = ((b**2/a**2-1)+(1+b**2/a**2)*(log(a)-log(b))).subs(dic_data)
σrr_sp = σrr.subs(Cab, Cab_val).subs(dic_data)
σθθ_sp = σθθ.subs(Cab, Cab_val).subs(dic_data)
σrθ_sp = σrθ.subs(Cab, Cab_val).subs(dic_data)

func1 = sp.lambdify(r, σrr_sp.subs({θ:   0.}),'numpy')
func2 = sp.lambdify(r, σrr_sp.subs({θ: np.pi/4.}),'numpy')
func3 = sp.lambdify(r, σθθ_sp.subs({θ:   0.}),'numpy')
func4 = sp.lambdify(r, σθθ_sp.subs({θ: np.pi/4.}),'numpy')
func5 = sp.lambdify(r, σrθ_sp.subs({θ: np.pi/2.}),'numpy')
func6 = sp.lambdify(r, σrθ_sp.subs({θ: np.pi/4.}),'numpy')
labels=[r'$σ_{rr}, θ=0$',r'$σ_{rr}, θ=\pi/4$', r'$σ_{θθ}, θ=0$',
        r'$σ_{θθ}, θ=\pi/4$',r'$σ_{rθ}, θ=\pi/2$',r'$σ_{rθ}, θ=\pi/4$']

r_data  = np.linspace(1, 2, 400)                # r values for the plot
x_data = [r_data,  r_data, r_data,  r_data, r_data, r_data]
y_data = [func1(x_data[0])*2, func2(x_data[1])*2, func3(x_data[2]),
          func4(x_data[3]), func5(x_data[4]), func6(x_data[5]) ]
# plt.ion()
# gr.plot_fs(x_data, y_data, labels, '$r$', 'stresses', \
#                                  'imagesSMp/CurveBeamS',ioff=True)
#                            # Use ioff=False to display the plot
# plt.show()
```

6.4.15 *Stress distribution plot*

The three stress components, σ_{rr}, $\sigma_{\theta\theta}$, and $\sigma_{r\theta}$, are plotted in Fig. 6.7 for various θ values. The dominant stress component is the normal stress $\sigma_{\theta\theta}$, which reaches its maximum and minimum values on the inner and outer surfaces of the beam at the $\theta = 0$ cross-section, respectively. Unlike in a straight beam, the distribution of normal stress along a cross-section is **nonlinear**. Notably, the value of $\sigma_{\theta\theta}$ on the inner surface of the semicircular beam is particularly large, as shown in Fig. 6.7, and this location is the most likely initiation point for yielding. These results again highlight the stress magnification effects caused by a concave boundary.

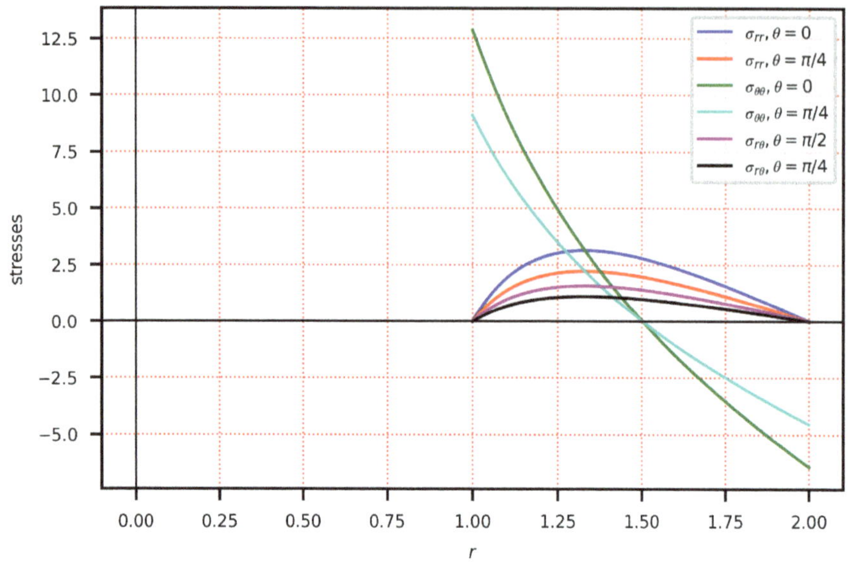

Figure 6.7. Distribution of the three stress components in the vertical, horizontal, and 45° radial directions.

For the selected θ values, there is a special coincidence between σ_{rr} and $\sigma_{r\theta}$:

$$\sigma_{rr}\Big|_{\theta=0} = \sigma_{r\theta}\Big|_{\theta=\pi/2}, \quad \sigma_{rr}\Big|_{\theta=\pi/4} = \sigma_{r\theta}\Big|_{\theta=\pi/4} \tag{6.23}$$

(see the derived stress formulas). To make the differences more visible, the value of σ_{rr} is intentionally magnified by a factor of two in the plots. In any case, the magnitudes of σ_{rr} and $\sigma_{r\theta}$ are much smaller compared to $\sigma_{\theta\theta}$, and hence they are not critical in this problem setting.

6.4.16 *Contour plot of displacements*

To generate contour plots of the displacements, we first express the displacements in Cartesian coordinates using Eq. (6.7). The resulting displacement expressions are then converted into NumPy functions. With these, we can use `gr.contour_plot()` to visualize the displacement fields over the beam domain.

The following code snippet illustrates the procedure:

```
# Contour of displacement component u_r:
x, y = sp.symbols('x, y', real=True)
a_val = 1; b_val = 2
dic_rt = {E:1,ν:0.3,P:1, a:a_val, b:b_val,
          r:sp.sqrt(x**2+y**2), θ:sp.atan(-x/y)}
Cab_val = ((b**2/a**2-1)+(1+b**2/a**2)*(log(a)-log(b))).subs(dic_rt)

ur_xy = ur.subs(Cab, Cab_val).subs(dic_rt)  # Displacements in Cartesian
uθ_xy = uθ.subs(Cab, Cab_val).subs(dic_rt)

σrr_xy = σrr.subs(Cab, Cab_val).subs(dic_rt)     # Stresses in Cartesian
σθθ_xy = σθθ.subs(Cab, Cab_val).subs(dic_rt)
σrθ_xy = σrθ.subs(Cab, Cab_val).subs(dic_rt)

display(Math(f" \\text{{Displacement in $x,y$ (as an example):}} "))
display(Math(f" \\text{{ $u_{{r}}(x, y)$}} =\small {latex(uθ_xy)}"))
```

Displacement in x, y (as an example):

$$u_r(x,y) = -\frac{\left(-\frac{5.2x}{y\sqrt{\frac{x^2}{y^2}+1}}+\left(x^2+y^2\right)\times\left(\frac{x\left(5.3x^2+5.3y^2+7.0\log\left(\sqrt{x^2+y^2}\right)+7.9\right)}{y\sqrt{\frac{x^2}{y^2}+1}}-\frac{20\operatorname{atan}\left(\frac{x}{y}\right)}{\sqrt{\frac{x^2}{y^2}+1}}\right)\right)}{2\left(3-5\log\left(2\right)\right)\left(x^2+y^2\right)}$$

```
# Create a mesh grid for a rectangular domain in x-y plane:
dX = 1.0*b_val;     dY = 1.0*b_val                  # plot 2dX × dY area
x_val = np.linspace(-dX, dX, 800)
y_val = np.linspace(dY*0.001, dY, 400)
X, Y = np.meshgrid(x_val, y_val)

# Define a ring beyond which the plot is masked:
#mask = (X/a_val)**2 + (Y/a_val)**2 <= 1
mask = np.logical_or((X / a_val)**2 + (Y / a_val)**2 <= 1,     # Inside
                     (X / b_val)**2 + (Y / b_val)**2 >= 1)    # Outside

np_ur_xy = lambdify((x,y), ur_xy, 'numpy')            # convert to numpy
Z = np_ur_xy(X, Y)                 # compute the values over the mesh grid
Z[mask] = np.nan  # mask the area beyond the beam, print nothing there

cntur = gr.contour_plot(X,Y, Z, x_lbl='x-axis',y_lbl='y-axis',
                title='Displacement $u_{{r}}$',lvls=50, ioff=True);

plt.savefig('imagesSMp/CurvedBeamCntur_ur.png', dpi=500);# save to file
#plt.show()
```

```
Minimum value in the domain: -0.393128
Maximum value in the domain: 33.7063
```

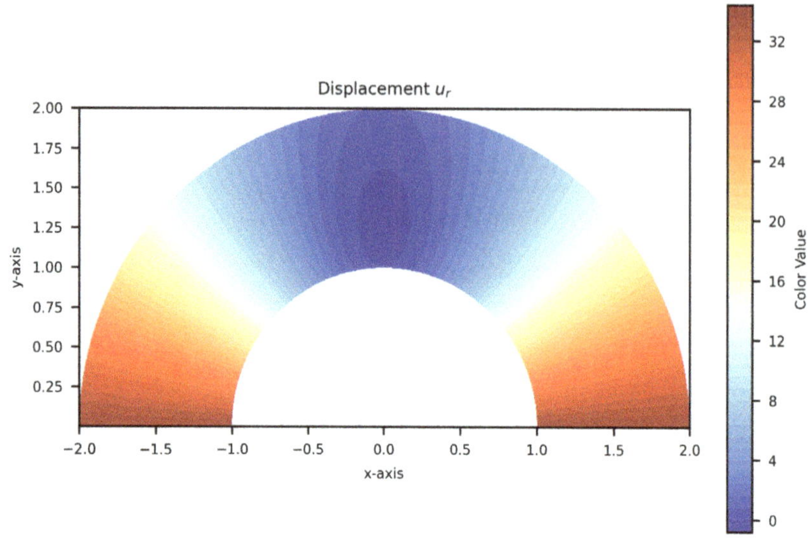

Figure 6.8. Contour plot of the radial displacement u_r in the semicircular beam subjected to a pair of horizontal shear forces at its two ends.

A contour plot of the radial displacement u_r within the semicircular beam subjected to a pair of horizontal shear forces at its two ends is shown in Fig. 6.8. The distribution of u_r exhibits mirror symmetry with respect to the y-axis.

The radial displacement varies little along the r-axis but depends on θ. Its maximum values occur at the two ends.

The following snippet plots the tangential displacement component, u_θ:

```
np_uθ_xy = lambdify((x,y), uθ_xy, 'numpy')            # convert to numpy
Z = np_uθ_xy(X, Y)                       # compute the value t over the mesh grid
Z[mask] = np.nan                                       # mask, print nothing there

cntur = gr.contour_plot(X,Y, Z, x_lbl='x-axis',y_lbl='y-axis',
                 title='Displacement $u_{{θ}}$',lvls=50, ioff=True);

plt.savefig('imagesSMp/CurvedBeamCntur_ut.png', dpi=500) # save to file
#plt.show()
```

```
Minimum value in the domain: -34.8808
Maximum value in the domain: 34.8808
```

A contour plot of the tangential displacement u_θ within the semicircular beam subjected to a pair of horizontal shear forces at its two ends is shown in Fig. 6.9. The distribution of u_θ exhibits mirror antisymmetry with respect

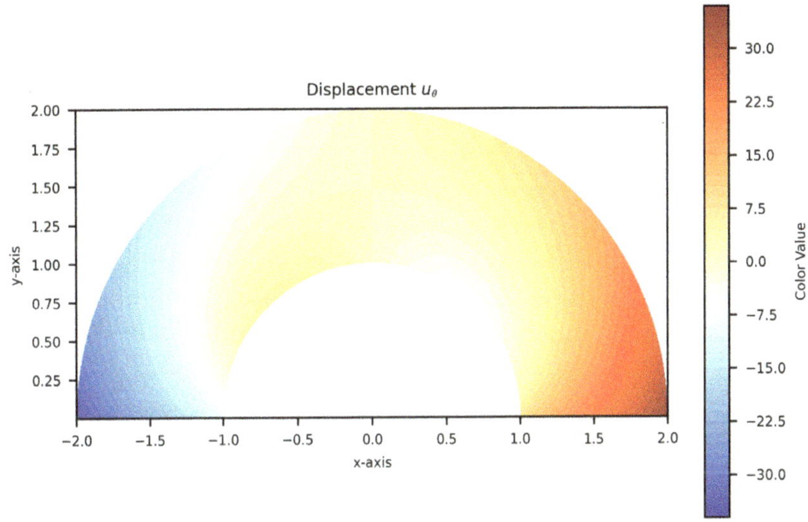

Figure 6.9. Contour plot of the tangential displacement u_θ in the semicircular beam subjected to a pair of horizontal shear forces at its two ends.

to the y-axis. The largest values occur at the two ends of the beam on the upper surface.

6.4.17 *Contour plot of stresses*

The following snippet plots the normal stress, σ_{rr}:

```
np_σrr_xy = lambdify((x,y), σrr_xy, 'numpy')          # convert to numpy
Z = np_σrr_xy(X, Y)                       # compute the values over the mesh grid
Z[mask] = np.nan                                # mask, print nothing there
cntur = gr.contour_plot(X,Y, Z, x_lbl='x-axis',y_lbl='y-axis',
                 title='Displacement $σ_{{rr}}$',lvls=50, ioff=True);

plt.savefig('imagesSMp/CurvedBeamCntSrr.png', dpi=500)    # save to file
#plt.show()
```

```
Minimum value in the domain: 1.3871e-05
Maximum value in the domain: 1.56562
```

A contour plot of the normal stress σ_{rr} within the semicircular beam subjected to a pair of horizontal shear forces at its two ends is shown in Fig. 6.10. The distribution exhibits mirror symmetry with respect to the y-axis. It is zero along all the edges of the beam, as required by the SBCs of this problem. The maximum normal stress occurs in the vertical cross-section along the y-axis near the inner circle.

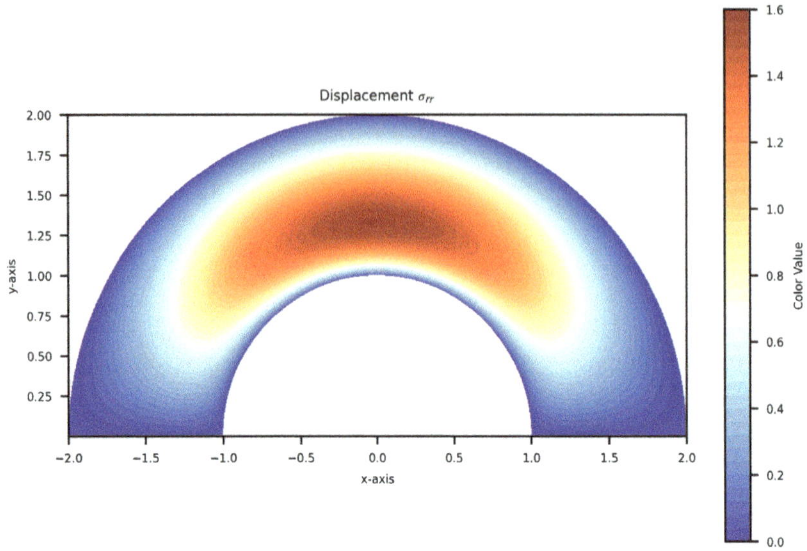

Figure 6.10. Contour plot of the normal stress σ_{rr} in the semicircular beam subjected to a pair of horizontal shear forces at its two ends.

The following snippet plots the same for the normal stress, $\sigma_{\theta\theta}$:

```
np_σθθ_xy = lambdify((x,y), σθθ_xy, 'numpy')          # convert to numpy
Z = np_σθθ_xy(X, Y)
Z[mask] = np.nan
cntur = gr.contour_plot(X,Y, Z, x_lbl='x-axis',y_lbl='y-axis',
                  title='Displacement $σ_{{θθ}}$',lvls=50, ioff=True);

plt.savefig('imagesSMp/CurvedBeamCnturStt.png', dpi=500) # save to file
#plt.show()
```

```
Minimum value in the domain: -6.42332
Maximum value in the domain: 12.8548
```

A contour plot of the normal stress $\sigma_{\theta\theta}$ within the semicircular beam subjected to a pair of horizontal shear forces at its two ends is shown in Fig. 6.11. The distribution of $\sigma_{\theta\theta}$ exhibits mirror symmetry with respect to the y-axis. The maximum and minimum normal stresses in the tangential direction occur in the vertical cross-section along the y-axis at the surfaces of the inner and outer circles, respectively. The inner circle, in particular, experiences high tension because (1) the boundary is concave and (2) the shear force load is directed outward.

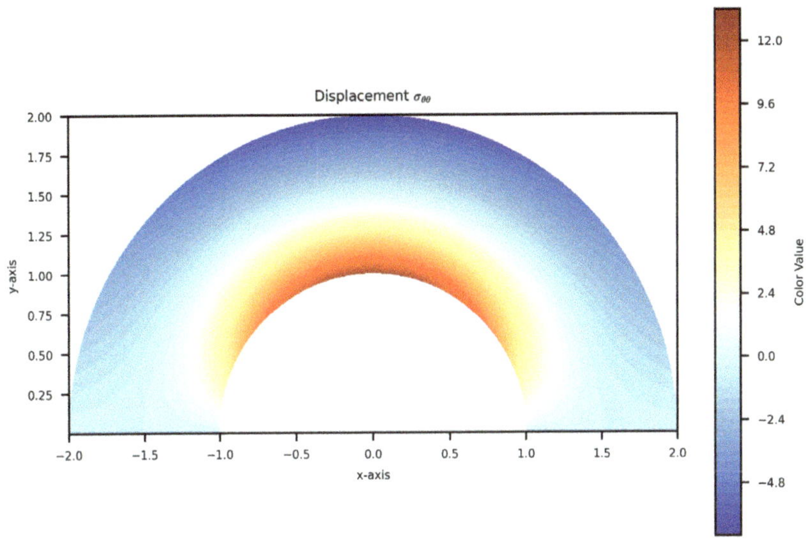

Figure 6.11. Contour plot of the normal stress $\sigma_{\theta\theta}$ in the semicircular beam subjected to a pair of horizontal shear forces at its two ends.

The following snippet plots the same for the shear stress, $\sigma_{r\theta}$:

```
np_σrθ_xy = lambdify((x,y), σrθ_xy, 'numpy')             # convert to numpy
Z = np_σrθ_xy(X, Y)
Z[mask] = np.nan
cntur = gr.contour_plot(X,Y, Z, x_lbl='x-axis',y_lbl='y-axis',
                  title='Displacement $σ_{{rθ}}$',lvls=50, ioff=True);

plt.savefig('imagesSMp/CurvedBeamCnturSrt.png', dpi=500) # save to file
#plt.show()
```

```
Minimum value in the domain: -1.56563
Maximum value in the domain: 1.56563
```

A contour plot of the shear stress $\sigma_{r\theta}$ within the semicircular beam subjected to a pair of horizontal shear forces at its two ends is shown in Fig. 6.12. The distribution of $\sigma_{r\theta}$ exhibits mirror antisymmetry with respect to the y-axis. The maximum and minimum shear stresses in the radial direction occur at the two ends near the inner circle. On the vertical cross-section, the shear stress is zero, as expected, since the shear forces are applied horizontally. It is also zero on both the inner and outer circular edges, as required by the boundary conditions of this problem.

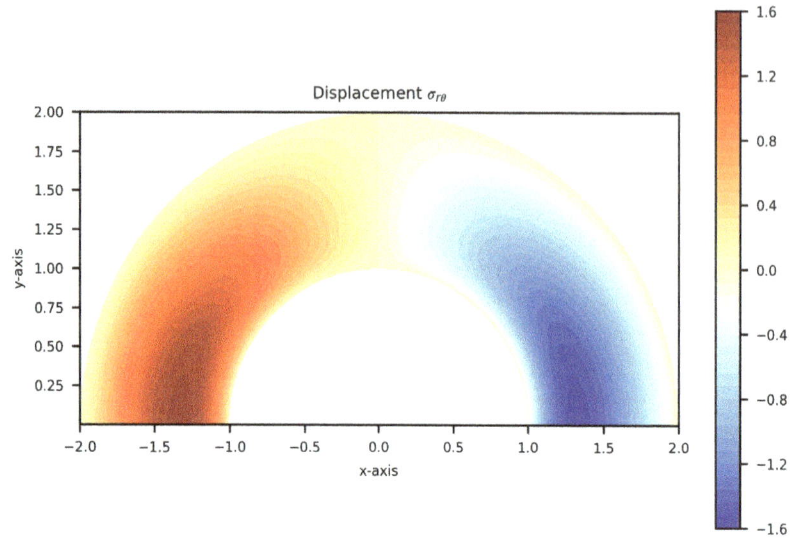

Figure 6.12. Contour plot of the shear stress $\sigma_{r\theta}$ in the semicircular beam subjected to a pair of horizontal shear forces at its two ends.

6.5 Review questions

1. Assume the displacements take the following form for the curved beam, and compute the general stresses by enforcing the differential equilibrium equations. Compare the resulting general expressions for the stresses with those given in Section 6.4.3:

$$\begin{aligned} u_r &= \frac{a_0 \cos\theta}{r^3} + \frac{a_1 \cos\theta}{r^2} + a_{10} r^2 \cos\theta + a_{11} r^3 \cos\theta \\ &\quad + a_3 \cos\theta + \frac{a_2 \cos\theta}{r} + a_4 \theta \sin\theta + a_5 r\theta \sin\theta + a_6 \\ &\quad + a_7 \log r \cos\theta + a_8 r \cos\theta + a_9 r \log r \cos\theta \\ u_\theta &= \frac{b_0 \sin\theta}{r^3} + \frac{b_1 \sin\theta}{r^2} + b_{10} r^2 \sin\theta + b_{11} r^3 \sin\theta \\ &\quad + b_3 \sin\theta + \frac{b_2 \sin\theta}{r} + b_4 \theta \cos\theta + b_5 r\theta \cos\theta \\ &\quad + b_6 r\theta + b_7 \log r \sin\theta + b_8 r \sin\theta + b_9 r \log r \sin\theta \end{aligned} \tag{6.24}$$

2. Consider a plate with a small hole subjected to a biaxial stress load at the far field, as shown in Fig. 5.8 and described in Section 5.2. Using a solution procedure similar to that presented in this chapter for the

uniaxial loading case, determine the complete set of solutions for the displacements, stresses, and strains.
3. Consider a plate with a small hole subjected to pure shear at the far field, as shown in Fig. 5.9 and described in Section 5.3. Using a solution procedure similar to that presented in this chapter for the uniaxial loading case, determine the complete set of solutions for the displacements, stresses, and strains.
4. Apply the displacement method to solve Review Problem 5 in Chapter 3: a semicircular beam subjected to a pair of vertical forces applied at its ends.

6.6 Remarks

This chapter presents displacement methods and applies them to analyze non-axisymmetric problems. The following conclusions can be drawn:

1. Two important cases were examined: stress concentration around a hole in a plate subjected to various types of loading, and a semicircular beam subjected to a pair of shear forces at its two ends.
2. Novel displacement methods were introduced in this chapter, including detailed techniques for properly assuming displacement fields.
3. Procedures and code were provided to obtain the complete set of solutions, including displacements, stresses, and strains.
4. The neutral axis, where the normal stress is zero, was identified for the semicircular beam. The neutral axis forms a circle, and a formula for computing its radius was derived.
5. The displacement method proposed is a powerful tool for solving such problems. However, the main challenge lies in selecting an appropriate set of basis functions that best suit the specific problem at hand.

The following chapter discusses stress methods for loaded wedges.

References

[1] G. R. Liu, *Solid Mechanics with Python: Cartesian Coordinates*, World Scientific, New Jersey, 2025.
[2] G. R. Liu, *Mechanics of Materials: Formulations and Solutions with Python*, World Scientific, New Jersey, 2024.
[3] G. R. Liu, *Numbers and Functions: Theory, Formulation, and Python Codes*, World Scientific, New Jersey, 2024.

Chapter 7

Wedges Subjected to Forces

Let us import the necessary external modules or dependencies for later use in this chapter:

```
# Place cursor in this cell, and press Ctrl+Enter to import dependencies.
import sys                              # for accessing the computer system
sys.path.append('../grbin/')              # Add in the path to your system

from commonImports import *            # Import dependencies from '../grbin/'
import grcodes as gr                        # Import the module of the author
importlib.reload(gr)                   # When grcodes is modified, reload it

from continuum_mechanics import vector
from continuum_mechanics.solids import sym_grad, strain_stress
init_printing(use_unicode=True)             # For latex-like quality printing
np.set_printoptions(precision=4,suppress=True)     # Digits in print-outs
```

Using the polar coordinate system, we solved axisymmetric problems in Chapter 3 using stress methods and various types of axisymmetric problems in Chapter 4 using displacement methods. Plates with holes were analyzed in Chapter 5 using stress methods, where stress concentration factors were determined. Novel displacement methods were developed in Chapter 6 for problems involving plates with holes and challenging curved beam configurations.

This chapter presents stress methods for solving another class of mechanics problems: **2D wedges** subjected to various types of loads, including point forces, point moments, and distributed pressures. All problems are assumed to be free of body forces and are treated under **plane stress** conditions. For plane strain problems, a simple conversion can be applied, as discussed in Ref. [2].

7.1 A point force on a wedge

Consider a two-dimensional (2D) wedge with an angle of $\alpha \in [0, \pi]$, as shown in Fig. 7.1. A concentrated force P is applied at its tip at an angle of $\beta \in [0, 2\pi]$. The problem is treated under plane stress conditions: the wedge extends infinitely in the x–y plane and has a unit thickness in the z-direction. The force P acts at the tip $(0, 0)$ in the x–y plane, viewed along the z-axis, and is uniformly distributed over the unit thickness. Hence, the dimension of P is **force per unit length**.

With this setup, our goal is to determine the **stress** and **displacement** distributions in the half-space as functions of x and y, or equivalently, in polar coordinates (r, θ).

We aim to derive the stress distribution within the wedge.

7.1.1 *Stress function construction*

For this type of problem, it is more convenient to work in polar coordinates (r, θ), which are defined in Fig. 7.1 relative to the Cartesian coordinates

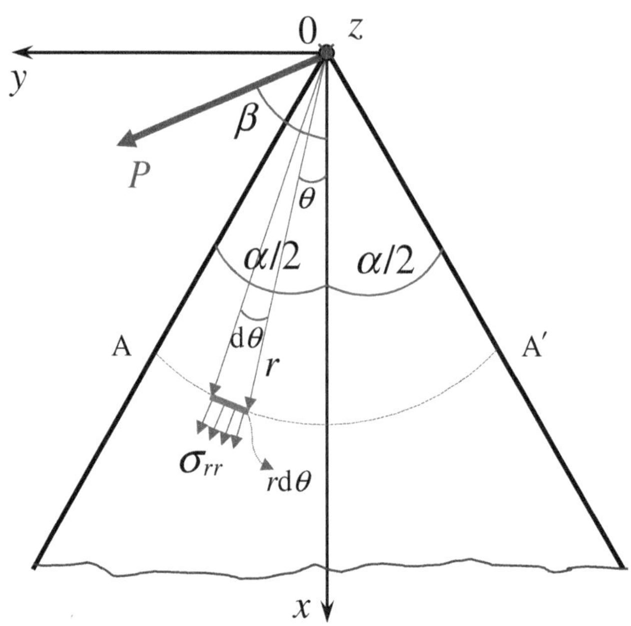

Figure 7.1. Schematic of a 2D wedge with an angle α, subjected to a vertical force P applied at an angle β.

(x, y). The transformation between these two coordinate systems is detailed in Section 4.1.

As we have done previously, when using the stress method, the first step is to identify the **stress function** $\phi(r, \theta)$, which is a 2D scalar function. An effective approach to constructing such a function is to express it as the product of two separate functions:

$$\phi(r, \theta) = g(r)f(\theta), \tag{7.1}$$

where $g(r)$ depends solely on r and $f(\theta)$ depends solely on θ. This approach is known as the **method of separation of variables**. For many problems, this separation can be achieved through a combination of dimensional reasoning and physical intuition.

For the present wedge problem, a simple **dimensional analysis** can be used to determine the form of $g(r)$. The dimensions of all parameters and coordinate variables involved are as follows:

- Applied force P: force per unit length
- Angles α and β: dimensionless
- Polar angle θ: dimensionless
- Polar radius r: length

The resulting stress must have dimensions consistent with P/r. Therefore, the stress function $\phi(r, \theta)$ must be constructed such that its second derivatives yield terms proportional to P/r.

Using Eq. (2.59) (neglecting body forces), the stress components can be expressed in terms of the second derivatives of ϕ with respect to r:

$$\begin{aligned}
\sigma_{rr} &= \frac{1}{r}\frac{\partial \phi}{\partial r} + \frac{1}{r^2}\frac{\partial^2 \phi}{\partial \theta^2} \\
\sigma_{\theta\theta} &= \frac{\partial^2 \phi}{\partial r^2} \\
\sigma_{r\theta} &= \frac{1}{r^2}\frac{\partial \phi}{\partial \theta} - \frac{1}{r}\frac{\partial^2 \phi}{\partial \theta \partial r} = -\frac{\partial}{\partial r}\left(\frac{1}{r}\frac{\partial \phi}{\partial \theta}\right)
\end{aligned} \tag{7.2}$$

Thus, ϕ should be two orders higher than the stresses in terms of r, and the stresses have a dimension of $\left[\frac{\phi}{r^2}\right]$. Therefore, it is reasonable to assume that ϕ takes the form

$$\phi = rf(\theta), \tag{7.3}$$

where $f(\theta)$ is an arbitrary function of θ. From this analysis, we find that $g(r) = r$. Consequently, determining ϕ reduces to solving for $f(\theta)$, which simplifies the problem to a one-dimensional (1D) form.

Since the body force in the wedge is zero, the stress function must satisfy $\nabla^4\phi = 0$. This equation can be solved for ϕ using the following SymPy code snippet:

```
# Solve ∇^4ϕ=0 for general solution:
r = sp.symbols('r', nonnegative=True)            # define symbolic variables
θ, α, β, P = sp.symbols('θ, α, β, P')             # involved in this problem

ϕ = sp.Function('ϕ')(r, θ)                      # define the stress function
f = sp.Function('f')(θ)                        # θ part of the stress function
ϕ = r*f                                          # form the stress function

lap = diff(ϕ,(r,2))  + (1/r)*diff(ϕ,r)  + (1/r**2)*diff(ϕ,(θ,2))
bih = diff(lap,(r,2))+ (1/r)*diff(lap,r) + (1/r**2)*diff(lap,(θ,2))

bih.simplify()                                      # bi harmonic operator
```

$$\frac{f(\theta) + 2\frac{d^2}{d\theta^2}f(\theta) + \frac{d^4}{d\theta^4}f(\theta)}{r^3}$$

For the foregoing expression to be zero, the numerator must be zero. This implies that $f(\theta)$ must satisfy the following differential equation (DE):

$$f(\theta) + 2\frac{d^2 f(\theta)}{d\theta^2} + \frac{d^4 f(\theta)}{d\theta^4} = 0 \tag{7.4}$$

This is a fourth-order, one-dimensional differential equation for $f(\theta)$. Its general solution can be found using the following:

```
# Define the 1D DE:
diff_eq = sp.Eq(f + 2*diff(f,(θ,2)) + diff(f,(θ,4)), 0)

g_sln = sp.dsolve(diff_eq, f).simplify()          # solve the DE for f(θ)

display(Math(f" \\text{{General solution for}} {latex(g_sln)}"))
display(Math(f" \\text{{General solution for $ϕ(r, θ)$}}=\
                                          {latex(r*(g_sln.args[1]))}"))
#g_sln.args # locate f(θ) in the solution
```

General solution for $f(\theta) = (C_1 + C_2\theta)\sin(\theta) + (C_3 + C_4\theta)\cos(\theta)$

General solution for $\phi(r,\theta) = r\left((C_1 + C_2\theta)\sin(\theta) + (C_3 + C_4\theta)\cos(\theta)\right)$

Since the terms $C_1 r\sin\theta + C_3 r\cos\theta = C_1 y + C_3 x$ form a linear function of x and y, they do not contribute to the stresses. This is because the stresses are derived from second-order partial derivatives of ϕ, as discussed in Chapter 2. This can also be confirmed using the following snippet:

```
# Contributions from terms with C1, C3 to the stresses:
C1, C3 = symbols('C1, C3')                                  # constants
ϕ13 =  C1*r*sp.sin(θ) + C3*r*sp.cos(θ)                 # the linear term
σ_rr = (diff(ϕ13,(θ,2))/r**2+diff(ϕ13,r)/r).expand()
display(Math(f" \\text{{Contribution to $σ_{{rr}}$}}={latex(σ_rr)}"))

σ_θθ = diff(ϕ13,(r,2)).expand()
display(Math(f" \\text{{Contribution to $σ_{{θθ}}$}}={latex(σ_θθ)}"))

σ_rθ = (-diff(diff(ϕ13,θ)/r,r)).expand()
display(Math(f" \\text{{Contribution to $σ_{{rθ}}$}}={latex(σ_rθ)}"))
```

Contribution to $\sigma_{rr} = 0$

Contribution to $\sigma_{\theta\theta} = 0$

Contribution to $\sigma_{r\theta} = 0$

Thus, the stress function simplifies to the following form:

$$\phi(r,\theta) = r\theta(c_1 \cos\theta + c_2 \sin\theta) = r(c_1\theta\cos\theta + c_2\theta\sin\theta) \tag{7.5}$$

Note that this stress function comprises two parts: a symmetric part and an antisymmetric part with respect to θ. If the problem setting is symmetric (for example, when $\beta = 0$ in Fig. 7.1), only the first term is required. If the problem is antisymmetric (for example, when $\beta = 90°$), only the second term is used. For all other cases, both terms must be included:

```
# Rewrite the simplified stress function for wedges under P in SymPy:
c0, c1 = symbols('c0, c1 ')                          # integration constants
ϕ = r*θ*(c0*sp.cos(θ) + c1*sp.sin(θ))
ϕ
```

$r\theta(c_0 \cos(\theta) + c_1 \sin(\theta))$

7.1.2 *General solution for stresses*

It is easy to confirm that the assumed stress function satisfies the compatibility condition: vanishing biharmonic. We can now use the stress function to compute the stresses, following Eq. (7.2):

```
σ_rrg = (diff(ϕ,(θ,2))/r**2+diff(ϕ,r)/r).expand()
display(Math(f" \\text{{General solution of $σ_{{rr}}$}}=\
                                        {latex(σ_rrg)}"))
σ_θθg = diff(ϕ,(r,2)).expand()
display(Math(f" \\text{{General solution of $σ_{{θθ}}$}}=\
                                        {latex(σ_θθg)}"))
σ_rθg = (-diff(diff(ϕ,θ)/r,r)).expand()
display(Math(f" \\text{{General solution of $σ_{{rθ}}$}}=\
                                        {latex(σ_rθg)}"))
```

General solution of $\sigma_{rr} = -\dfrac{2c_0 \sin(\theta)}{r} + \dfrac{2c_1 \cos(\theta)}{r}$

General solution of $\sigma_{\theta\theta} = 0$

General solution of $\sigma_{r\theta} = 0$

It is clear that for this problem, the only nonzero stress component in the entire domain is σ_{rr}. We write the general solution for these stresses in the following equation for future reference:

$$\begin{aligned} \sigma_{rr} &= \frac{2}{r}(-c_0 \sin\theta + c_1 \cos\theta) \\ \sigma_{\theta\theta} &= 0; \quad \sigma_{r\theta} = 0 \end{aligned} \tag{7.6}$$

The general solution for the stress σ_{rr} contains two undetermined constants. This solution applies to all wedges subjected to a point force P. To fully determine these constants, we need to establish two boundary conditions based on the specific problem setup.

7.1.3 *Stress boundary conditions*

For the problem defined in Fig. 7.1, the two surfaces of the wedge are stress-free, which gives the following conditions:

$$\sigma_{\theta\theta} = 0 \quad \text{and} \quad \sigma_{r\theta} = 0 \quad \text{at } \theta = \pm\alpha/2 \tag{7.7}$$

Clearly, these stress boundary conditions (SBCs) are already satisfied by Eq. (7.6), regardless of the undetermined constants. Therefore, we must use either the integral equilibrium condition or the force boundary conditions (FBCs) to determine them.

It is important to note that the stress given in Eq. (7.6) exhibits an r^{-1} singularity: σ_{rr} approaches infinity as $r \to 0$ at a rate of r^{-1}. Consequently, when applying the integral equilibrium condition, we must evaluate it at a distance away from $r = 0$. This is a common approach in problems involving point forces.

7.1.4 *Integral equilibrium condition*

The externally applied force P must be balanced by the integrated internal stresses. This means that the sum of all forces in both the x- and y-directions must vanish along an arc A–A′ at an arbitrary nonzero r, as shown in Fig. 7.1. The equilibrium condition along the arc A–A′ is expressed as

$$\begin{aligned}
\sum F_x = 0 : \int_{-\alpha/2}^{\alpha/2} \sigma_{rr} r d\theta \cos\theta + P\cos\beta = 0 \\
\sum F_y = 0 : \int_{-\alpha/2}^{\alpha/2} \sigma_{rr} r d\theta \sin\theta + P\sin\beta = 0
\end{aligned} \tag{7.8}$$

Here, we first compute the forces using the stress and then project them onto the x-direction, which gives the first equation in Eq. (7.8). Its projection onto the y-direction gives the second equation in Eq. (7.8). These two equations can be used to determine the two constants, c_0 and c_1. This is done by substituting Eq. (7.6) into Eq. (7.8) and then carrying out the integrals. We use the following snippet to perform this calculation:

```
θ = sp.symbols('θ')
eq1=(integrate(σ_rrg*r*sp.cos(θ),(θ,-α/2,α/2)) + P*sp.cos(β)).simplify()
print(eq1)

eq2=(integrate(σ_rrg*r*sp.sin(θ),(θ,-α/2,α/2)) + P*sp.sin(β)).simplify()
print(eq2)
```

```
P*cos(β) + c1*α + c1*sin(α)
P*sin(β) - c0*α + c0*sin(α)
```

After obtaining these two equations, we can solve them for the two constants c_0 and c_1:

```
cs = sp.solve([eq1, eq2],[c0, c1])
cs
```

$$\left\{c_0 : -\frac{P\sin(\beta)}{-\alpha+\sin(\alpha)},\ c_1 : -\frac{P\cos(\beta)}{\alpha+\sin(\alpha)}\right\}$$

As shown, the first one accounts for the contribution from the antisymmetric part of the load, and the **second** for the contribution from the symmetric part of the load. In other words, if the force is vertical ($\beta = 0$), c_0 will be zero, and if the force is horizontal ($\beta = \pi/2$), c_1 will be zero.

7.1.5 *Final stress solution*

We substitute these constants back into the expressions for the general stresses:

```
σ_rr = σ_rrg.subs(cs)
#display(Math(f"\\text{{Final solution of $σ_{{rr}}$}}={latex(σ_rr)}"))
```

Let's rewrite the final solution for these stresses as follows for future reference:

$$\sigma_{rr} = -\frac{2P}{r}\left(\frac{\cos(\beta)\cos(\theta)}{(\alpha+\sin(\alpha))} + \frac{\sin(\beta)\sin(\theta)}{(\sin(\alpha)-\alpha)}\right)$$
$$\sigma_{\theta\theta} = 0; \quad \sigma_{r\theta} = 0 \tag{7.9}$$

The stress solution exhibits an r^{-1} singularity due to the concentrated force. Note that θ is confined: $-\alpha/2 \le \theta \le \alpha/2$. The distribution of σ_{rr} is a simple sine, cosine, or a combination of both, and hence is smooth in θ.

We redefine σ_{rr} explicitly in Python for later use, so that we do not need to rerun all the code above:

```
r = sp.symbols('r', nonnegative=True)          # define symbolic variables
θ, α, β, P = sp.symbols('θ, α, β, P')        # parameters for this problem

σ_rr_ = -2*P*sp.cos(β)*sp.cos(θ)/(r*( α + sp.sin(α)))\
        +2*P*sp.sin(β)*sp.sin(θ)/(r*(-α + sp.sin(α)))
(σ_rr - σ_rr_).simplify()  # check the re-defined formula, should be 0.
```

0

7.1.6 *Plot the distribution of the stress*

Finally, we can plot the distribution of the stresses using Python. Note that the wedge angle can be in the range $0 \le \alpha \le \pi$, as shown in Fig. 7.2(a). When $\alpha = \pi$, the wedge becomes a half-space, and the formulas developed in this chapter are still applicable. When $\alpha > \pi$, the domain becomes concave with a sharp corner, which is no longer a wedge. The "tip" of the wedge becomes a geometrically singular point, and hence the formulas developed in this chapter are not applicable. Alternative approaches that can handle singularities should be used (e.g., Ref. [3]).

All the formulas derived above apply to all wedges with $0 \le \alpha \le \pi$ and $0 \le \beta \le 2\pi$. The following Python function converts SymPy formulas

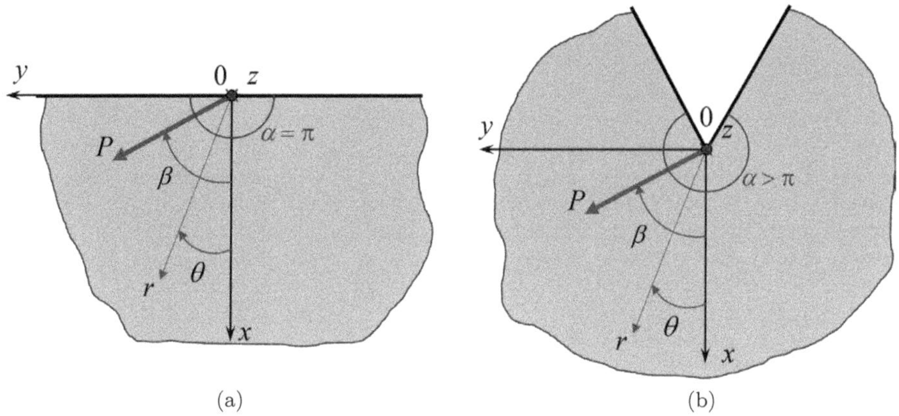

Figure 7.2. Schematic view of a 2D wedge with large angles of α, subjected to a vertical force P at an angle β. (a) $\alpha = \pi$, where the formulas developed in this chapter are still applicable; (b) $\alpha > \pi$, where the formulas are not applicable.

to NumPy functions and then uses `plt.plot()` to produce plots of stress variation with coordinates. Using this function, we can plot multiple figures on one page:

```
def plot_fij(fig_s, x, y, labels, xlbl, ylbl,fij):
    '''plot multiple x-y curves in multiple figures.
    fig_s: a plt.figure() object
    x: x data points, 1D numpy array
    y: y data, list of 1D numpy array
    labels: labels for these curves
    xlbl: for ax.set_xlabel()
    ylbl: for ax.set_ylabel()
    fij: figure grids. First 2 indexes gives the grid, third is for
         figure number. The last one controls where plot xlabel.
    '''
    colors = ['b', 'r', 'g', 'c', 'm', 'k']

    ax = fig_s.add_subplot(fij[0],fij[1],fij[2])  #(1,1, 1)
    for i in range(len(y)):
        ax.plot(x, y[i], label=labels[i], color=colors[i])

    ax.grid(c='r', linestyle=':', linewidth=0.5)
    #ax.axvline(x=0, c="k", lw=0.6);ax.axhline(y=0, c="k", lw=0.6)
    if fij[3] == 'xl': ax.set_xlabel(xlbl)
    ax.set_ylabel(ylbl)
    ax.legend() #loc='center right', bbox_to_anchor=(1, 0.5))

    #plt.title('x-y curves'+p_name+'')
    #plt.savefig('imagesSMp/'+p_name[0]+'.png',dpi=500,bbox_inches='tight')
    #plt.show()
```

```
# SymPy expressions to Numpy functions:

alfa_ = [sp.pi/4, sp.pi/2, 3*sp.pi/4, sp.pi]     # define α(<=pi) cases
nf = len(alfa_)                                         # number of figures

f_ij=[[1, 1, 1, "xl"]]                               # define figure grids
if nf == 2: f_ij=[[1, 2, 1, "xl"],[1, 2, 2, "xl"]]
if nf >= 3: f_ij=[[2,2,1,"no"],[2,2,2,"no"],[2,2,3,"xl"],[2,2,4,"xl"]]
#if nf >= 3: f_ij=[[4,1,1,"no"],[4,1,2,"no"],[4,1,3,"no"],[4,1,4,"xl"]]

plt.ioff()
#plt.ion()                           # uncomment this to display the plots
#plt.rcParams.update({'font.size': 5})            # settings for plots
fig_s = plt.figure()#figsize=(4,2.5))

for k in range(nf):
    fij=f_ij[k]
    alfa = alfa_[k]
    a_np = np.array(alfa.evalf(),dtype=float)  # convert to numpy float
    degree = a_np*180/np.pi

    func1=sp.lambdify(θ, σ_rr.subs({P:1,r:1,α:alfa,β:0}),       'numpy')
    func2=sp.lambdify(θ, σ_rr.subs({P:1,r:1,α:alfa,β:sp.pi/2}),'numpy')
    func3=sp.lambdify(θ, σ_rr.subs({P:1,r:1,α:alfa,β:sp.pi/4}),'numpy')

    labels=[r'$σ_{rr}, α='+f'{degree:.0f}'+r'$∘$'+', β=0$',
            r'$σ_{rr}, α='+f'{degree:.0f}'+r'$∘$'+', β=\pi/2$',
            r'$σ_{rr}, α='+f'{degree:.0f}'+r'$∘$'+', β=\pi/4$']

    x_data = np.linspace(-a_np/2, a_np/2, 400)    # x-axis for θ in plot
    y_data = [func1(x_data), func2(x_data), func3(x_data)]  # fs values

    plot_fij(fig_s, x_data, y_data, labels, '$θ$', '$σ_{rr}$', fij)

plt.savefig('imagesSMp/PonWedge.png',dpi=500,bbox_inches='tight')
#plt.show()                          # uncomment this to display the plots
```

Figure 7.3 shows the distribution of stress σ_{rr} with respect to θ along the arc of $r = 1$. The concentrated force is set to unity: $P = 1$. Therefore, the figure gives the profile of the solutions.

We found the following:

1. As α increases, the stress level decreases, which is expected.
2. When $\beta = 0$, the force is vertical; σ_{rr} is a cosine function in terms of θ in $[-\alpha/2, \alpha/2]$ and is symmetric.
3. When $\beta = 90°$, the force is horizontal; σ_{rr} is a sine function in terms of θ in $[-\alpha/2, \alpha/2]$ and is antisymmetric. In this case, the bending effect dominates.

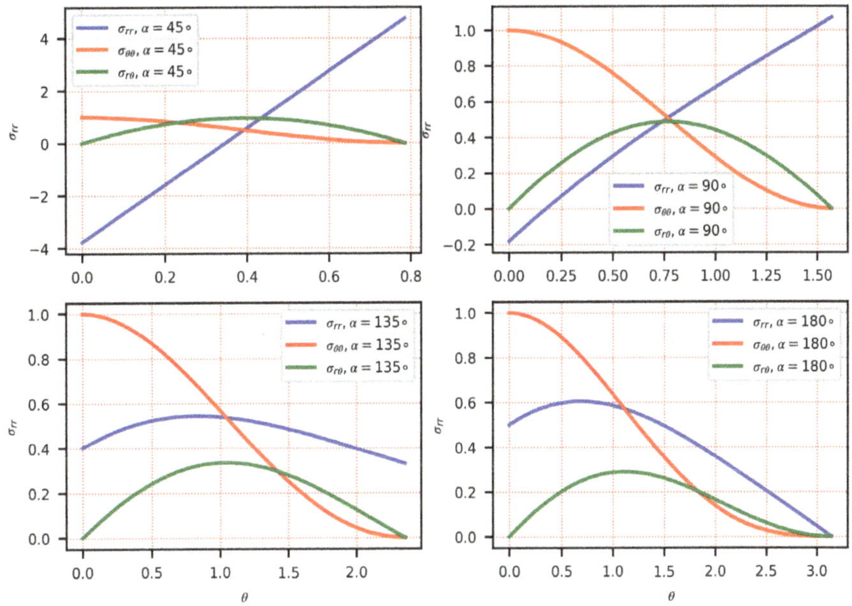

Figure 7.3. Distribution of stress σ_{rr} along the arc of $r = 1$ over wedges with various α values. The load is set as $P = 1$.

4. For small α, the range of θ is also small and hence the distribution of σ_{rr} is almost linear.
5. The last case shown in the preceding figure is for $\alpha = 180°$, which corresponds to a half-space as shown in Fig. 7.2(a). The half-space case will be studied further in the following chapter.

Readers may explore further by simply changing the variable `alfa_` for wedges with different angles and rerunning the code cell.

The following code plots the stress distributions for different r values:

```
# SymPy expressions to numpy functions:

alfa = sp.pi/2                                        # define α cases
r_ = [0.1, 0.4, 0.7, 1.0]                             # define r cases
nf = len(r_)                                          # number of figures

f_ij=[[1, 1, 1, "xl"]]                                # define figure grids
if nf == 2: f_ij=[[1, 2, 1, "xl"],[1, 2, 2, "xl"]]
if nf >= 3: f_ij=[[2,2,1,"no"],[2,2,2,"no"],[2,2,3,"xl"],[2,2,4,"xl"]]
#if nf >= 3: f_ij=[[4,1,1,"no"],[4,1,2,"no"],[4,1,3,"no"],[4,1,4,"xl"]]

plt.ioff()
#plt.ion()                        # uncomment this to display the plots
#plt.rcParams.update({'font.size': 5})              # settings for plots
fig_s = plt.figure()#figsize=(4,2.5))
```

```
for k in range(nf):
    fij=f_ij[k]
    rk = r_[k]
    a_np = np.array(alfa.evalf(),dtype=float) # convert to numpy float
    degree = a_np*180/np.pi

    func1=sp.lambdify(θ, σ_rr.subs({P:1,r:rk,:αalfa,β:0}),       'numpy')
    func2=sp.lambdify(θ, σ_rr.subs({P:1,r:rk,:αalfa,β:sp.pi/2}),'numpy')
    func3=sp.lambdify(θ, σ_rr.subs({P:1,r:rk,:αalfa,β:sp.pi/4}),'numpy')

    labels=[r'$σ_{rr}, r='+f'{rk:.1f}'+', β=0$',
            r'$σ_{rr}, r='+f'{rk:.1f}'+', β=\pi/2$',
            r'$σ_{rr}, r='+f'{rk:.1f}'+', β=\pi/4$']

    x_data = np.linspace(-a_np/2, a_np/2, 400)   # x-axis for θ in plot
    y_data = [func1(x_data), func2(x_data), func3(x_data)] # fs values

    plot_fij(fig_s, x_data, y_data, labels, '$θ$', '$σ_{rr}$', fij)

plt.savefig('imagesSMp/stressOnWedge_r.png',dpi=500,bbox_inches='tight')
#plt.show()                        # uncomment this to display the plots
```

Figure 7.4 shows the distribution of stress σ_{rr} with respect to θ at various r values, where the concentrated force is set to $P = 1$. The wedge angle is set to $\pi/2$, which corresponds to a quarter-space.

It can be seen that the distribution patterns are the same, but the values of σ_{rr} increase rapidly as r decreases. This means that r controls the magnitude of the stress level at a rate of r^{-1}. High stress is always observed at the tip of the wedge. In the analytical solution we obtained, the stress is singular at the tip. However, for real wedge structures, the stress remains finite (though it can be very high) in regions away from the tip. The reasons are as follows:

1. The point force P is an idealized case. In reality, a force load is never applied at a single point. It is always distributed over an area, however small. Therefore, the analytical solution remains valid in the wedge at some finite distance r away.
2. In regions of high stress, the material can undergo plastic deformation, causing it to deviate from purely elastic behavior. Therefore, the load on the elastic part of the wedge is effectively applied at some finite r and is distributed along the interface between the plastic and elastic zones.

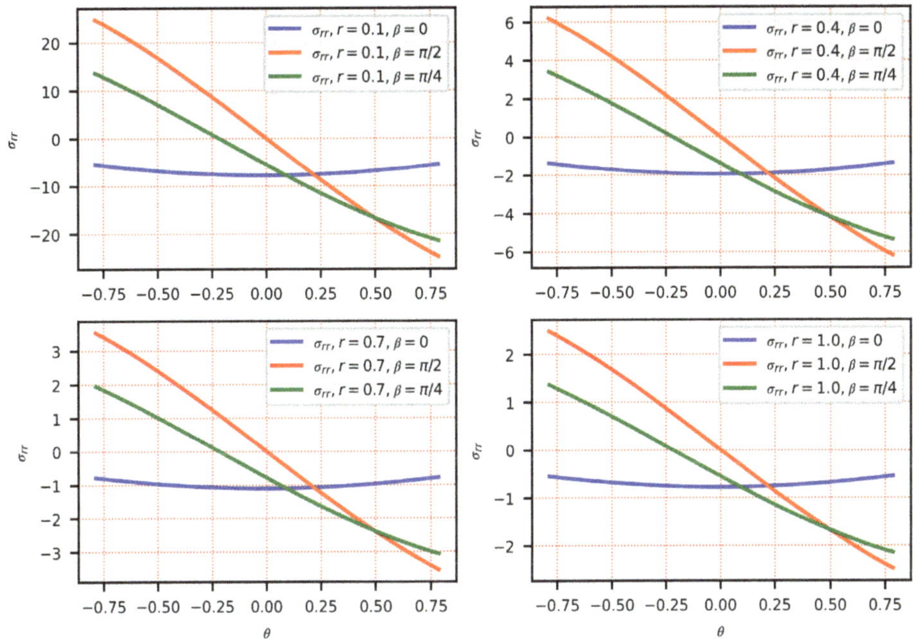

Figure 7.4. Distribution of stress σ_{rr} along θ over a quarter-space ($\alpha = \pi/2$) with various r values. The load is set as $P = 1$.

7.2 A point moment on wedges

Consider again a two-dimensional (2D) wedge with an angle α, as shown in Fig. 7.5. This time, the load is a concentrated moment M applied at the wedge tip. We aim to find the stress distribution in the wedge.

The problem setup is the same as that shown in Fig. 7.1, with the only difference being that the load is now a moment M about the z-axis.

7.2.1 *Stress function construction*

To find the stress solution to this problem, we first need to determine the stress function ϕ through dimensional analysis, as we did for the previous case. Since the problem is 2D, the unit of the moment M should be (force/length) $\times$ length $=$ force. The other physical parameter that affects the solution is the angle α, which is dimensionless. In the coordinate system, the radius r has the dimension of length, while the polar angle θ is dimensionless.

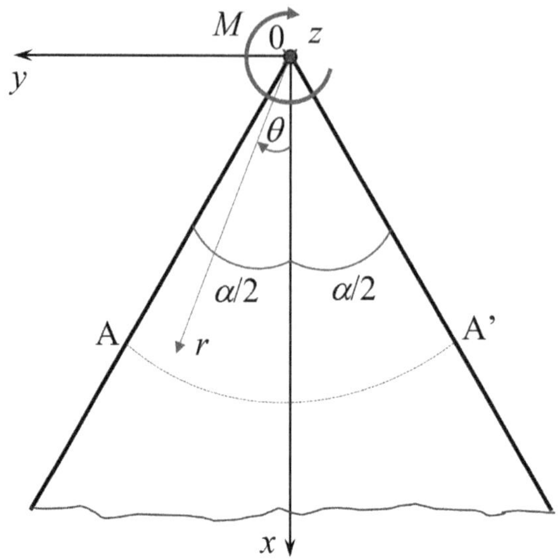

Figure 7.5. Schematic view of a 2D wedge with an angle α, subjected to a concentrated moment M at its tip.

The stress we seek must be related to M/r^2 to have the correct units. Using Eq. (7.2), the stresses are obtained from the second derivatives of ϕ with respect to r. Therefore, it is reasonable to assume that ϕ has the form:

$$\phi = \phi(\theta) \tag{7.10}$$

where $\phi(\theta)$ is an arbitrary function of θ only, reducing the problem to one dimension. For zero body force, the stress function should satisfy $\nabla^4\phi = 0$. This equation can be solved using the following snippet:

```
# Solve ∇^4ϕ=0 for general solution:
r = sp.symbols('r', nonnegative=True)          # define symbolic variables
θ, α, M = sp.symbols('θ, α, M')                 # involved in this problem

ϕ = sp.Function('ϕ')(θ)                         # define the stress function

lap = diff(ϕ,(r,2))  + (1/r)*diff(ϕ,r)  + (1/r**2)*diff(ϕ,(θ,2))
bih = diff(lap,(r,2))+ (1/r)*diff(lap,r) + (1/r**2)*diff(lap,(θ,2))

bih.simplify()                                  # bi harmonic operator
```

$$\frac{4\frac{d^2}{d\theta^2}\phi(\theta) + \frac{d^4}{d\theta^4}\phi(\theta)}{r^4}$$

For this expression to be zero, we need to find a $\phi(\theta)$ that satisfies

$$4\frac{d^2\phi(\theta)}{d\theta^2} + \frac{d^4\phi(\theta)}{d\theta^4} = 0 \tag{7.11}$$

This is a fourth-order 1D differential equation for $\phi(\theta)$. Its general solution can be found using the following code:

```
# define differential equation
diff_eq = sp.Eq(4*diff(ϕ,(θ,2)) + diff(ϕ,(θ,4)), 0)

g_sln = sp.dsolve(diff_eq, ϕ).simplify()  # solve differential equation
display(Math(f" \\text{{General solution of the stress function: }} "))
display(Math(f" \small {latex(g_sln)}"))
```

General solution of the stress function:

$\phi(\theta) = C_1 + C_2\theta + C_3 \sin(2\theta) + C_4 \cos(2\theta)$

As shown in Fig. 7.5, our problem is antisymmetric with respect to θ. Therefore, φ must be antisymmetric in θ, and the terms $C_1 + C_4 r \cos 2\theta$ should be removed. The stress function can be simplified to the following form:

$$\varphi(\theta) = c_0 \sin 2\theta + c_1\theta \tag{7.12}$$

where c_0 and c_1 are arbitrary constants. The following snippet defines this general stress function for wedges subjected to a moment at the tip:

```
# Rewrite the simplified stress function in SymPy:
c0, c1 = symbols('c0, c1 ')                          # integration constant
ϕ = c0*sp.sin(2*θ) + c1*θ
display(Math(f" \\text{{ The final general form of the stress \
                                      function: ϕ(θ)}}={latex(ϕ)}"))
```

The final general form of the stress function: $\phi(\theta) = c_0 \sin(2\theta) + c_1\theta$

7.2.2 *General solution for stresses*

We can now compute the stresses using the stress function, as shown in Eq. (7.2):

```
σ_rrg = (diff(ϕ,(θ,2))/r**2+diff(ϕ,r)/r).expand()
display(Math(f" \\text{{General solution of $σ_{{rr}}$}}=\
                                            {latex(σ_rrg)}"))
σ_θθg = diff(ϕ,(r,2)).expand()
display(Math(f" \\text{{General solution of $σ_{{θθ}}$}}=\
                                            {latex(σ_θθg)}"))
σ_rθg = (-diff(diff(ϕ,θ)/r,r)).expand()
display(Math(f" \\text{{General solution of $σ_{{rθ}}$}}=\
                                            {latex(σ_rθg)}"))
```

General solution of $\sigma_{rr} = -\dfrac{4c_0 \sin(2\theta)}{r^2}$

General solution of $\sigma_{\theta\theta} = 0$

General solution of $\sigma_{r\theta} = \frac{2c_0 \cos(2\theta)}{r^2} + \frac{c_1}{r^2}$

It is clear that the stress component $\sigma_{\theta\theta}$ is zero throughout the entire domain for this problem. The general solution for the stresses involves two constants. We create two boundary conditions to determine them.

7.2.3 *Stress boundary conditions*

Since the two surfaces of the wedge are stress-free, the boundary conditions should be the same as those given in Eq. (7.7). Clearly, the first condition in Eq. (7.7) is already satisfied regardless of the undetermined constants. To satisfy the second condition in Eq. (7.7), we have

$$\frac{2c_0 \cos\alpha}{r^2} + \frac{c_1}{r^2} = 0, \quad \text{or} \quad c_1 = -2c_0 \cos\alpha \tag{7.13}$$

With this relation, all these nonzero stresses can be written with only one unknown constant c_0. These expressions can be found using the following snippet:

```
σ_rrg = σ_rrg.subs(c1, -2*c0*cos(α))
display(Math(f" \\text{{General solution of $σ_{{rr}}$}}=\
                                                {latex(σ_rrg)}"))
σ_rθg = σ_rθg.subs(c1, -2* c0*cos(α))
display(Math(f" \\text{{General solution of $σ_{{rθ}}$}}=\
                                                {latex(σ_rθg)}"))
```

General solution of $\sigma_{rr} = -\frac{4c_0 \sin(2\theta)}{r^2}$

General solution of $\sigma_{r\theta} = -\frac{2c_0 \cos(\alpha)}{r^2} + \frac{2c_0 \cos(2\theta)}{r^2}$

There is still one unknown constant to be determined, for which we need to use the integral equilibrium condition.

7.2.4 *Integral equilibrium condition*

Note that the stress has an $r{-}2$ singularity: σ_{rr} becomes infinite as $r \to 0$ at a rate of $r{-}2$. Therefore, any application of the boundary conditions must be made at a distance away from $r = 0$.

The externally applied moment M must be balanced by that generated by the internal stresses along the arc A–A′ shown in Fig. 7.5. The equation should be

$$\sum M_0 = 0 : \int_{-\alpha/2}^{\alpha/2} \sigma_{r\theta} r^2 d\theta + M = 0 \tag{7.14}$$

This integral equation can be used to find the constant c_0:

```
θ = sp.symbols('θ')
eq1 = (integrate(σ_rθg*r**2,(θ,-α/2,α/2)) + M).simplify()
display(Math(f" \\text{{ Equilibrium equation to solve for $c_0$: }}\
                                                    {latex(eq1)}=0"))
# Solve eq1 for constant $c_0$:
cs = sp.solve(eq1, c0)[0]
display(Math(f" \\text{{ Constant found: $c_{{0}}$}}={latex(cs)}"))
```

Equilibrium equation to solve for c_0: $M - 2c_0\alpha\cos(\alpha) + 2c_0\sin(\alpha) = 0$

Constant found: $c_0 = \dfrac{M}{2(\alpha\cos(\alpha) - \sin(\alpha))}$

7.2.5 *Final stress solution*

We substitute the constant c_0 back into the expressions for the stresses to obtain the final solutions:

```
σ_rr = σ_rrg.subs(c0, cs)
σ_rθ = σ_rθg.subs(c0, cs).simplify()
#display(Math(f" \\text{{Final solution of $σ_{{rθ}}$}}=\
#                                   {latex(σ_rθ)}"))
```

We rewrite the final solution for these stresses as follows for future reference:

$$\begin{aligned}
\sigma_{rr} &= -\frac{2M\sin 2\theta}{r^2(\alpha\cos\alpha - \sin\alpha)} \\
\sigma_{\theta\theta} &= 0 \\
\sigma_{r\theta} &= \frac{M(-\cos\alpha + \cos 2\theta)}{r^2(\alpha\cos\alpha - \sin\alpha)}
\end{aligned} \tag{7.15}$$

Note that θ is confined: $-\alpha/2 \le \theta \le \alpha/2$. We redefine these stresses explicitly in Python for future use:

```
r = sp.symbols('r', nonnegative=True)        # define symbolic variables
θ, α, M = sp.symbols('θ, α, M')             # parameters for this problem

σ_rr = -2*M*sin(2*θ)/(r**2*(α*cos(α) - sin(α)))
σ_rθ = M*(-cos(α) + cos(2*θ))/(r**2*(α*cos(α) - sin(α)))
```

Using the final stress solution, we can also check whether these stresses satisfy the integral equilibrium equations for forces in both the x- and y-directions along the arc A–A′ shown in Fig. 7.5. These conditions are as follows:

$$\begin{aligned}\sum F_x = 0 : \int_{-\alpha/2}^{\alpha/2} (\sigma_{rr}\cos\theta - \sigma_{r\theta}\sin\theta) r d\theta = 0 \\ \sum F_y = 0 : \int_{-\alpha/2}^{\alpha/2} (\sigma_{rr}\sin\theta + \sigma_{r\theta}\cos\theta) r d\theta = 0\end{aligned} \tag{7.16}$$

Here, we project the forces resulting from the stresses onto both the x- and y-directions.

The snippet to check these two equations is follows:

```
Fx=(integrate(σ_rr*r*sp.cos(θ)-σ_rθ*r*sp.sin(θ),(θ,-α/2,α/2)) ).simplify()
display(Math(f" \\text{{ $∑F_x$ = 0? {Fx==0}}}"))

Fy=(integrate(σ_rr*r*sp.sin(θ)+σ_rθ*r*sp.cos(θ),(θ,-α/2,α/2)) ).simplify()
display(Math(f" \\text{{ $∑F_y$ = 0? {Fy==0}}}"))
```

$\sum F_x = 0?$ True

$\sum F_y = 0?$ True

Finally, we can plot and examine the distribution of stresses over the wedge, using the following snippet:

```
# SymPy expressions to Numpy functions:

alfa_ = [sp.pi/4, sp.pi/2, 3*sp.pi/4, sp.pi]              # define α cases
nf = len(alfa_)                                        # number of figures

f_ij=[[1, 1, 1, "xl"]]                              # define figure grids
if nf == 2: f_ij=[[1, 2, 1, "xl"],[1, 2, 2, "xl"]]
if nf >= 3: f_ij=[[2,2,1,"no"],[2,2,2,"no"],[2,2,3,"xl"],[2,2,4,"xl"]]

plt.ioff()
#plt.ion()                              # uncomment this to display the plots
#plt.rcParams.update({'font.size': 5})   # settings for plots, if needed
fig_s = plt.figure()#figsize=(4,2.5))
```

```python
for k in range(nf):
    fij=f_ij[k]
    alfa = alfa_[k]
    a_np = np.array(alfa.evalf(),dtype=float)  # convert to numpy float
    degree = a_np*180/np.pi

    func1=sp.lambdify(θ, σ_rr.subs({M:1,r:1,α:alfa}),'numpy')
    func2=sp.lambdify(θ, σ_rθ.subs({M:1,r:1,α:alfa}),'numpy')

    labels=[r'$σ_{rr}, α=$'+f'{degree:.0f}'+r'$∘$',
            r'$σ_{rθ}, α=$'+f'{degree:.0f}'+r'$∘$']

    x_data = np.linspace(-a_np/2, a_np/2, 400)   # x-axis for θ in plot
    y_data = [func1(x_data), func2(x_data)]                   # fs values

    plot_fij(fig_s, x_data, y_data, labels, '$θ$', '$σ_{rr}$', fij)

plt.savefig('imagesSMp/MonWedge.png',dpi=500,bbox_inches='tight')
#plt.show()                        # uncomment this to display the plots
```

Figure 7.6 shows the distribution of both stress components σ_{rr} and $\sigma_{r\theta}$ with respect to θ along the arc of $r = 1$, when the concentrated moment is set to $M = 1$. The wedge has four different angles. The pattern of the

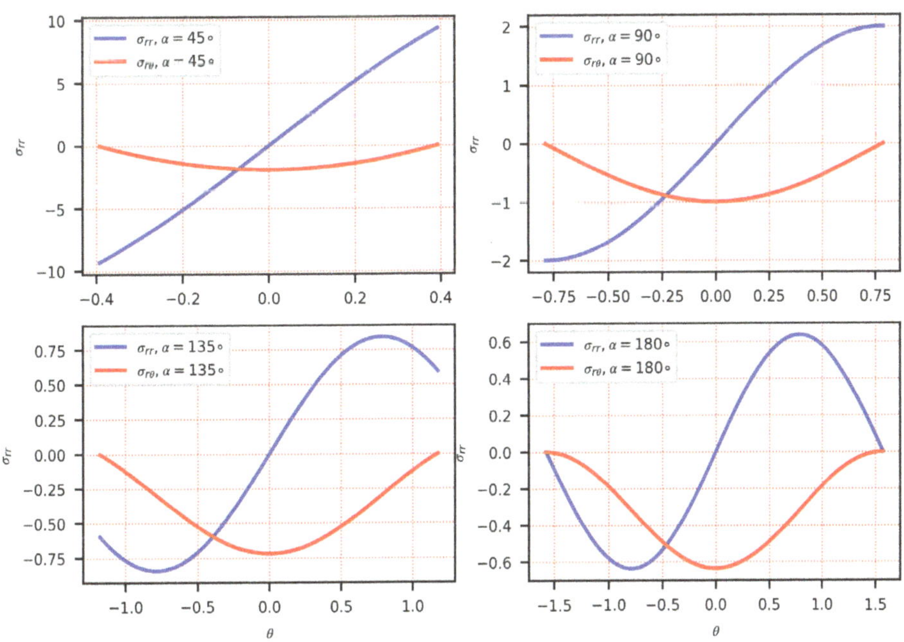

Figure 7.6. Distribution of stress components along θ over wedges along the arc of $r = 1$. The load is set as $M = 1$.

distributions does not change with r, and the amplitude is proportional to $1/r^2$. In other words, the stress decays very rapidly (quadratically) in the radial direction.

It is also seen that the stress σ_{rr} is always antisymmetric and $\sigma_{r\theta}$ is symmetric due to the antisymmetric nature of the moment load. The antisymmetric normal stress σ_{rr} results from the applied moment M. All the shear stresses are zero on the two surfaces of the wedge due to the boundary condition settings.

7.3 Distributed pressure on wedges

Consider a two-dimensional (2D) wedge with an angle α, as shown in Fig. 7.7. A uniformly distributed pressure p is applied on its right edge. We aim to find the stress distribution in the wedge.

7.3.1 *Stress function construction*

The stress function ϕ can again be assumed using dimensional analysis. First, all the variables involved in this problem are the wedge angle α, the polar angle θ, the radial coordinate r, and the pressure p. Since the pressure has

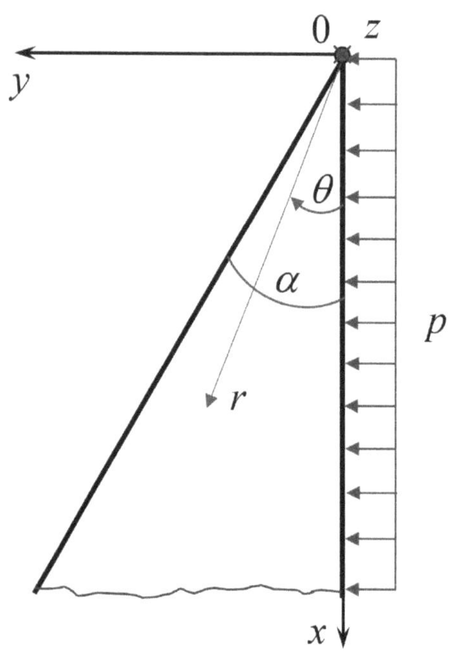

Figure 7.7. Schematic view of a 2D wedge with an angle α, subjected to a uniformly distributed pressure p on the right vertical edge.

the same units as the stresses, the other variables that might appear in the stress expression are these dimensionless angles. In other words, r will not appear in the stress expressions. Since the stresses come from the second derivatives of ϕ with respect to r, the stress function should include r^2 as a factor. Hence, it can be written in the following form:

$$\phi = r^2 f(\theta) \tag{7.17}$$

where $f(\theta)$ is an arbitrary function of θ. Since the body force in the wedge is zero, we have $\nabla^4\phi = 0$, where $\nabla^4\phi$ can be computed using the following snippet:

```
# Solve ∇^4ϕ=0 for general solution:
r = sp.symbols('r', nonnegative=True)          # define symbolic variables
θ, α, p = sp.symbols('θ, α, p')                 # involved in this problem

ϕ = sp.Function('ϕ')(r, θ)                     # define the stress function
f = sp.Function('f')(θ)                       # θ part of the stress function
ϕ = r**2*f                                     # form the stress function

lap = diff(ϕ,(r,2))  + (1/r)*diff(ϕ,r)  + (1/r**2)*diff(ϕ,(θ,2))
bih = diff(lap,(r,2))+ (1/r)*diff(lap,r) + (1/r**2)*diff(lap,(θ,2))

bih.simplify()                                          # biharmonic operator
display(Math(f"\\text{{Equation for}} f(θ): {latex(bih.simplify())}=0"))
```

Equation for $f(\theta)$: $\dfrac{4\frac{d^2}{d\theta^2}f(\theta) + \frac{d^4}{d\theta^4}f(\theta)}{r^2} = 0$

Therefore, we need to find a function $f(\theta)$ that makes the numerator of the above equation vanish. Such a function was found earlier and is recomputed using the following code:

```
# define differential equation
diff_eq = sp.Eq(f + 2*diff(f,(θ,2)) + diff(f,(θ,4)), 0)

g_sln = sp.dsolve(diff_eq, f).simplify()  # solve differential equation
display(Math(f" \\text{{General solution:}}{latex(g_sln)}"))
#print(g_sln.expand())                 # to generate the Sympy expression
```

General solution: $f(\theta) = (C_1 + C_2\theta)\sin(\theta) + (C_3 + C_4\theta)\cos(\theta)$

We can now form the stress function as follows:

$$\phi = r^2(c_0 \sin\theta + c_1\theta\sin\theta + c_2\cos\theta + c_3\theta\cos\theta) \tag{7.18}$$

This time, we have no reason to remove any of these terms. Hence, we keep all of them. The SymPy expression for the stress function ϕ is follows:

```
# Redefine stress function ϕ:
c0, c1, c2, c3 = sp.symbols('c0, c1, c2, c3', real=True)
ϕ = r**2*(c0*sin(θ) + c1*θ*sin(θ) + c2*cos(θ) + c3*θ*cos(θ)).expand()
```

7.3.2 *General solution for stresses*

Using the stress function obtained above, we can now compute the stresses using Eq. (7.2):

```
# Compute stresses:
σ_rrg = (diff(ϕ,(θ,2))/r**2+diff(ϕ,r)/r).expand()
display(Math(f" \\text{{ $σ_{{rr}}$}}=\small {latex(σ_rrg)}"))

σ_θθg = diff(ϕ,(r,2)).expand()
display(Math(f" \\text{{ $σ_{{θθ}}$}}=\small {latex(σ_θθg)}"))

σ_rθg = (-diff(diff(ϕ,θ)/r,r)).expand()
display(Math(f" \\text{{ $σ_{{rθ}}$}}=\small {latex(σ_rθg)}"))
```

$$\sigma_{rr} = c_0 \sin(\theta) + c_1\theta \sin(\theta) + 2c_1 \cos(\theta) + c_2 \cos(\theta) + c_3\theta \cos(\theta) - 2c_3 \sin(\theta)$$

$$\sigma_{\theta\theta} = 2c_0 \sin(\theta) + 2c_1\theta \sin(\theta) + 2c_2 \cos(\theta) + 2c_3\theta \cos(\theta)$$

$$\sigma_{r\theta} = -c_0 \cos(\theta) - c_1\theta \cos(\theta) - c_1 \sin(\theta) + c_2 \sin(\theta) + c_3\theta \sin(\theta) - c_3 \cos(\theta)$$

These are the general solutions for the stresses. For this problem, all these stresses are generally nonzero. There are a total of four constants, and we need to create four boundary conditions to determine them.

7.3.3 *Stress boundary conditions*

Since the two surfaces of the wedge are stress-free, we have

$$\begin{aligned} &\sigma_{\theta\theta} = 0 \quad \text{and} \quad \sigma_{r\theta} = 0 \quad \text{at} \quad \theta = \alpha \\ &\sigma_{\theta\theta} = p \quad \text{and} \quad \sigma_{r\theta} = 0 \quad \text{at} \quad \theta = 0 \end{aligned} \tag{7.19}$$

Using these four boundary conditions, we can determine the four constants. The snippet is as follows:

```
# Define BC equations and use these to solve for the unknown constants:
eq1 = σ_θθg.subs(θ, α)
eq2 = σ_rθg.subs(θ, α)
eq3 = σ_θθg.subs(θ, 0) + p
eq4 = σ_rθg.subs(θ, 0)
cs = [c0, c1, c2, c3]
C_sln_dict = sp.solve([eq1, eq2, eq3, eq4], cs)
C_sln = {k: C_sln_dict[k].simplify() for k, v in C_sln_dict.items()}
# solution for constants [c_0, c_1, c_2, c_3]

display(Math(f" \\text{{Constants found: }}"))
display(Math(f" \scriptsize {latex(C_sln)}"))
```

Constants found:

$$\left\{c_0 : -\frac{p\left(\alpha + \frac{\sin(2\alpha)}{2}\right)}{2\alpha^2 - 2\sin^2(\alpha)},\ c_1 : \frac{p\sin^2(\alpha)}{2\left(\alpha^2 - \sin^2(\alpha)\right)},\ c_2 : -\frac{p}{2},\ c_3 : \frac{p\left(\alpha + \frac{\sin(2\alpha)}{2}\right)}{2\left(\alpha^2 - \sin^2(\alpha)\right)}\right\}$$

7.3.4 *Final stress solution*

We substitute the constants found back into the expressions for the general stresses, which gives the stress solutions for this particular problem:

```
σ_rr = σ_rrg.subs(C_sln_dict).simplify().collect(cos(θ)).collect(sin(α))
display(Math(f" \\text{{ $σ_{{rr}}$}}=\scriptsize {latex(σ_rr)}"))

σ_θθ = σ_θθg.subs(C_sln_dict).simplify().collect(cos(θ)).collect(sin(α))
display(Math(f" \\text{{ $σ_{{θθ}}$}}=\scriptsize {latex(σ_θθ)}"))

σ_rθ = σ_rθg.subs(C_sln_dict).simplify().collect(cos(θ)).collect(sin(α))
display(Math(f" \\text{{ $σ_{{rθ}}$}}=\scriptsize{latex(σ_rθ)}"))
```

$$\sigma_{rr} = \frac{p\left(-3\alpha\sin(\theta) + \left(-\alpha^2 + \alpha\theta\right)\cos(\theta) + \left(\theta\cos(\alpha-\theta) + 3\sin(\alpha-\theta)\right)\sin(\alpha)\right)}{2(\alpha^2 - \sin^2(\alpha))}$$

$$\sigma_{\theta\theta} = \frac{p\left(\alpha\sin(\theta) + \left(-\alpha^2 + \alpha\theta\right)\cos(\theta) + \left(\theta\cos(\alpha-\theta) + \sin(\alpha-\theta)\right)\sin(\alpha)\right)}{\alpha^2 - \sin^2(\alpha)}$$

$$\sigma_{r\theta} = \frac{p\left(-\alpha^2\sin(\theta) + \alpha\theta\sin(\theta) - \theta\sin(\alpha)\sin(\alpha-\theta)\right)}{2\left(\alpha^2 - \sin^2(\alpha)\right)}$$

This is the final solution for all the stresses. For this problem, all the stress components are generally nonzero and independent of r.

We redefine these stresses explicitly in Python for later use, without re-running all the code above:

```
r = sp.symbols('r', nonnegative=True)        # define symbolic variables
θ, α, p = sp.symbols('θ, α, p')             # parameters for this problem

σ_rr =-p*(-3*α*sin(θ) + (-α**2 + α*θ)*cos(θ) + (θ*cos(α - θ)\
     + 3*sin(α - θ))*sin(α))/(2*(α**2 - sin(α)**2))

σ_θθ =-p*(-α*sin(θ) + (-α**2 + α*θ)*cos(θ) + (θ*cos(α - θ)\
     + sin(α - θ))*sin(α))/(α**2 - sin(α)**2)

σ_rθ =-p*(-α**2*sin(θ) + α*θ*sin(θ) \
     - θ*sin(α)*sin(α - θ))/(2*(α**2 - sin(α)**2))
```

7.3.5 *Stress in 2D half-space*

Let $\alpha = \pi$, making the wedge a half-space, as shown in Fig. 7.8(a). A uniformly distributed pressure p is applied on the surface on the right. Let us find the stress distribution in the half-space.

The stress formulas obtained above still apply by substituting $\alpha = \pi$ and can be computed using the following snippet:

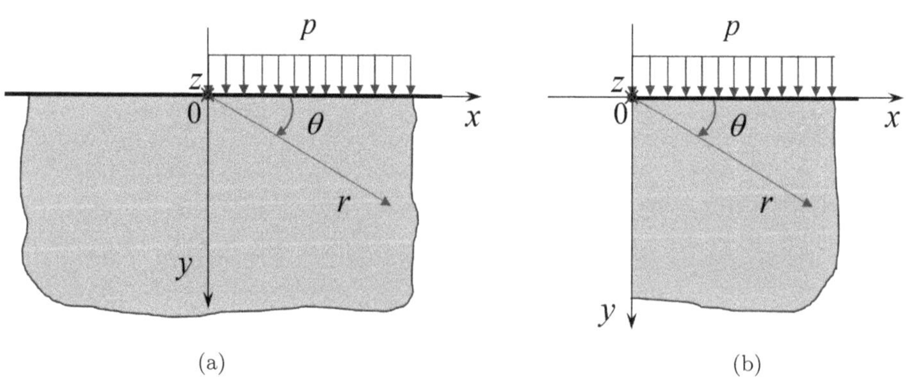

Figure 7.8. Schematic view of (a) a 2D half-space subjected to a uniformly distributed pressure p on the right surface; (b) a 2D quarter-space subjected to a uniformly distributed pressure p on the top surface.

```
# half-space with uniform pressure on half of the top surface:
x, y = sp.symbols('x, y', real=True)
dict_p = {α:sp.pi}
σ_rr_hs = σ_rr.subs(dict_p).simplify()
display(Math(f" \\text{{Stress $σ_{{rr}}$ in half-space}} \
                                                ={latex(σ_rr_hs)}"))
σ_θθ_hs = σ_θθ.subs(dict_p).simplify()
display(Math(f" \\text{{Stress $σ_{{θθ}}$ in half-space}} \
                                                ={latex(σ_θθ_hs)}"))
σ_rθ_hs = σ_rθ.subs(dict_p).simplify()
display(Math(f" \\text{{Stress $σ_{{rθ}}$ in half-space}} \
                                                ={latex(σ_rθ_hs)}"))
```

7.3.6 *Contour plots for stresses*

We can now plot the stress contours. To do this, we first convert the formulas into Cartesian coordinates:

```
# Compute the stress components in the Cartesian coordinates:

x, y = sp.symbols('x, y', real=True)
dict_p = {α:sp.pi, p:1}                                # set parameter
dic_rt = {r:sp.sqrt(x**2+y**2), θ:sp.atan2(y,x)}     # must use atan2()

σ_rrxy_hs = σ_rr_hs.subs(dict_p).subs(dic_rt).simplify()
display(Math(f" \\text{{Stress $σ_{{rr}}$ written in $(x, y)$}} \
                                                ={latex(σ_rrxy_hs)}"))
σ_θθxy_hs = σ_θθ_hs.subs(dict_p).subs(dic_rt).simplify()
display(Math(f" \\text{{Stress $σ_{{θθ}}$ written in $(x, y)$}} \
                                                ={latex(σ_θθxy_hs)}"))
σ_rθxy_hs = σ_rθ_hs.subs(dict_p).subs(dic_rt).simplify()
display(Math(f" \\text{{Stress $σ_{{rθ}}$ written in $(x, y)$}} \
                                                ={latex(σ_rθxy_hs)}"))
```

$$\text{Stress } \sigma_{rr} \text{ written in } (x,y) = \frac{-x\left(\operatorname{atan}_2(y,x) - \pi\right) + 3y}{2\pi\sqrt{x^2+y^2}}$$

$$\text{Stress } \sigma_{\theta\theta} \text{ written in } (x,y) = \frac{-x\left(\operatorname{atan}_2(y,x) - \pi\right) + y}{\pi\sqrt{x^2+y^2}}$$

$$\text{Stress } \sigma_{r\theta} \text{ written in } (x,y) = \frac{y\left(\pi - \operatorname{atan}_2(y,x)\right)}{2\pi\sqrt{x^2+y^2}}$$

```
# Create a mesh grid for a rectangular domain in x-y plane:

dX = 2.0;     dY = 2.0                              # plot dX × dY area
x_val = np.linspace(-dX+0.001, dX, 1000)
y_val = np.linspace(0, dY, 500)
X, Y = np.meshgrid(x_val, y_val)

np_σ_rθxy_hs = lambdify((x,y), σ_rθxy_hs, 'numpy')  # convert to numpy
Z = np_σ_rθxy_hs(X, Y)            # compute the stress over the mesh grid

gr.contour_plot(X,Y, Z, x_lbl='x-axis',y_lbl='y-axis',
       title=r'shear stress $σ_{{xy}}$',lvls=80, flip='y', ioff=True);

plt.savefig('imagesSMp/halfSpaceShear.png', dpi=500)      # save to file
```

```
Minimum value in the domain: 0
Maximum value in the domain: 0.289615
```

A contour plot of the shear stress σ_{xy} within the half-space is shown in Fig. 7.9. It is clear that the shear stress σ_{xy} is zero on the surface because the surface is loaded only with pressure.

This contour plot may appear unusual: all the contour lines are straight radial rays, as shown in Fig. 7.9. This occurs because, for this problem, all the

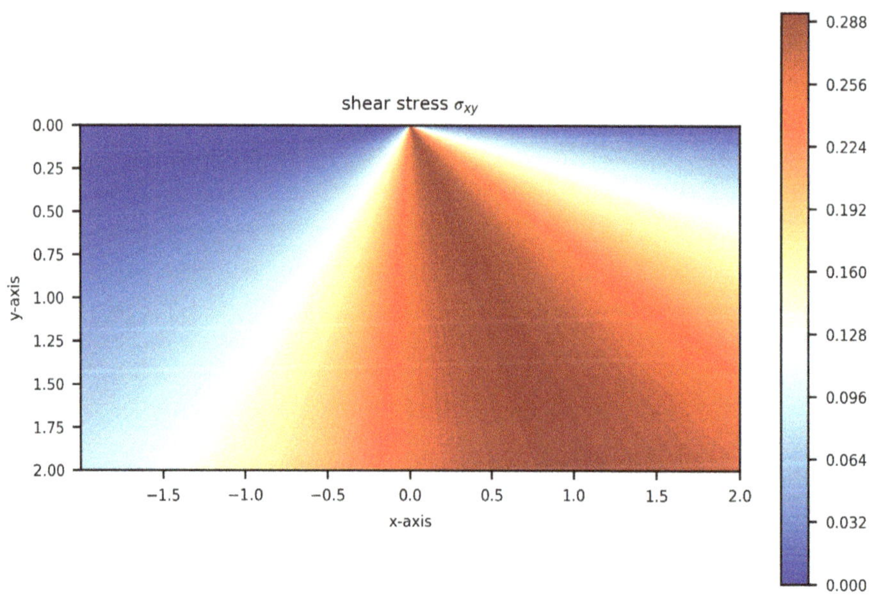

Figure 7.9. Contour plot of the shear stress in the half-space subjected to a unit pressure on the right side of the top surface.

stress components are independent of the radius r. Therefore, the contour lines correspond to lines of constant θ, which are radial lines. It represents a typical radial stress field [3].

Readers can plot the contour curves for other stress components using the same code.

7.3.7 *Stress in 2D quarter-space*

Let $\alpha = \pi/2$, so the wedge becomes a quarter-space, as shown in Fig. 7.8(b). A uniformly distributed pressure p is applied on the surface. We aim to determine the stress distribution in the quarter-space.

The stress formulas obtained earlier still apply by substituting $\alpha = \pi/2$, and the stresses can be computed using the following code snippet:

```
# A quarter-space with uniform pressure on its top surface:
dict_p = {α:sp.pi}
σ_rr_hs = σ_rr.subs(dict_p).simplify()
display(Math(f" \\text{{Stress $σ_{{rr}}$ in quarter-space}} \
                                            ={latex(σ_rr_hs)}"))
σ_θθ_hs = σ_θθ.subs(dict_p).simplify()
display(Math(f" \\text{{Stress $σ_{{θθ}}$ in quarter-space}} \
                                            ={latex(σ_θθ_hs)}"))
σ_rθ_hs = σ_rθ.subs(dict_p).simplify()
display(Math(f" \\text{{Stress $σ_{{rθ}}$ in quarter-space}} \
                                            ={latex(σ_rθ_hs)}"))
```

$$\text{Stress } \sigma_{rr} \text{ in quarter-space} = -\frac{p\left(\left(\theta - \pi\right)\cos\left(\theta\right) - 3\sin\left(\theta\right)\right)}{2\pi}$$

$$\text{Stress } \sigma_{\theta\theta} \text{ in quarter-space} = -\frac{p\left(\left(\theta - \pi\right)\cos\left(\theta\right) - \sin\left(\theta\right)\right)}{\pi}$$

$$\text{Stress } \sigma_{r\theta} \text{ in quarter-space} = \frac{p(\pi - \theta)\sin\left(\theta\right)}{2\pi}$$

7.3.8 *Stresses plots*

We can use the following snippet to plot the stresses in wedges with different angles, including the quarter-space and half-space:

```
# SymPy expressions to Python functions:

alfa_ = [sp.pi/4, sp.pi/2, 3*sp.pi/4, sp.pi]          # define α cases
nf = len(alfa_)                                      # number of figures

f_ij=[[1, 1, 1, "xl"]]                              # define figure grids
if nf == 2: f_ij=[[1, 2, 1, "xl"],[1, 2, 2, "xl"]]
if nf >= 3: f_ij=[[2,2,1,"no"],[2,2,2,"no"],[2,2,3,"xl"],[2,2,4,"xl"]]

plt.ioff()
#plt.ion()                          # uncomment this to display the plots
#plt.rcParams.update({'font.size': 5})              # settings for plots
fig_s = plt.figure()#figsize=(4,2.5))

for k in range(nf):
    fij=f_ij[k]
    alfa = alfa_[k]
    dict_p = {r:1, α:alfa, p:1}                         # set parameters
    a_np = np.array(alfa.evalf(),dtype=float)  # convert to numpy float
    degree = a_np*180/np.pi

    func1=sp.lambdify(θ, σ_rr.subs(dict_p),'numpy')
    func2=sp.lambdify(θ, σ_θθ.subs(dict_p),'numpy')
    func3=sp.lambdify(θ, σ_rθ.subs(dict_p),'numpy')

    labels=[r'$σ_{rr}, α=$'+f'{degree:.0f}'+r'$∘$',
            r'$σ_{θθ}, α=$'+f'{degree:.0f}'+r'$∘$',
            r'$σ_{rθ}, α=$'+f'{degree:.0f}'+r'$∘$']

    x_data = np.linspace(0, a_np, 400)    # x-axis for θ in plot
    y_data = [func1(x_data), func2(x_data), func3(x_data)]  # fs values

    plot_fij(fig_s, x_data, y_data, labels, '$θ$', '$σ_{rr}$', fij)

plt.savefig('imagesSMp/pOnWedge.png',dpi=500,bbox_inches='tight')
#plt.show()                         # uncomment this to display the plots
```

Figure 7.10 shows the distribution of all three stress components with respect to θ for any r, when the pressure is set to $p = 1$. The wedge has four different angles.

It can be observed that the stresses $\sigma_{\theta\theta}$ and $\sigma_{r\theta}$ satisfy the boundary conditions on the two surfaces of the wedge. The level of σ_{rr} dominates the stress when α is small, with both the normal stress $\sigma_{\theta\theta}$ and the shear stress $\sigma_{r\theta}$ being relatively small. In this case, the wedge behaves more like a thin beam [2]. As α increases, $\sigma_{\theta\theta}$ can become larger than the other components, and the shear stress $\sigma_{r\theta}$ also increases. All these stresses must be considered

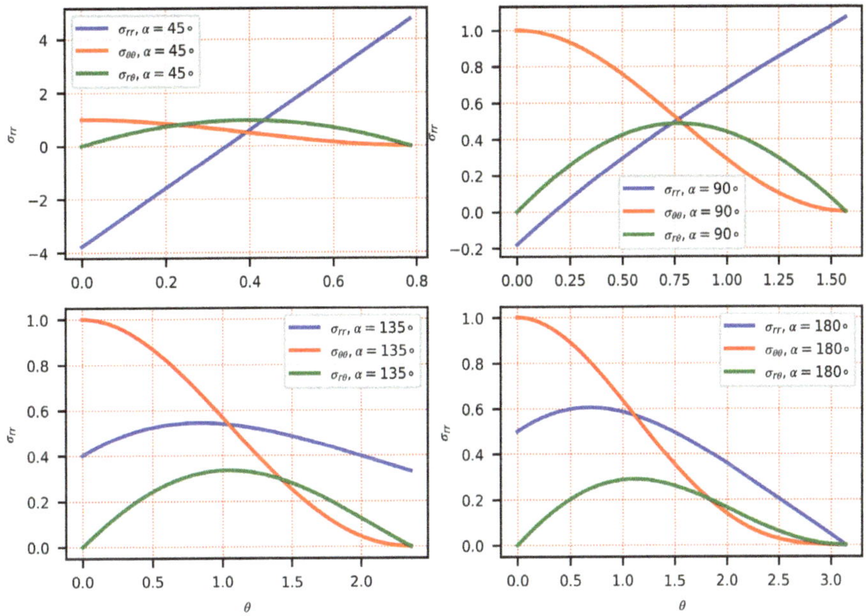

Figure 7.10. Distribution of stress components along θ over wedges. The load is set as $p = 1$.

when designing such a structure. It is important to note that these stresses do not depend on r.

7.3.9 *Contour plot of von Mises stress*

We now write the following snippet to plot a contour of the von Mises stress, with $p = 1$. For this, we need to convert the solutions in polar coordinates to those in Cartesian coordinates:

```
# Compute the stress components in the Cartesian coordinates:

x, y = sp.symbols('x, y', real=True)
dict_p = {α:sp.pi, p:1}                                    # set parameters
dic_rt = {r:sp.sqrt(x**2+y**2), θ:sp.atan2(y,x)}
σ_rrxy = σ_rr.subs(dict_p).subs(dic_rt).simplify()
display(Math(f" \\text{{Stress $σ_{{rr}}$ written in $(x, y)$}} \
                                              ={latex(σ_rrxy)}"))
σ_θθxy = σ_θθ.subs(dict_p).subs(dic_rt).simplify()
display(Math(f" \\text{{Stress $σ_{{θθ}}$ written in $(x, y)$}} \
                                              ={latex(σ_θθxy)}"))
σ_rθxy = σ_rθ.subs(dict_p).subs(dic_rt).simplify()
display(Math(f" \\text{{Stress $σ_{{rθ}}$ written in $(x, y)$}} \
                                              ={latex(σ_rθxy)}"))
```

$$\text{Stress } \sigma_{rr} \text{ written in } (x, y) = \frac{-x\left(\operatorname{atan}_2(y, x) - \pi\right) + 3y}{2\pi\sqrt{x^2 + y^2}}$$

$$\text{Stress } \sigma_{\theta\theta} \text{ written in } (x, y) = \frac{-x\left(\operatorname{atan}_2(y, x) - \pi\right) + y}{\pi\sqrt{x^2 + y^2}}$$

$$\text{Stress } \sigma_{r\theta} \text{ written in } (x, y) = \frac{y\left(\pi - \operatorname{atan}_2(y, x)\right)}{2\pi\sqrt{x^2 + y^2}}$$

The von Mises stress can be computed using the stress components, as described in Section 3.10 of Ref. [1]:

$$\sigma_{vm} = \sqrt{\frac{1}{2}\left[(\sigma_{11} - \sigma_{22})^2 + (\sigma_{22} - \sigma_{33})^2 + (\sigma_{33} - \sigma_{11})^2 + 6(\sigma_{12}^2 - \sigma_{13}^2 - \sigma_{23}^2)\right]} \tag{7.20}$$

For 2D plane stress problems, it is simplified as

$$\sigma_{vm} = \sqrt{\frac{1}{2}\left[(\sigma_{11} - \sigma_{22})^2 + \sigma_{22}^2 + \sigma_{11}^2 + 6\sigma_{12}^2\right]} \tag{7.21}$$

We can now write the following snippet to generate a contour plot of the von Mises stress:

```
# Create a mesh grid for a rectangular domain in x-y plane:
Mises=sqrt(((σ_rrxy-σ_θθxy)**2 + σ_θθxy**2 + σ_rrxy**2 + 6*σ_rθxy**2)/2)

dX = 3.0;     dY = 2.0                                # plot dX × dY area
x_val = np.linspace(0.001, dX, 500)
y_val = np.linspace(0, dY, 500)
X, Y = np.meshgrid(x_val, y_val)

np_Mises = lambdify((x,y), Mises, 'numpy')              # convert to numpy
Z = np_Mises(X, Y)                      # compute the stress over the mesh grid

gr.contour_plot(X,Y, Z, x_lbl='x-axis',y_lbl='y-axis',
                title='Mises',lvls=80, flip='y', ioff=True);
plt.savefig('imagesSMp/quadSpaceMises.png', dpi=500)       # save to file
```

```
Minimum value in the domain: 0.604194
Maximum value in the domain: 0.866976
```

A contour plot of the von Mises stress σ_{vm} within the quarter-space is shown in Fig. 7.11.

The contour lines appear as straight radial rays, as shown in Fig. 7.11. This occurs because the von Mises stress is independent of the radius r. The contour lines correspond to $\theta =$ constants, which is characteristic of a typical radial stress field.

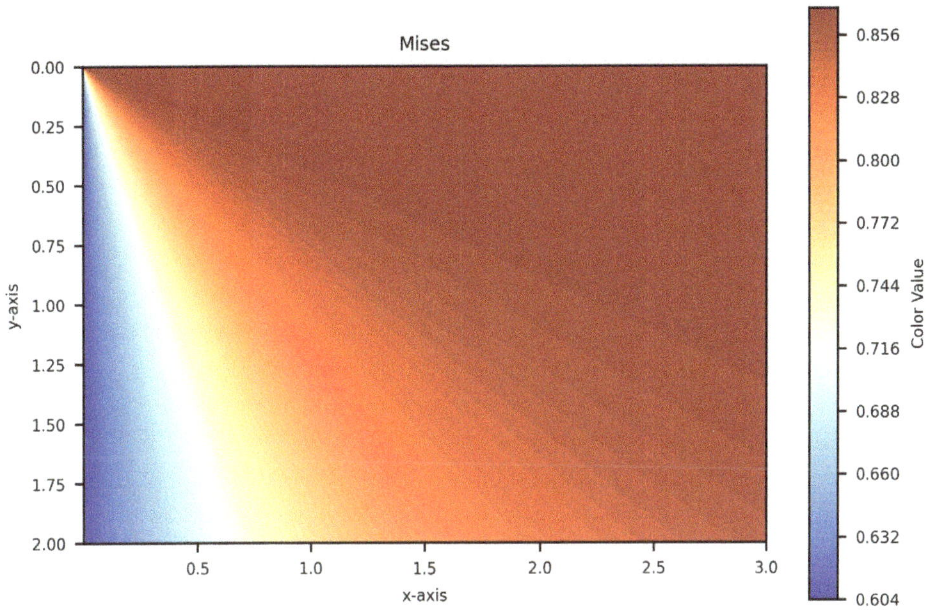

Figure 7.11. Contour plot of the von Mises stress in the quarter-space subjected to a unit pressure on the top surface.

It is also evident that the largest stress occurs on the top surface where the pressure is applied. Readers may plot the contour curves for other stress components using the same code.

7.4 Review questions

1. Consider a wedge subjected to uniformly distributed shear forces q on its two edges, as shown in Fig. 7.12(a).

   (a) Derive the formulas for the stresses.
   (b) Plot the stress distributions with respect to θ for $q = 1$.
   (c) Discuss the results obtained.

   [Hint: Use Eq. (7.18)]

2. Consider a quarter-space subjected to a point force P at its tip, as shown in Fig. 7.12(b).

   (a) Derive the formulas for the stresses.
   (b) Plot the stress distributions with respect to θ for $P = 1$, $r = 1$.
   (c) Discuss the results obtained.

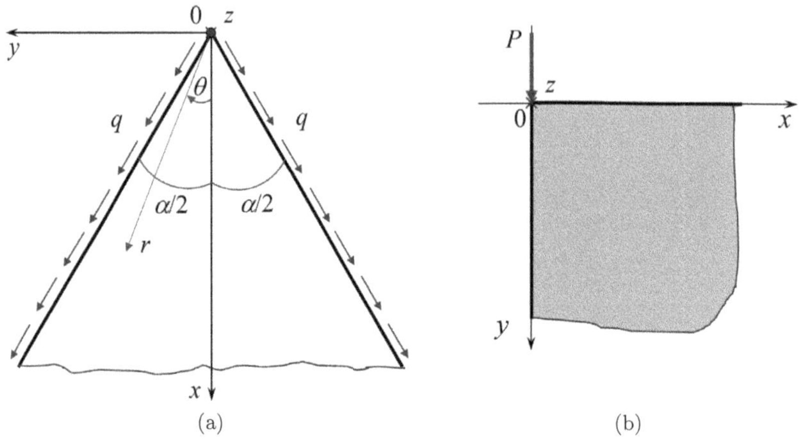

Figure 7.12. (a) A wedge subjected to uniformly distributed shear force. (b) A quarter-space subjected to a point force.

7.5 Remarks

This chapter discusses the solution process for 2D wedges subjected to various types of loads, including point forces, point moments, and distributed pressure. The analysis covers wedges of different sizes, including quarter-space and half-space configurations. Stress methods are employed to compute the stresses in these wedges, and a set of Python codes is provided for implementation.

The following chapter will focus on half-space problems, using the stress method to provide a complete set of solutions, including displacements, stresses, and strains.

References

[1] G. R. Liu, *Mechanics of Materials: Formulations and Solutions with Python*, World Scientific, New Jersey, 2024.

[2] G. R. Liu and S. S. Quek, *The Finite Element Method: A Practical Course*, Butterworth-Heinemann, New York, 2013.

[3] G. R. Liu, *Vector Calculus: Formulation, Applications, and Python Codes*, World Scientific, New Jersey, 2025.

Chapter 8

Half-Space Subjected to Surface Forces

Let us import the necessary external modules or dependencies for later use in this chapter:

```
# Place cursor in this cell, and press Ctrl+Enter to import dependencies.
import sys                                  # for accessing the computer system
sys.path.append('../grbin/')                  # Add in the path to your system

from commonImports import *              # Import dependencies from '../grbin/'
import grcodes as gr                         # Import the module of the author
importlib.reload(gr)                       # When grcodes is modified, reload it

from continuum_mechanics import vector
from continuum_mechanics.solids import sym_grad, strain_stress
init_printing(use_unicode=True)              # For latex-like quality printing
np.set_printoptions(precision=4,suppress=True)     # Digits in print-outs
```

8.1 Setting of the problem

Consider a two-dimensional (2D) half-space that is free from body forces. We assume the problem is under a plane stress condition: the domain is infinite in the x–y plane, but with unit thickness in the z-direction. It is subjected to a vertical force P, as shown in Fig. 8.1. The force is applied at point $(0, 0)$ in the x–y plane, viewed along the z-direction. Along the unit thickness, it is uniformly distributed. Therefore, the dimension of the force should be force/length. With this setting, we aim to find the stress and displacement distribution in the half-space as functions of x and y.

Note that the half-space is, in fact, a special case of the wedge discussed in the previous chapter, but with $\alpha = \pi/2$. Hence, readers can obtain the

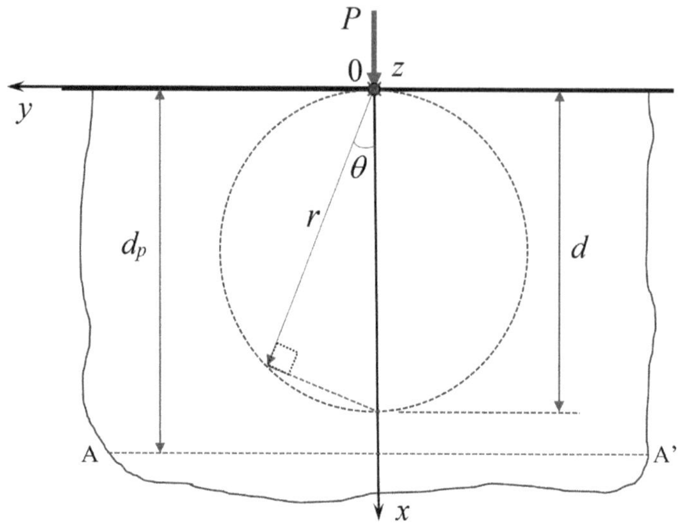

Figure 8.1. Schematic view of a 2D half-space subjected to a vertical force P.

stress solutions using the formulas derived in the previous section. We discuss it further in this chapter because it has some special properties in both formulas and solutions, and it also has wide applications. In addition, we derive displacement solutions, which provide an ideal example of how the displacement field can be obtained for such problems.

8.2 Construction of the stress function

We can perform a similar dimensional analysis and procedure as we did for the wedges and then form the stress function for the half-space problem. Here, we use a shortcut. From Eq. (7.5), we see that the expression contains two parts: symmetric and antisymmetric components (with respect to θ). For the half-space problem with a vertical load, the problem is symmetric. Therefore, we omit the term with $\theta \cos\theta$, which is antisymmetric, and write the stress function directly:

$$\phi(r, \theta) = c_0 r\, \theta \, \sin\theta \tag{8.1}$$

We can use the following snippets to quickly create the above stress function, compute the stress in the half-space using it, and then verify whether its biharmonic vanishes:

```
r = sp.symbols('r', nonnegative=True)
c0 = symbols('c0')                              # integration constant
a, dp, θ, P = symbols('a, d_p, θ, P')                    # dp: depth
ϕ = sp.Function('ϕ')(r, θ)
ϕ = c0*r*θ*sp.sin(θ)                      # define the stress function

# Compute the second derivative:
ϕ2 = (diff(ϕ,(r,2)) + diff(ϕ,r)/r + diff(ϕ,(θ,2))/r/r).expand()

# Compute the fourth derivative:
ϕ4 = (diff(diff(ϕ2,r),r)+diff(ϕ2,r)/r+diff(diff(ϕ2,θ),θ)/r/r).expand()
ϕ4                                                # this should be zero
```

As seen, the compatibility condition, or the biharmonic equation, is satisfied by the assumed ϕ. We can now compute the stresses using Eq. (7.2).

8.3 General solution for stresses

```
σ_rr = (diff(ϕ,(θ,2))/r/r+diff(ϕ,r)/r).expand()
display(Math(f" \\text{{General solution for $σ_{{rr}}$}}=\
                                        {latex(σ_rr)}"))
σ_θθ = diff(ϕ,(r,2)).expand()
display(Math(f" \\text{{General solution for $σ_{{θθ}}$}}=\
                                        {latex(σ_θθ)}"))
σ_rθ = (-diff(diff(ϕ,θ)/r,r)).expand()
display(Math(f" \\text{{General solution for $σ_{{rθ}}$}}=\
                                        {latex(σ_rθ)}"))
```

General solution for $\sigma_{rr} = \dfrac{2c_0 \cos(\theta)}{r}$
General solution for $\sigma_{\theta\theta} = 0$
General solution for $\sigma_{r\theta} = 0$

8.3.1 *Stress solution*

It is clear that, for this problem, the only nonzero stress component in the entire domain is σ_{rr}. We write the general solution for these stresses in the following equation for future reference:

$$\sigma_{rr} = \frac{2c_0 \cos(\theta)}{r}$$
$$\sigma_{\theta\theta} = 0; \quad \sigma_{r\theta} = 0 \tag{8.2}$$

This general solution for stress σ_{rr} has only one constant, c_0.

8.3.2 *Stress boundary conditions (SBCs)*

Since the surface of the half-space is free from stresses (except at the origin), we have the following SBCs:

$$\sigma_{\theta\theta} = 0 \quad \text{and} \quad \sigma_{r\theta} = 0 \quad \text{at } \theta = \pm\pi/2 \tag{8.3}$$

Clearly, these SBCs are already satisfied by Eq. (8.2) regardless of the undetermined constant, except at the singular point $r = 0$. We shall use the integral equilibrium condition or the force boundary conditions (FBCs) to determine c_0.

8.3.3 *Force boundary conditions*

To avoid the singular point, we consider the total resultant force from the stresses acting on a cylindrical surface of nonzero radius r in the half-space. This force must balance the applied force P, as shown in Fig. 8.2.

The integral formula is given as follows:

$$\begin{aligned} \sum F_x = 0&: \int_{-\pi/2}^{\pi/2} \sigma_{rr} \cos\theta r d\theta + P = 0 \\ \sum F_y = 0&: \int_{-\pi/2}^{\pi/2} \sigma_{rr} \sin\theta r d\theta = 0 \end{aligned} \tag{8.4}$$

The first integral in Eq. (8.4) can be evaluated using the following snippet. Since the integrand is an even function of θ, we integrate it over $[0, \pi/2]$ and then double the result:

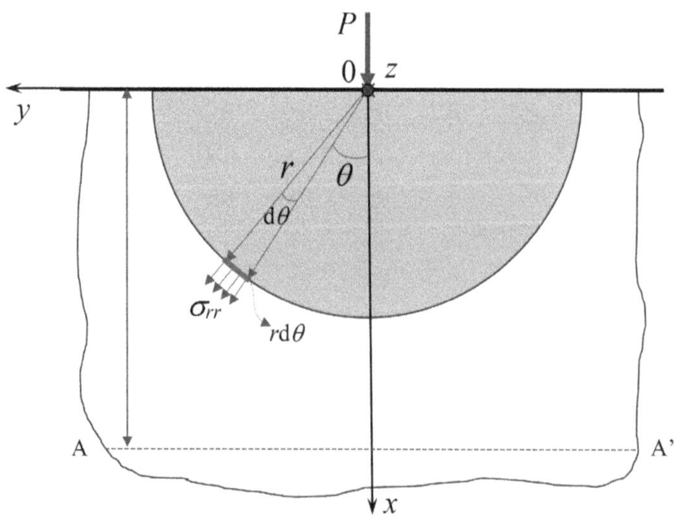

Figure 8.2. Integration of the stress σ_{rr} along a cylindrical surface of radius r in a 2D half-space, balanced by the vertical force P.

```
eqn = sp.integrate(2*σ_rr*sp.cos(θ)*r, (θ, 0, pi/2)) + P     # first Eq.
display(Math(f" \\text{{ Equilibrium equation to solve for $c_0$: }}\
                                              {latex(eqn)}=0"))

# Solve eqn for constant $c_0$:
cs = sp.solve(eqn, c0)[0]
display(Math(f" \\text{{ Constant found: $c_{{0}}$}}={latex(cs)}"))
```

Equilibrium equation to solve for c_0: $P + \pi c_0 = 0$

Constant found: $c_0 = -\frac{P}{\pi}$

We obtained c_0. The second equation in Eq. (8.4) clearly holds because the integrand is an odd function of θ, and the integral domain is symmetric. Nevertheless, let us confirm it:

```
eqn = sp.integrate(σ_rr*sp.sin(θ)*r, (θ, -pi/2, pi/2)) #     second Eq.
display(Math(f" \\text{{ Does the second Eq. hold? }} {latex(eqn==0)}"))
```

Does the "second Eq." hold? True

Therefore, all the integral equilibrium conditions are satisfied.

8.3.4 *Final stress solution*

Substituting $c_0 = -P/\pi$ back into the general solution for the stress σ_{rr}, we obtain the final stress solution:

```
σ_rr = σ_rr.subs({c0:-P/sp.pi})
display(Math(f" \\text{{Final solution for stresses: $σ_{{rr}}$}}=\
                                              {latex(σ_rr)}"))
```

Final solution for stresses: $\sigma_{rr} = -\frac{2P\cos(\theta)}{\pi r}$.

This is the same as the equation for the wedge, obtained by setting $\beta = 0$ and $\alpha = \pi$. It exhibits an r^{-1} singularity.

For a given depth ($x = d$), we can define a circular path, where $r = d\cos\theta$, represented by the dashed line in Fig. 8.1. The normal stress σ_{rr} along this circle is given as follows:

```
# on the circle where r = d cosθ:
σ_rrd = σ_rr.subs({r:dp*cos(θ)})
display(Math(f" \\text{{ On the circle where $r=d\cosθ$, $σ_{{rr}}$}}=\
                                              {latex(σ_rrd)}"))
```

On the circle where $r = d\cos\theta$, $\sigma_{rr} = -\frac{2P}{\pi d_p}$

It remains constant.

An interesting result is that along the circle with diameter d, where $r = d\cos\theta$, the normal stress σ_{rr} is constant.

8.3.5 *Stress solution in Cartesian coordinates*

We now switch to Cartesian coordinates for further analysis, requiring only an inverse coordinate transformation for stresses (a second-order tensor). Using the transformation matrix $\mathbf{T}$ and Eq. (2.5), the stresses in Cartesian coordinates can be computed with the following snippet:

```
T = Matrix([[ sp.cos(θ), sp.sin(θ)],                # T: Cartesian to Polar
            [-sp.sin(θ), sp.cos(θ)]])

σ_polar = Matrix([[σ_rr, 0],                        # stress tensor in Polar
                  [ 0,   0]])

σ_Carte_rt = sp.simplify(T.T@σ_polar@T)             # inverse transformation
display(Math(f" \\text{{Stress tensor in $(x,y)$ \
                    written in $(r,θ)$}}: {latex(σ_Carte_rt)}"))
```

Stress tensor in (x, y) written in (r, θ): $\left[\begin{matrix}- \frac{2 P \cos^{3}{\left(\theta \right)}}{\pi r} & - \frac{2 P \sin{\left(\theta \right)} \cos^{2}{\left(\theta \right)}}{\pi r}\\- \frac{2 P \sin{\left(\theta \right)} \cos^{2}{\left(\theta \right)}}{\pi r} & - \frac{2 P \sin^{2}{\left(\theta \right)} \cos{\left(\theta \right)}}{\pi r}\end{matrix}\right]$

The stress tensor is still expressed in polar coordinates. Converting it to Cartesian coordinates is straightforward and requires only a coordinate substitution:

```
x, y = sp.symbols('x, y', real=True)          # define Carte. coordinates
dic_rt = {r:sp.sqrt(x**2+y**2), θ:sp.atan(y/x)}              # substitution

σ_Carte_xy = sp.simplify(σ_Carte_rt.subs(dic_rt))
display(Math(f" \\text{{Stress in $(x,y)$ \
                   written in $(x,y)$}}={latex(σ_Carte_xy)}"))
```

Stress in (x, y) written in $(x, y) = \left[\begin{matrix}- \frac{2 P x^{2} \left|{x}\right|}{\pi \left(x^{2} + y^{2}\right)^{2}} & - \frac{2 P x y \left|{x}\right|}{\pi \left(x^{2} + y^{2}\right)^{2}}\\- \frac{2 P x y \left|{x}\right|}{\pi \left(x^{2} + y^{2}\right)^{2}} & - \frac{2 P y^{2} \left|{x}\right|}{\pi \left(x^{2} + y^{2}\right)^{2}}\end{matrix}\right]$

```
# Stress tensor at depth a in the Cartesian coordinates:
a = sp.symbols('a', positive=True)                        # at depth x = a
σ_Carte_xy.subs(x,a)
```

$$\begin{bmatrix} -\frac{2Pa^3}{\pi(a^2+y^2)^2} & -\frac{2Pa^2y}{\pi(a^2+y^2)^2} \\ -\frac{2Pa^2y}{\pi(a^2+y^2)^2} & -\frac{2Pay^2}{\pi(a^2+y^2)^2} \end{bmatrix}$$

8.3.6 *Plots of the stress solution*

We can now plot the distribution of the stress components in the x–y plane of the half-space. The snippet is as follows:

```
# SymPy expressions to Numpy functions:

dp_ = [0.1, 0.4, 0.7, 1.0]                    # define cases for the depth
nf = len(dp_)                                          # number of figures

f_ij=[[1, 1, 1, "xl", "yl"]]                         # define figure grids
if nf == 2: f_ij=[[1, 2, 1, "xl", "yl"],[1, 2, 2, "xl", "no"]]
if nf >= 3: f_ij=[[2, 2, 1, "no", "yl"],[2, 2, 2, "no", "no"],
                  [2, 2, 3, "xl", "yl"],[2, 2, 4, "xl", "no"]]
#if nf >=3: f_ij=[[4, 1, 1, "no", "yl"],[4, 1, 2, "no", "yl"], #vertical
#                  [4, 1, 3, "no", "yl"],[4, 1, 4, "xl", "yl"]]

plt.ioff()
#plt.ion()                         # uncomment this to display the plots
#plt.rcParams.update({'font.size': 5})             # settings for plots
fig_s = plt.figure()#figsize=(4,2.5))

for k in range(nf):
    fij=f_ij[k]
    func1=sp.lambdify(y, σ_Carte_xy[0,0].subs({P:1, x:dp_[k]}),'numpy')
    func2=sp.lambdify(y, σ_Carte_xy[1,1].subs({P:1, x:dp_[k]}),'numpy')
    func3=sp.lambdify(y, σ_Carte_xy[0,1].subs({P:1, x:dp_[k]}),'numpy')

    labels=[r'$σ_{xx}$, depth='+f'{dp_[k]:.1f}',
            r'$σ_{yy}$, depth='+f'{dp_[k]:.1f}',
            r'$σ_{xy}$, depth='+f'{dp_[k]:.1f}']

    y_axis = np.linspace(-2*dp_[-1], 2*dp_[-1], 400)   # y-axis in plot
    y_data = [func1(y_axis), func2(y_axis), func3(y_axis)]  # fs values

    gr.plot_fij(fig_s, y_axis, y_data, labels, '$y$', '$σ_{ij}$', fij)
plt.savefig('imagesSMp/stressInHalfSpace.png',dpi=500,bbox_inches='tight')
#plt.show()                        # uncomment this to display the plots
```

Figure 8.3 shows the distribution of stress σ_{ij} with respect to y at various depth levels. The concentrated force is set to unity, $P = 1$.

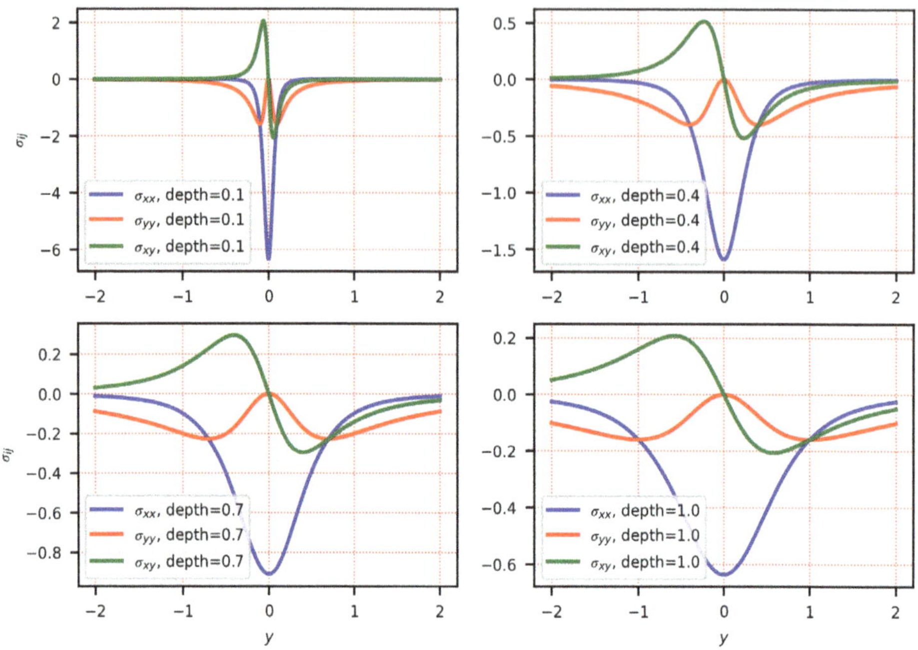

Figure 8.3. Distribution of stress σ_{ij} along y over the half-space at various depth levels.

8.4 Displacements in half-space

8.4.1 *General procedure*

This section provides a procedure and code to derive the displacement solution using the stress formulas obtained in the previous section. Detailed procedures, results, and analysis are provided for this half-space problem.

The strategy is as follows:

1. Use the constitutive equations to express strains in terms of stresses (only one stress component, σ_{rr}, is nonzero for this problem).
2. Use the strain–displacement relations to express the derivatives of the displacement components in terms of the known stresses.
3. Integrate the stress–displacement relations to find the solution for each displacement component, ensuring strain compatibility.

We now write the code by first defining the necessary variables and functions:

```
E, v = sp.symbols('E, v ', nonnegative=True)        # material constants

r = sp.symbols('r', nonnegative=True)               # polar coordinate r
θ = sp.symbols('θ')                                 # polar coordinate θ
P = sp.symbols('P')                                           # force

f = sp.Function('f')(θ)               # integration constants (functions)
g = sp.Function('g')(r)
```

Define the stress formula in SymPy obtained earlier:

```
σ_rr = -2*P*sp.cos(θ)/(sp.pi*r)
display(Math(f" \\text{{ Given stress $σ_{{rr}}$}}={latex(σ_rr)}"))
```

Given stress $\sigma_{rr} = -\dfrac{2P\cos(\theta)}{\pi r}$

8.4.2 *Stress–displacement relations*

To obtain the stress–displacement relations, we use the constitutive equation in polar coordinates, which has the same form as in Cartesian coordinates because it depends only on material properties. For a plane stress problem, using the constitutive equation Eq. (5.74) in Ref. [1], we have

$$\begin{aligned}
\varepsilon_{rr} &= \frac{1}{E}(\sigma_{rr} - \nu\sigma_{\theta\theta}) \\
\varepsilon_{\theta\theta} &= \frac{1}{E}(\sigma_{\theta\theta} - \nu\sigma_{rr}) \\
\varepsilon_{r\theta} &= \frac{1}{2G}\sigma_{r\theta} = \frac{1+\nu}{E}\sigma_{r\theta}
\end{aligned} \tag{8.5}$$

where E is Young's modulus and ν is Poisson's ratio. For our problem, $\sigma_{\theta\theta} = \sigma_{r\theta} = 0$, and using Eq. (2.28), Eq. (8.5) becomes

$$\begin{aligned}
\frac{\partial u_r}{\partial r} &= \frac{1}{E}\sigma_{rr} \\
\frac{u_r}{r} + \frac{\partial u_\theta}{r\partial\theta} &= -\frac{\nu}{E}\sigma_{rr} \\
\frac{\partial u_r}{r\partial\theta} + \frac{\partial u_\theta}{\partial r} - \frac{u_\theta}{r} &= 0
\end{aligned} \tag{8.6}$$

Since σ_{rr} is known, the above equations can be solved for the displacements u_r and u_θ sequentially by performing indefinite integrations.

8.4.3 *General form for component u_r*

Using the first equation in Eq. (8.6) and the given stress σ_{rr}, we obtain the displacement u_r by integration with respect to r:

```
ur_with_f = sp.integrate(σ_rr/E, r) + f  # f: integration function of θ
ur_with_f                                # should be f - 2 P Log(r) cos(θ)/(πE)
```

$$f(\theta) - \frac{2P\log(r)\cos(\theta)}{\pi E}$$

Since the expression for u_r is obtained by integration with respect to r, it includes an arbitrary function of θ, $f(\theta)$. We therefore need to determine $f(\theta)$ before u_r can be fully specified.

8.4.4 *General form for component u_θ*

Using the second equation in Eq. (8.6), the known stress σ_{rr}, and the expression for u_r including $f(\theta)$, we obtain the displacement u_θ by integration with respect to θ:

```
uθ_with_g=(integrate(-r*v*σ_rr/E - ur_with_f, (θ)).doit() + g).expand()
uθ_with_g
```

$$g(r) - \int f(\theta)d\theta + \frac{2P\nu\sin(\theta)}{\pi E} + \frac{2P\log(r)\sin(\theta)}{\pi E}$$

Since the expression for u_θ is obtained by integration with respect to θ, it includes an arbitrary function of r, $g(r)$. We therefore need to determine $g(r)$ before u_θ can be fully specified.

8.4.5 *Techniques for equation splitting*

Substituting u and v into the third equation in Eq. (8.6) gives as follows:

```
expr3 = (diff(ur_with_f, θ)+r*diff(uθ_with_g, r)-uθ_with_g).expand()
expr3
```

$$r\frac{d}{dr}g(r) - g(r) + \frac{d}{d\theta}f(\theta) + \int f(\theta)d\theta - \frac{2P\nu\sin(\theta)}{\pi E} + \frac{2P\sin(\theta)}{\pi E}$$

This expression must be zero. It is rearranged as

$$r\frac{d}{dr}g(r) - g(r) = -\frac{d}{d\theta}f(\theta) - \int f(\theta)\,d\theta + \frac{2P(\nu-1)\sin(\theta)}{\pi E} \tag{8.7}$$

The left-hand side of the above equation is a function of r only, while the right-hand side is a function of θ only. This equality can hold only if both sides are equal to the same constant. Let us denote this common constant as C. Hence, Eq. (8.7) gives two separate equations:

$$\begin{aligned} r\frac{d}{dr}g(r) - g(r) &= C \\ -\frac{d}{d\theta}f(\theta) - \int f(\theta)d\theta + \frac{2P(\nu-1)\sin(\theta)}{\pi E} &= C \end{aligned} \tag{8.8}$$

These two equations can be solved separately. We need to find $f(\theta)$ and $g(r)$ in `expr3` to form these two equations. Let us use the following code to find these terms:

```
expr3.args
```

$$\left(-g(r),\ r\frac{d}{dr}g(r),\ \frac{2P\sin(\theta)}{\pi E},\ -\frac{2P\nu\sin(\theta)}{\pi E},\ \frac{d}{d\theta}f(\theta),\ \int f(\theta)\,d\theta\right)$$

Now, the first equation in Eq. (8.8) is formed as follows:

```
C = symbols('C')                                # the common constant
Fr = expr3.args[0] + expr3.args[1] - C           # function of only r
Fr
```

$$-C + r\frac{d}{dr}g(r) - g(r)$$

This is an expression in only r and must be zero. It can be solved for $g(r)$:

```
# Solve it for g(r):
g_ = sp.dsolve(Fr,g)
g_.args                                          # Check the terms in g_
```

$$(g(r), -C + C_1 r)$$

Let us redefine the expression for $g(r)$, so that it does not interfere with the others:

```
cg = symbols('c_g')                  # define the constant for function g
g_r = -C + cg*r
```

Next, the function of only θ is formed as follows:

```
Fθ = -expr3.args[2] - expr3.args[3] -expr3.args[4] - expr3.args[5] - C
Fθ
```

$$-C - \frac{d}{d\theta}f(\theta) - \int f(\theta)\, d\theta + \frac{2P\nu\sin(\theta)}{\pi E} - \frac{2P\sin(\theta)}{\pi E}$$

Since this equation involves both differentiation and integration of the unknown function $f(\theta)$, we convert it into a differential equation. This is done by differentiating and then solving it for $f(\theta)$:

```
Fθ = diff(Fθ, θ)
Fθ
```

$$-f(\theta) - \frac{d^2}{d\theta^2}f(\theta) + \frac{2P\nu\cos(\theta)}{\pi E} - \frac{2P\cos(\theta)}{\pi E}$$

```
# Solve it for f(θ):
f_ = sp.dsolve(Fθ,f).expand()
f_.args
```

$$\left(f(\theta),\ C_1\sin(\theta) + C_2\cos(\theta) + \frac{P\nu\theta\sin(\theta)}{\pi E} - \frac{P\theta\sin(\theta)}{\pi E}\right)$$

Let us rewrite the expression for $f(\theta)$ to avoid it interfering with other variables:

```
cf1, cf2 = symbols('c_f1, c_f2')
fθ = cf1*sp.sin(θ) + cf2*sp.cos(θ) + P*(v-1)/(E*sp.pi)*θ*sp.sin(θ)
fθ
```

$$c_{f1}\sin(\theta) + c_{f2}\cos(\theta) + \frac{P\theta(\nu-1)\sin(\theta)}{\pi E}$$

Using the second equation in Eq. (8.8), the term $\int f(\theta)\, d\theta$ can be found as follows:

```
f_int = - C - diff(fθ, θ) + 2*P*(v-1)/(E*sp.pi)*sp.sin(θ)
f_int
```

$$-C - c_{f1}\cos(\theta) + c_{f2}\sin(\theta) - \frac{P\theta(\nu-1)\cos(\theta)}{\pi E} + \frac{P(\nu-1)\sin(\theta)}{\pi E}$$

8.4.6 *General solutions for displacements*

Let us first find the terms in expression `ur_with_f`:

```
ur_with_f.args                          # check what is currently in ur_with_f
```

$$\left(-\frac{2P\log(r)\cos(\theta)}{\pi E},\ f(\theta)\right)$$

Using **ur_with_f** and **fθ** obtained earlier, the displacement u_r can be written as follows:

```
ur = ur_with_f.args[0] + fθ              # form the final expression for ur
ur.simplify().expand()
```

$$c_{f1}\sin(\theta) + c_{f2}\cos(\theta) + \frac{P\nu\theta\sin(\theta)}{\pi E} - \frac{P\theta\sin(\theta)}{\pi E} - \frac{2P\log(r)\cos(\theta)}{\pi E}$$

```
uθ_with_g.args                        # check what is currently in uθ_with_g
```

$$\left(-\int f(\theta)\,d\theta,\ \frac{2P\nu\sin(\theta)}{\pi E},\ \frac{2P\log(r)\sin(\theta)}{\pi E},\ g(r)\right)$$

```
# Form uθ:
uθ = uθ_with_g.args[1] + uθ_with_g.args[2] + g_r - f_int     # form uθ
#display(Math(f" \\text{{$u_θ$}}= {latex(uθ.expand())}"))   #Lengthy
```

Finally, by simply regrouping, we obtain the expressions for the displacements:

$$\begin{aligned} u_r &= c_{f1}\sin(\theta) + c_{f2}\cos(\theta) + \frac{P(\nu-1)}{\pi E}\theta\sin(\theta) \\ &\quad - \frac{2P\log(r)\cos(\theta)}{\pi E} \\ u_\theta &= c_{f1}\cos(\theta) - c_{f2}\sin(\theta) + c_g r + \frac{P(\nu-1)}{\pi E}\theta\cos(\theta) \\ &\quad + \frac{P(\nu+1)}{\pi E}\sin(\theta) + \frac{2P}{\pi E}\log(r)\sin(\theta) \end{aligned} \tag{8.9}$$

There are a total of three unknown constants: c_{f1}, c_{f2}, and c_g. These need to be determined using displacement boundary conditions (DBCs).

8.4.7 *Final solution for displacements*

Assuming $u_\theta = 0$ at $\theta = 0$, meaning that the horizontal displacement along the vertical x-axis is zero. This condition leads to $c_{f1} = 0$, $c_g = 0$.

Next, at $\theta = 0$, we have

$$u_r\Big|_{\theta=0} = -\frac{2P\log(r)}{\pi E} + c_{f2} \tag{8.10}$$

If we set $\theta = 0$, $x = d_0$ as the reference point for measuring the displacement u_r: $u_r\big|_{\theta=0,x=d_0} = 0$, we obtain

$$c_{f2} = \frac{2P\log(d_0)}{\pi E} \tag{8.11}$$

We have now determined all the constants using these displacement boundary conditions. Let us substitute these constants into the general solution of displacements:

```
d0 = symbols("d_0")
ur = ur.subs({cf1:0, cf2:2*P/sp.pi/E*sp.log(d0), cg:0})
uθ = (uθ.subs({cf1:0, cf2:2*P/sp.pi/E*sp.log(d0), cg:0}))
#print(f"ur = {ur}"); print(f"ur = {uθ}")                    # Lengthy
```

The above cell produces the final expressions for the displacements, which are quite lengthy. By regrouping the terms, we obtain the final, more readable expressions for the displacements:

$$\begin{aligned} u_r &= \frac{P}{\pi E}\left((\nu-1)\theta\sin\theta + 2\log\left(\frac{d_0}{r}\right)\cos\theta\right) \\ u_\theta &= \frac{P}{\pi E}\left((\nu-1)\theta\cos\theta + (\nu+1)\sin\theta + 2\log\left(\frac{r}{d_0}\right)\sin\theta\right) \end{aligned} \tag{8.12}$$

It is seen that u_r is symmetric and u_θ is antisymmetric with respect to θ.

The following cell defines these displacements in Python for later use without rerunning the code above:

```
# redefine the solution for future use:
ur = P/(sp.pi*E)*( θ*(v - 1)*sp.sin(θ) + 2*sp.log(d0)*sp.cos(θ)\
  - 2*sp.log(r)*sp.cos(θ) )

vθ = P/(sp.pi*E)*( θ*(v - 1)*sp.cos(θ) + (v + 1)*sp.sin(θ) - \
    2*sp.log(d0)*sp.sin(θ) + 2*sp.log(r)*sp.sin(θ) )
```

8.4.8 *Plots of u_r and u_θ*

The following code plots u_r and u_θ, using `gr.plot_fij()`:

```
# SymPy expressions to Numpy functions:

dp_ = [0.1, 0.4, 0.7, 1.0]                  # define cases for the depth
d0_= 0.1                                      # reference depth for ur
nf = len(dp_)                                     # number of figures

f_ij=[[1, 1, 1, "xl", "yl"]]                          # define figure grids
if nf == 2: f_ij=[[1, 2, 1, "xl", "yl"],[1, 2, 2, "xl", "no"]]
if nf >= 3: f_ij=[[2, 2, 1, "no", "yl"],[2, 2, 2, "no", "no"],
                  [2, 2, 3, "xl", "yl"],[2, 2, 4, "xl", "no"]]
```

```
plt.ioff()
#plt.ion()                          # uncomment this to display the plots
#plt.rcParams.update({'font.size': 5})           # settings for plots
fig_s = plt.figure()#figsize=(4,2.5))

for k in range(nf):
    fij=f_ij[k]

    func1=sp.lambdify((r,θ), ur.subs({P:1,E:1, v:.3,d0:d0_}),'numpy')
    func2=sp.lambdify((r,θ), vθ.subs({P:1,E:1, v:.3,d0:d0_}),'numpy')

    labels=[r'$u_{r}$, depth='+f'{dp_[k]:.1f}',
            r'$u_{θ}$, depth='+f'{dp_[k]:.1f}']

    y_data = np.linspace(-2*dp_[-1], 2*dp_[-1], 400)   # y-axis in plot
    ri = np.sqrt(y_data**2 + dp_[k]**2)          # compute r along y-axis
    ti = np.arctan(y_data/dp_[k])                # compute θ along y-axis

    f_data = [func1(ri,ti), func2(ri,ti)]                    # fs values

    gr.plot_fij(fig_s, y_data, f_data, labels, '$y$', '$u_{i}$', fij)

plt.savefig('imagesSMp/dispInHalfSpace_rt.png',dpi=500,bbox_inches='tight')
#plt.show()                         # uncomment this to display the plots
```

The displacement components u_r and u_θ are plotted in Fig. 8.4 with respect to y for various depths. The concentrated force is set to $P = 1$, $E = 1$, $\nu = 0.3$, and $d_0 = 0.1$.

8.4.9 *Plots of u_x and u_y*

It is often more intuitive to plot the displacements u_x and u_y in Cartesian coordinates. The following code does this, again using `gr.plot_fij()`. We simply need to include the coordinate transformation for the displacement vector. The formula is given in Eq. (2.8):

```
# SymPy expressions to Numpy functions:

x_ = [0.1, 0.4, 0.7, 1.0]                  # define cases for the depth
d0_= 0.1                                      # reference depth for ur
nf = len(x_)                                       # number of figures

f_ij=[[1, 1, 1, "xl", "yl"]]                       # define figure grids
if nf == 2: f_ij=[[1, 2, 1, "xl", "yl"],[1, 2, 2, "xl", "no"]]
if nf >= 3: f_ij=[[2, 2, 1, "no", "yl"],[2, 2, 2, "no", "no"],
                  [2, 2, 3, "xl", "yl"],[2, 2, 4, "xl", "no"]]
```

```
plt.ioff()
#plt.ion()                      # uncomment this to display the plots
#plt.rcParams.update({'font.size': 5})          # settings for plots
fig_s = plt.figure()#figsize=(4,2.5))

for k in range(nf):
    fij=f_ij[k]

    func1=sp.lambdify((r,θ), ur.subs({P:1,E:1, v:.3,d0:d0_}),'numpy')
    func2=sp.lambdify((r,θ), vθ.subs({P:1,E:1, v:.3,d0:d0_}),'numpy')

    labels=[r'$u_{x}$, depth='+f'{dp_[k]:.1f}',
            r'$u_{y}$, depth='+f'{dp_[k]:.1f}']

    y_data = np.linspace(-2*x_[-1], 2*x_[-1], 400)     # y-axis in plot
    ri = np.sqrt(y_data**2 + x_[k]**2)           # compute r along y-axis
    ti = np.arctan(y_data/x_[k])                 # compute θ along y-axis

    # perform coordinate transformation:
    ux = func1(ri,ti)*np.cos(ti) - func2(ri,ti)*np.sin(ti)
    uy = func1(ri,ti)*np.sin(ti) + func2(ri,ti)*np.cos(ti)

    f_data = [ux, uy]                                      # fs values

    gr.plot_fij(fig_s, y_data, f_data, labels, '$y$', '$u_{i}$', fij)

plt.savefig('imagesSMp/dispInHalfSpace_xy.png',dpi=500,bbox_inches='tight')
#plt.show()                     # uncomment this to display the plots
```

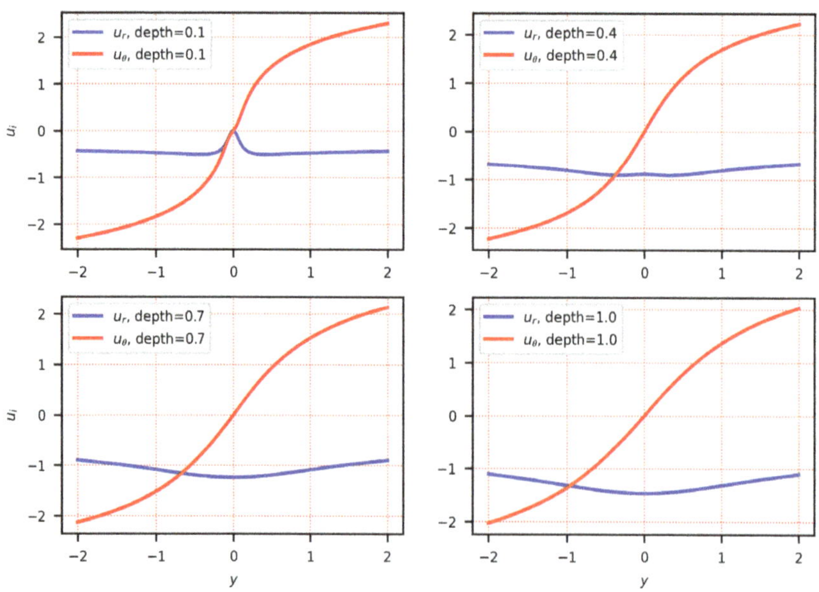

Figure 8.4. Distribution of displacement components u_r and u_θ along y over the half-space for various depths.

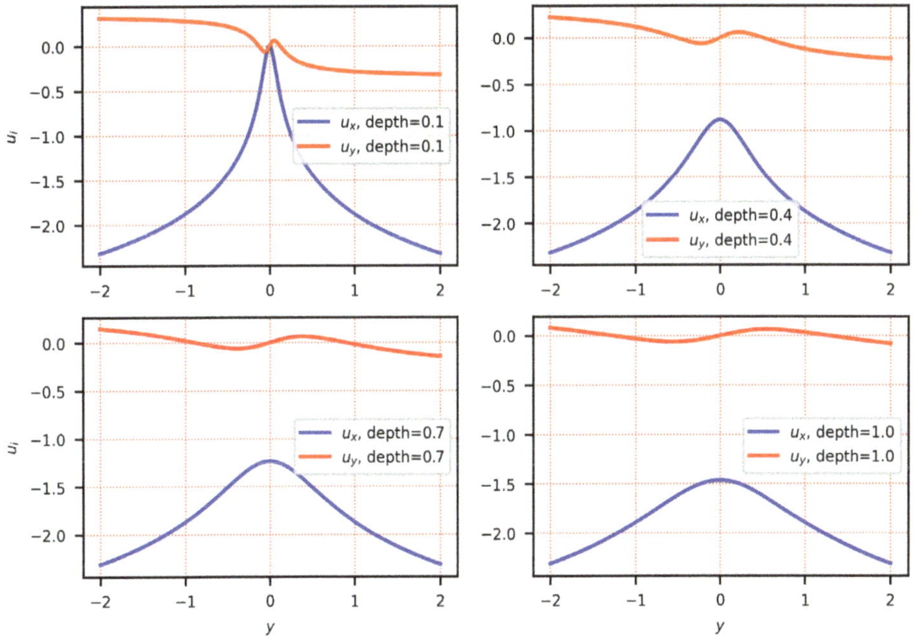

Figure 8.5. Distribution of displacement components u_x and u_y along y over the half-space for various depths.

Displacement components u_x and u_y are plotted in Fig. 8.5 with respect to y for various depths. The concentrated force is set to $P = 1$, $E = 1$, $\nu = 0.3$, and $d_0 = 0.1$.

8.4.10 *Special cases*

The displacement u_r on the surface of the half-space, where $\theta = \pm\pi/2$, is given by

```
ur.subs(θ,sp.pi/2)                              # or -sp.pi/2 # symmetric.
```

$$\frac{P(\nu-1)}{2E}$$

which is constant and negative, independent of r. This indicates that the material moves symmetrically toward the applied force P. The displacement v on the surface of the half-space is given as follows:

```
vs = vθ.subs(θ, sp.pi/2)      # =-v.subs(θ,-sp.pi/2)    # antisymmetric.
vs
```

$$\frac{P(\nu - 2\log(d_0) + 2\log(r) + 1)}{\pi E}$$

It exhibits a $\log(r)$ singularity.

8.5 Distributed loads on the half-space surface

Once a solution (stress or displacement) is known for a point load, solutions for any type of distributed load can be obtained by superposition. The point load solution serves as a Green's function, providing a fundamental solution. Solutions for other types of distributed loads can then be found by integration. If the product of the load function and the Green's function is integrable, the solution can be found analytically; otherwise, numerical integration is required. This section demonstrates the analytical procedure for displacements, which can also be used to compute stress solutions.

8.5.1 *Solution at points outside the distributed load*

Consider a distributed load applied on the half-space surface. The vertical displacement (deflection) at an observation point O on the surface can be computed using the point load solution. Let dv_s denote the deflection at point O due to an infinitesimal force $P = q\,dr$, as illustrated in Fig. 8.6.

Using the second equation in Eq. (8.12) with $\theta = \pi/2$, we have

$$dv_s = \frac{q\,dr}{\pi E}\left((\nu+1) + 2\log\left(\frac{r}{d_0}\right)\right) \tag{8.13}$$

The total deflection at point O, measured from the reference point at $x = d_0$, is

$$v_s = \frac{1}{\pi E}\int_a^{l+a} q\left((\nu+1) + 2\log\left(\frac{r}{d_0}\right)\right)dr \tag{8.14}$$

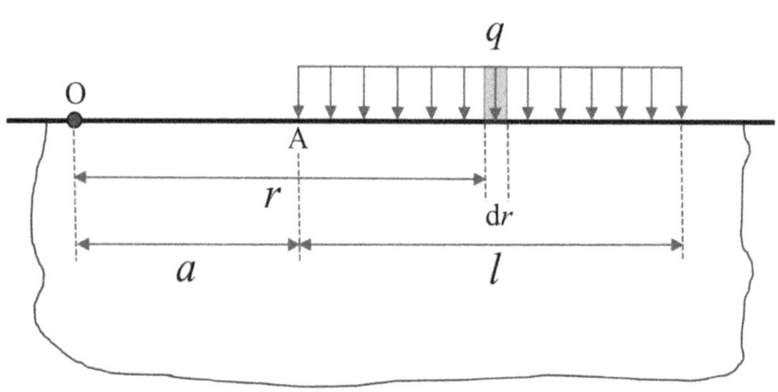

Figure 8.6. Distributed load on the surface of a half-space. The observation point is to the left of point A.

This integral can be evaluated using SymPy. For a uniformly distributed load, q is constant, and the result is given as follows:

```
# Define the integrand:
q, l, a = sp.symbols('q, l, a')
vs = q/(sp.pi*E)*(1+v - 2*sp.log(d0) + 2*sp.log(r))
display(Math(f"\\text{{ The integrand, $v_{{s}}$}}={latex(vs)}"))

# Integration (when O is at the Left side of A):
vs_O = integrate(vs,(r, a, a+l)).expand().factor(E)

gr.printM(vs_O, 'Deflection at O (in reference to that at x=d_0) by q:')
```

The integrand, $v_s = \dfrac{q(\nu - 2\log(d_0) + 2\log(r) + 1)}{\pi E}$

```
Deflection at O (in reference to that at x=d_0) by q:
```

$$\frac{q(-2a\log(a) + 2a\log(a+l) + l\nu - 2l\log(d_0) + 2l\log(a+l) - l)}{\pi E}$$

8.5.2 *Solution at points within the distributed load*

When the observation point O is located to the right of point A, where the distributed load begins, as shown in Fig. 8.7, the same formula still applies, but a should be replaced by $-a$. The corresponding Python snippet is as follows:

```
# Integration (when O is at the right side of A):
vs_A = integrate(vs,(r, -a, -a+l)).expand().factor(E)

gr.printM(vs_A, 'Deflection at O (in reference to that at x=d_0) by q:')
```

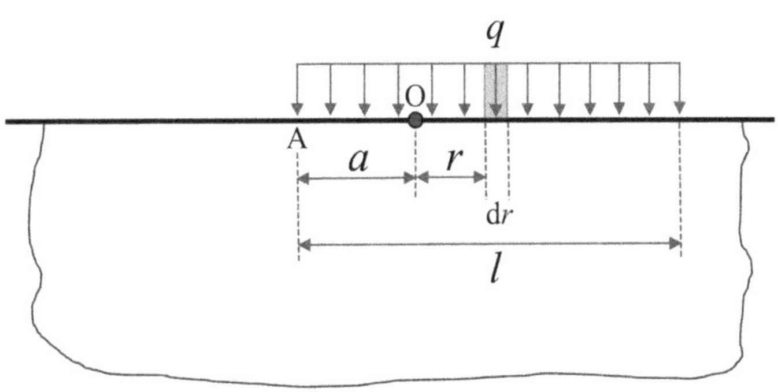

Figure 8.7. Distributed load on the surface of a half-space. The observation point is located to the right of point A.

```
Deflection at O (in reference to that at x=d_0) by q:
```

$$\frac{q(2a\log{(-a)} - 2a\log{(-a+l)} + l\nu - 2l\log{(d_0)} + 2l\log{(-a+l)} - l)}{\pi E}$$

If q is not uniform but given as a function that is analytically integrable, the solution can be obtained in the same manner using the same code. In this case, q is included as part of the integrand and integrated accordingly. If the function is not analytically integrable, numerical integration can be used. The solution can always be obtained, provided that the combined integrand is Riemann integrable [1].

8.6 Review questions

1. Consider distributed forces on a half-space. Use the stress function $\phi = Cr^2\theta$.
   - 1.1. Derive the formulas for the stresses.
   - 1.2. Plot the stresses on the surface of the half-space.
   - 1.3. Shift the origin to the right by a length l, and repeat items 1 and 2.

   [Hint: Use superposition]
2. Consider a uniformly distributed force q applied on the half-space over a finite length l.
   - 2.1. Using the solution obtained in question 1, derive the expressions for the stresses.
   - 2.2. Plot the stress distributions below the force q over a range of $2l$ at a depth of $d_p = 1$, for $q = 1$.

   [Hint: Use superposition]
3. Consider a full-space subjected to a concentrated force P.
   - 3.1. Construct the stress function.
   - 3.2. Derive the expressions for the stresses.
   - 3.3. Plot the stresses in the space for $P = 1$.
   - 3.4. Derive the formulas for the displacements.
   - 3.5. Plot the displacements in the space for $P = 1$.

8.7 Remarks

This chapter discusses the solution process for half-spaces subjected to various loads. Key points are as follows:

1. Fundamental solutions are obtained for a half-space subjected to a point force. Stress methods are employed to compute stresses, and displacements are then derived via indefinite integration.
2. Solutions for distributed forces are obtained through superposition, performed using definite integration. Python codes are provided to implement this superposition technique efficiently.

The following chapter focuses on loaded wedges, including the half-space, using the displacement method to obtain a complete set of solutions, including displacements, stresses, and strains.

Reference

[1] G. R. Liu, *Calculus: A Practical Course with Python*, World Scientific, New Jersey, 2025.

Chapter 9

Displacement Methods for Wedges

Let us import the necessary external modules or dependencies for later use in this chapter:

```
# Place cursor in this cell, and press Ctrl+Enter to import dependencies.
import sys                              # for accessing the computer system
sys.path.append('../grbin/')               # Add in the path to your system

from commonImports import *          # Import dependencies from '../grbin/'
import grcodes as gr                        # Import the module of the author
importlib.reload(gr)                    # When grcodes is modified, reload it

from continuum_mechanics import vector
from continuum_mechanics.solids import sym_grad, strain_stress
init_printing(use_unicode=True)          # For latex-like quality printing
np.set_printoptions(precision=4,suppress=True)    # Digits in print-outs
```

In Chapter 7, stress methods were employed to solve two-dimensional wedge problems subjected to various types of loads. While these methods provided stress solutions, displacement solutions were not obtained. This chapter introduces a displacement method for the same class of problems, which naturally yields both displacements and stresses. Building on the approaches developed in the previous chapters, the method is systematically implemented using SymPy, making the exposition here concise and code-focused. These codes enable the automation of the solution procedure for similar problems using the displacement method.

9.1 Wedges under a vertical point load

The problem setup is similar to that studied in Chapter 7. Consider a wedge subjected to a vertical point force P applied at its tip, as shown in Fig. 9.1.

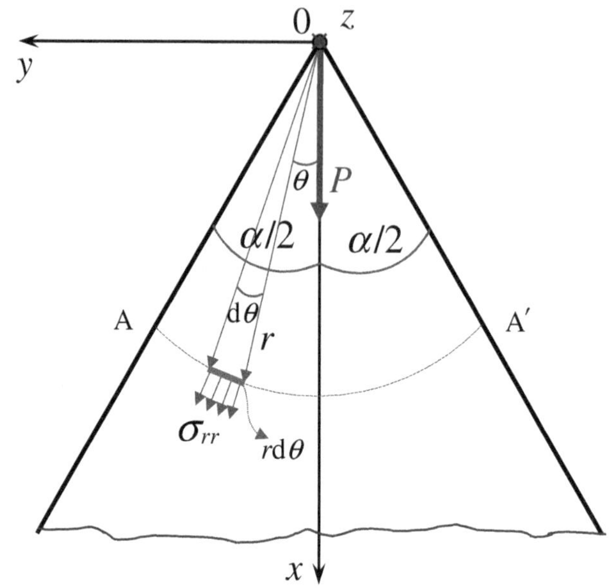

Figure 9.1. A wedge with angle α subjected to a vertical point force P at its tip.

A polar coordinate system is employed, with $\theta = 0$ aligned along the vertical axis of symmetry of the wedge.

Based on the dimensional analysis in Section 7.1.1 for wedges under a point load, the stress must scale as P/r to maintain proper units.

9.1.1 *General expressions for displacements*

In a displacement method, we assume first the expressions for the displacements. For this particular problem, the displacements can be expressed in the following general form:

$$\begin{aligned} u_r(r,\theta) &= a_0 \log(r)\cos\theta + a_1\theta\sin\theta + a_2\cos\theta \\ u_\theta(r,\theta) &= b_0\log(r)\sin\theta + b_1\theta\cos\theta + b_2\sin\theta + b_3\theta \end{aligned} \tag{9.1}$$

where a_i and b_i are arbitrary constants. The rationale for including these terms is as follows:

1. Since the radial stress σ_{rr} varies with r as $1/r$, and σ_{rr} is related to the derivative of the radial displacement u_r, the term $\log(r)$ should be included in the displacement expressions because it is the antiderivative of $1/r$.
2. The problem is symmetric with respect to θ, so u_r must also be symmetric. This means that u_r should have the same value for θ and $-\theta$, and hence

only basis terms that are even functions of θ are included. Higher-order terms are omitted for now but can be added later if needed.

3. By the same symmetry argument, the tangential displacement u_θ must be antisymmetric with respect to θ. Therefore, only basis terms that are odd functions of θ are included in the expression for u_θ in Eq. (9.1).

The following code snippet generates the displacement expressions based on these considerations:

```
# Define variables for displacement method for symmetric wedges:
E, ν = symbols('E, ν', positive=True)          # given material constants
P = symbols('P', real=True)                   # given vertical point force
r = sp.symbols('r', positive=True)              # variable for radial axis
θ = sp.symbols('θ', real=True)              # variable for tangential axis
α = sp.symbols('α')                                          # wedge angle
d0 = sp.symbols('d0', positive=True)                    # depth from tip

#ur_rbs_=[log(r)*cos(θ),θ*sin(θ),cos(θ),log(r)*sin(θ),θ*cos(θ),sin(θ)]
ur_rbs_ = [log(r)*cos(θ), θ*sin(θ), cos(θ)]
ur_rbs = sp.Matrix(ur_rbs_)

uθ_rbs_ = [log(r)*sin(θ), θ*cos(θ), sin(θ), θ]
uθ_rbs = sp.Matrix(uθ_rbs_)

# unknown constants for displacements:
ai = sp.symbols(f'a:{len(ur_rbs)}')                    # unknown constants
display(Math(f" \\text{{Unknown constants for $u$}}={latex(ai)}"))
ai = Matrix([ai[i] for i in range(len(ur_rbs))])# form coeffs in vector

bi = sp.symbols(f'b:{len(uθ_rbs)}')                    # unknown constants
display(Math(f" \\text{{Unknown constants for $v$}}={latex(bi)}"))
bi = Matrix([bi[i] for i in range(len(uθ_rbs))])          # form a vector

cs = ai.col_join(bi)                # form a single vector of constants

urg = (ai.T@Matrix(ur_rbs))[0]         # Form the general displacements
uθg = (bi.T@Matrix(uθ_rbs))[0]
display(Math(f"\\text{{Assumed displacements, $u_r(r,θ)$}}=\
                                                    {latex(urg)}"))
display(Math(f"\\hspace{{4.5cm}}\\text{{$u_θ(r,θ)$}}= {latex(uθg)}"))
```

Unknown constants for $u = (a_0, \ a_1, \ a_2)$

Unknown constants for $v = (b_0, \ b_1, \ b_2, \ b_3)$

Assumed displacements, $u_r(r, \theta) = a_0 \log(r) \cos(\theta) + a_1 \theta \sin(\theta) + a_2 \cos(\theta)$

$$u_\theta(r, \theta) = b_0 \log(r) \sin(\theta) + b_1 \theta \cos(\theta) + b_2 \sin(\theta) + b_3 \theta$$

We have constructed the displacement expressions, which involve a total of seven unknown constants. These constants must now be determined by enforcing equilibrium and stress boundary conditions. To do so, explicit stress expressions are required.

9.1.2 *Stress expressions*

Next, we compute the strain components from the displacement field and then calculate the corresponding stresses. Alternatively, the strains can also be computed directly from the stresses using the constitutive equations. The Python function to perform these computations is given in the following:

```
def strainStressFrDisp(urg, uθg, r, θ, E, ν):
    '''Compute 2D strains and stresses for given displacements urg and
       uθg in polar coordinates r and θ for material with E and ν.
       return: εvg, σvg (strain and stress in vector form)
    '''
    εrr  = urg.diff(r)
    εθθ  = urg/r + uθg.diff(θ)/r
    εrθ  = (urg.diff(θ)/r + uθg.diff(r) - uθg/r)/2
    εvg  = sp.Matrix([εrr, εθθ, εrθ])          # define strains as a vector

    # Compute stress vector using the strain vector:
    σvg  = sp.simplify(sp.Matrix(gr.strain2stress2D(E,ν,εvg,\
                                                    Ptype='stress')))
    σvg  = σvg.applyfunc(lambda expr:sp.collect(expr,sin(θ))\
                                                .collect(cos(θ)))
    εvg  = εvg.applyfunc(lambda expr:sp.collect(expr,sin(θ))\
                                                .collect(cos(θ)))
    return εvg, σvg

εvg, σvg = strainStressFrDisp(urg, uθg, r, θ, E,ν)

display(Math(f" \\text{{General expression for stresses:}} "))
display(Math(f" \\text{{$σ_{{v}}$=}} {latex(σvg)}"))
display(Math(f" \\text{{$ε_{{v}}$=}} {latex(εvg)}"))
```

General expression for stresses:

$$\sigma_v = \begin{bmatrix} -\frac{E(a_0\cos(\theta)+(b_3+(a_1\theta-b_1\theta)\sin(\theta)+(a_0\log(r)+a_2+b_0\log(r)+b_1+b_2)\cos(\theta)))}{r(^2-1)} \\ \frac{E(-b_3+(-a_1\theta+b_1\theta)\sin(\theta)+(-a_0-a_0\log(r)-a_2-b_0\log(r)-b_1-b_2)\cos(\theta))}{r(^2-1)} \\ \frac{E(-b_3\theta+(a_1\theta-b_1\theta)\cos(\theta)+(-a_0\log(r)+a_1-a_2-b_0\log(r)+b_0-b_2)\sin(\theta))}{2r(+1)} \end{bmatrix}$$

$$\varepsilon_v = \begin{bmatrix} \frac{a_0 \cos(\theta)}{r} \\ \frac{a_1\theta \sin(\theta) + (a_0 \log(r) + a_2)\cos(\theta)}{r} + \frac{-b_1\theta \sin(\theta) + b_3 + (b_0 \log(r) + b_1 + b_2)\cos(\theta)}{r} \\ \frac{b_0 \sin(\theta)}{2r} + \frac{a_1\theta\cos(\theta) + (-a_0 \log(r) + a_1 - a_2)\sin(\theta)}{2r} - \frac{b_1\theta\cos(\theta) + b_3\theta + (b_0\log(r) + b_2)\sin(\theta)}{2r} \end{bmatrix}$$

These general stress expressions contain unknown constants, which must be determined through the following procedure:

1. **Differential equilibrium equations**: These apply at every point in the domain and relate the stresses to body forces (which are zero in this problem).
2. **Stress boundary conditions (SBCs)**: These are applied on the wedge surfaces and relate the stresses to the surface tractions (which are zero here).
3. **Integral equilibrium conditions**: These enforce force balance over an arbitrary cross-section, ensuring consistency with the applied point load P.
4. **Displacement boundary conditions (DBCs)**: These eliminate rigid-body motions and define the reference for displacement measurements.

Throughout this process, we frequently solve systems of expressions derived from the above conditions to determine the unknown constants. For this purpose, we employ the Python function `gr.constantUpdate()` introduced in Chapter 4.

9.1.3 *Enforcing differential equilibrium equations*

From the numerators of the general stress expressions, we observe that there are four fundamental basis functions:

$$\log(r),\ \sin\theta,\ \cos\theta,\ \theta \tag{9.2}$$

When an expression involving these stresses is required to vanish, all coefficients corresponding to the possible combinations of these basis functions must also be zero. For the present problem, the general form of these combinations can be expressed as

$$(\log(r))^i, (\sin\theta)^j, (\cos\theta)^k, \theta^l \tag{9.3}$$

where $i + j + k + l \leq 2$. The following code snippet generates all possible combinations of these terms:

```
bases = sp.Matrix([log(r), sin(θ), cos(θ), θ])

order = 1
# compute all possible powers of the 4D bases:
ijkl = [(i, j, k, l) for i in range(order + 1)
                     for j in range(order + 1)
                     for k in range(order + 1)
                     for l in range(order + 1) if i+j+k+l <= order+1]
display(ijkl[:6])
display(ijkl[6:])
```

$$[(0,\ 0,\ 0,\ 0),\ (0,\ 0,\ 0,\ 1),\ (0,\ 0,\ 1,\ 0),\ (0,\ 0,\ 1,\ 1),$$

$$(0,\ 1,\ 0,\ 0),\ (0,\ 1,\ 0,\ 1)]$$

$$[(0,\ 1,\ 1,\ 0),\ (1,\ 0,\ 0,\ 0),\ (1,\ 0,\ 0,\ 1),\ (1,\ 0,\ 1,\ 0),\ (1,\ 1,\ 0,\ 0)]$$

For example, the tuple $(0, 1, 1, 0)$ corresponds to $((\log r)^0, (\sin\theta)^1, (\cos\theta)^1, \theta^0)$, which produces the combined basis term $\sin\theta\cos\theta$.

The following code first generates the two equilibrium equations based on Eq. (2.48) and then automatically extracts the coefficient expressions corresponding to all possible combined basis terms:

```
def applyEquilibrium(σvg, r, θ, br, bθ, bases, ijkl, cs):
    '''Apply differential equilibrium equations to a given
    stress vector σv over a wedge body.
    '''
    # Define equilibrium equations that need to be satisfied:
    equi1=(σvg[0].diff(r)+σvg[2].diff(θ)/r+(σvg[0]-σvg[1])/r+br)\
                                                  .simplify().factor()
    equi1=sp.numer(equi1).expand()

    equi2=(σvg[2].diff(r)+σvg[1].diff(θ)/r+2*σvg[2]/r+bθ)\
                                                  .simplify().factor()
    equi2=sp.numer(equi2).expand()

    # Find expressions of the coefficient of all possible combined bases:
    exprs = []
    for i in range(len(ijkl)):
        exprs.append(equi1.coeff(bases[0],ijkl[i][0]) # from equation 1
                          .coeff(bases[1],ijkl[i][1])
                          .coeff(bases[2],ijkl[i][2])
                          .coeff(bases[3],ijkl[i][3]))
        exprs.append(equi2.coeff(bases[0],ijkl[i][0]) # from equation 2
                          .coeff(bases[1],ijkl[i][1])
                          .coeff(bases[2],ijkl[i][2])
                          .coeff(bases[3],ijkl[i][3]))
```

```
    exprs = gr.remove_zero_exprs(exprs)        # remove zero expression
    # Find possible expressions that do not contain unknown constants:
    for i, expr in enumerate(exprs):
        if len(set(cs) & expr.free_symbols)==0:
            print(f"Alert:expression cannot be zero:i={i},expr={expr}")

    sln_equi = sp.solve(exprs, cs)
    if len(sln_equi)==0: print(f"Alert: no solution were found.")

    display(sln_equi)

    # Update unknown constants:
    sln_cs = {**sln_equi}
    sln_cs = {key: value.subs(sln_cs).simplify().factor() for key, \
                                                  value in sln_cs.items()}
    print(f"Current len of sln_cs={len(sln_cs)}; Total cs={len(cs)}")
    #display(sln_cs)

    # Check how many constants are found and how many unknowns remain:
    list_keys = list(sln_cs.keys())
    cs_Eql = list(set(list(cs))-set(list_keys))
    print(f"Constants yet to be found: {cs_Eql}")

    return sln_cs, sln_equi
```

The operation above can generate a large number of expressions, which are then solved for the unknown constants. Note that during the automatic generation process, many expressions may evaluate to zero. Such expressions are removed when present. We also check for expressions that do not contain any unknown constants, since these can never be satisfied regardless of the constant values. This situation indicates that the assumed displacement functions cannot fulfill certain equilibrium conditions. Therefore, it is good practice to include an alarm for this case. Another alarm is set for situations where no solution is found for any of the unknown constants, indicating that no combination of constants can satisfy the equilibrium equations. When either of these alarms is triggered, a careful reexamination of the assumed displacement functions is necessary.

Let us now apply the equilibrium equations to the assumed stresses and determine some of the unknown constants:

```
br = 0; bθ = 0                                 # Consider zero body forces
sln_cs, sln_equi = applyEquilirium(σvg, r, θ, br, bθ, bases, ijkl, cs)
```

```
Total number of expressions = 22
Number of nonzero expressions = 8
```

$$\left\{a_0 : -b_0,\ a_1 : b_1,\ a_2 : \frac{b_0(-\nu - 1)}{\nu - 3} + \frac{b_1(\nu + 1)}{\nu - 3} - b_2,\ b_3 : 0\right\}$$

```
Current len of sln_cs=4; Total cs=7
Constants yet to be found: [b1, b2, b0]
```

No alarm is triggered in the foregoing process, and several relationships among the unknown constants are established. Consequently, only three constants remain to be determined.

9.1.4 *Enforcing stress boundary conditions*

Since the two surfaces of the wedge are traction-free, the stress boundary conditions (SBCs) are given by

$$\sigma_{\theta\theta} = 0 \text{ and } \sigma_{r\theta} = 0, \quad \text{at } \theta = \pm\alpha/2 \tag{9.4}$$

The following Python function applies these SBCs. Note that it is sufficient to set the numerators of the SBC expressions to zero. Since these expressions involve $\log(r)$, both the coefficients of $(\log(r))^1$ and $(\log(r))^0$ must vanish. The function `gr.constantUpdate()` is then invoked to perform this task automatically:

```
def applySBCs(σv, r, θ, αL, αR, trL, tθL, trR, tθR, sln_cs, cs):
    '''Apply stress boundary conditions or SBCs to a given
    stress vector σv on two sides of a wedge.
    '''
    # Define stress boundary conditions (SBCs) need to be satisfied:
    SBC1=sp.simplify(σv[1].subs(θ, αL)-tθL).simplify().factor()
    SBC1=sp.numer(SBC1).expand()
    SBC2=sp.simplify(σv[2].subs(θ, αL)-trL).simplify().factor()
    SBC2=sp.numer(SBC2).expand()

    SBC3=sp.simplify(σv[1].subs(θ, αR)+tθR).simplify().factor()
    SBC3=sp.numer(SBC3).expand()
    SBC4=sp.simplify(σv[2].subs(θ, αR)+trR).simplify().factor()
    SBC4=sp.numer(SBC4).expand()

    sln_SBCs = gr.constantUpdate([SBC1,SBC2,SBC3,SBC4],\
                                              log(r),1,cs,prnt=False)
    display(sln_SBCs)
```

```
    # Update unknown constants:
    sln_cs.update(sln_SBCs)# = {**sln_equi, **sln_SBCs}
    sln_cs = {key: value.subs(sln_cs).simplify().factor() for key,\
                                            value in sln_cs.items()}
    print(f" Current len of sln_cs={len(sln_cs)}; Total cs={len(cs)}")
    display(sln_cs)

    # Check how many unknown constants remain:
    list_keys = list(sln_cs.keys())
    cs_Eql = list(set(list(cs))-set(list_keys))
    print(f"Constants yet to be found: {cs_Eql}")

    return sln_cs, sln_SBCs
```

Let us apply these SBCs to the current stresses:

```
σv = σvg.subs(sln_cs)
sln_cs, sln_SBCs = applySBCs(σv, r, θ, α/2, -α/2, 0,0,0,0, sln_cs, cs)
```

```
Total number of expressions = 4
Number of nonzero expressions = 4
```

$$\left\{b_0 : \frac{2b_1}{\nu - 1}\right\}$$

```
Current len of sln_cs=5; Total cs=7
```

$$\left\{a_0 : -\frac{2b_1}{\nu - 1},\ a_1 : b_1,\ a_2 : \frac{b_1\nu + b_1 - b_2\nu + b_2}{\nu - 1},\ b_0 : \frac{2b_1}{\nu - 1},\ b_3 : 0\right\}$$

```
Constants yet to be found: [b1, b2]
```

We have determined one of the three constants, leaving two yet to be found.

9.1.5 *Enforcing integral equilibrium conditions*

To apply these conditions, we introduce an arbitrary circular section A–A′, as illustrated in Fig. 9.1. Along this section, the following equilibrium condition must be satisfied:

$$\begin{aligned} \sum F_x = 0 : \int_{-\alpha/2}^{\alpha/2} \sigma_{rr} r d\theta \cos\theta + P = 0 \\ \sum F_y = 0 : \int_{-\alpha/2}^{\alpha/2} \sigma_{rr} r d\theta \sin\theta = 0 \end{aligned} \tag{9.5}$$

Here, we first compute the resultant forces by projecting the stresses onto the x-direction, which must balance the applied point force P, as expressed in the first equation in Eq. (9.5). Similarly, the total projected stress in the y-direction must vanish, as stated in the second equation in Eq. (9.5). These two equilibrium equations can then be used to determine the remaining unknown constants.

The following Python function is written to perform this task, which involves integrating the stress components over the specified surface:

```
def applyIntegralEqui(σv, r, θ, αL, αR, sln_cs, cs, Px, Py):
    '''Apply integral equilibrium conditions or FBCs to a given
    stress vector σv over an arbitrary cut at r.
    '''
    # Define the integral equilibrium equations need to be satisfied:
    equI1=(integrate(σv[0]*r*sp.cos(θ),(θ, αR, αL)) + Px)\
                                              .simplify().factor()
    equI1=sp.numer(equI1).expand()

    equI2=(integrate(σv[0]*r*sp.sin(θ),(θ, αR, αL)) + Py)\
                                              .simplify().factor()
    equI2=sp.numer(equI2).expand()

    sln_equI = gr.constantUpdate([equI1, equI2],log(r),1,cs,prnt=False)

    # Update unknown constants:
    sln_cs.update(sln_equI)    # = {**sln_SBCs, **sln_equi, **sln_equI}
    sln_cs = {key: value.subs(sln_cs).simplify() for
                                        key, value in sln_cs.items()}
    print(f" Current len of sln_cs={len(sln_cs)}; Total cs={len(cs)}")

    # Check how many unknown constants remain:
    list_keys = list(sln_cs.keys())
    cs_Eql = list(set(list(cs))-set(list_keys))
    print(f"Constants yet to be found: {cs_Eql}")
    return sln_cs, sln_equI
```

Now, apply these FBCs to the current stresses:

```
σv = σvg.subs(sln_cs)                                # update stresses
Px = P; Py = 0
sln_cs, sln_equI = applyIntegralEqui(σv, r,θ, α/2,-α/2,sln_cs,cs,Px,Py)

sln_cs_= gr.split_dict_n(sln_cs, 2)
display(sln_cs_[0]); display(sln_cs_[1])

σv = sp.simplify(σv.subs(sln_cs))                    # update stresses
display(Math(f" \\text{{ Current stresses $σ_{{v}}$=}} \
                                  {latex(sp.simplify(σv).T)}"))
```

```
 Total number of expressions = 2
 Number of nonzero expressions = 1
 Current len of sln_cs=6; Total cs=7
Constants yet to be found: [b2]
```

$$\left\{a_0 : -\frac{2P}{E\left(\alpha+\sin\left(\alpha\right)\right)},\ a_1 : \frac{P\left(\nu-1\right)}{E\left(\alpha+\sin\left(\alpha\right)\right)},\right.$$
$$\left. a_2 : \frac{-Eb_2\alpha - Eb_2\sin\left(\alpha\right)+P\nu+P}{E\left(\alpha+\sin\left(\alpha\right)\right)}\right\}$$

$$\left\{b_0 : \frac{2P}{E\left(\alpha+\sin\left(\alpha\right)\right)},\ b_1 : \frac{P\left(\nu-1\right)}{E\left(\alpha+\sin\left(\alpha\right)\right)},\ b_3 : 0\right\}$$

Current stresses $\sigma_v = \begin{bmatrix} -\frac{2P\cos\left(\theta\right)}{r\left(\alpha+\sin\left(\alpha\right)\right)} & 0 & 0\end{bmatrix}$

We have successfully determined all the unknown constants required to evaluate the stresses and updated them in the expressions. It is found that the only nonzero stress component is σ_{rr}, which coincides with the symmetric portion of the stress obtained earlier in Eq. (7.9) using the stress method. The following code snippet verifies this result:

```
σrr_  = -2*P*cos(θ)/(r*(α + sin(α)))                    # by displacement method
σ_rr_strssMthd = -2*P*sp.cos(θ)/(r*( α + sp.sin(α)))   # symmetric part

# check the differences:
(σrr_ - σ_rr_strssMthd).simplify()                                   # should be 0
```

0

9.1.6 *Validation of solution*

It is always good practice to validate the obtained solution by confirming that it satisfies all the prescribed conditions, including equilibrium equations, boundary conditions, and integral constraints. The following Python function performs this verification systematically:

```
def checkConditions(σv, r, θ, br, bθ, αL, αR, Px, Py):
    '''Check for satisfaction of diff and integral equilibriums and SBCs
       by a given stress vector σv.'''
    # Check the differential equilibrium equations:
    equi1=(σv[0].diff(r)+σv[2].diff(θ)/r+(σv[0]-σv[1])/r+br).simplify()
    equi1=sp.numer(equi1).expand()
    equi2=(σv[2].diff(r)+σv[1].diff(θ)/r+2*σv[2]/r+bθ).simplify()
    equi2=sp.numer(equi2).expand()
    display(Math(f"\\text{{Are the equilibrium equations satisfied? }}\
                                  {latex((equi1**2+equi2*2).simplify()==0)}"))
```

```
    # Check the stress boundary conditions:
    SBC1=sp.simplify(σv[1].subs(θ, αL)).simplify().factor()
    SBC1=sp.numer(SBC1).expand()
    SBC2=sp.simplify(σv[2].subs(θ, αL)).simplify().factor()
    SBC2=sp.numer(SBC2).expand()

    SBC3=sp.simplify(σv[1].subs(θ, αR)).simplify().factor()
    SBC3=sp.numer(SBC3).expand()
    SBC4=sp.simplify(σv[2].subs(θ, αR)).simplify().factor()
    SBC4=sp.numer(SBC4).expand()
    display(Math(f" \\text{{Are the SBCs satisfied? }} \
            {latex((SBC1**2+SBC2**2+SBC3**2+SBC4**2).simplify()==0)}"))

    # Check the integral equilibrium conditions:
    equI1=(integrate(σv[0]*r*sp.cos(θ),(θ, αR,αL)) + Px).simplify()
    equI1=sp.numer(equI1).expand()
    equI2=(integrate(σv[0]*r*sp.sin(θ),(θ, αR,αL)) + Py).simplify()
    equI2=sp.numer(equI2).expand()
    display(Math(f" \\text{{Are the FBCs satisfied? }} \
                          {latex((equI1**2+equI1**2).simplify()==0)}"))
```

```
# Validation of Solutions:
br = 0; bθ = 0
Px = P; Py = 0
checkConditions(σv, r, θ, br, bθ, α/2, -α/2, Px, Py)
```

Are the equilibrium equations satisfied? True

Are the SBCs satisfied? True

Are the FBCs satisfied? True

It is confirmed that the stress solution obtained is correct. We can now update the strains using the constitutive equation and the displacements using the constants found so far:

```
# Update the strain vector using the constants found so far:
εv = sp.simplify(εvg.subs(sln_cs))

display(Math(f" \\text{{ Current strains $ε_{{v}}$=}} \
                                    {latex(sp.simplify(εv).T)}"))

εrr_  = -2*P*cos(θ)/(E*r*(α + sin(α)))         # re-define for later use
εθθ_  = 2*P*ν*cos(θ)/(E*r*(α + sin(α)))
```

Current strains $\varepsilon_v = \begin{bmatrix} -\frac{2P\cos(\theta)}{Er(\alpha+\sin(\alpha))} & \frac{2P\nu\cos(\theta)}{Er(\alpha+\sin(\alpha))} & 0 \end{bmatrix}$

```
# Update the displacements using the constants found so far:
ur = urg.subs(sln_cs).expand().simplify().collect(P).collect(E)\
                                     .collect(sin(θ)).collect(cos(θ))
display(Math(f"\\text{{Current displacements: }}"))
display(Math(f"\\text{{$u_r$ = }}\\small {latex(ur)}"))

uθ = uθg.subs(sln_cs).expand().simplify().collect(P).collect(E)
display(Math(f"\\text{{$u_θ$ = }}\\small {latex(uθ)}"))

# Check how many unknown constants remain:
list_keys = list(sln_cs.keys())
cs_Eql = list(set(list(cs))-set(list_keys))
print(f"Constants yet to be found: {cs_Eql}")
```

Current displacements:

$$u_r = \frac{E\left(-b_2\alpha - b_2\sin\left(\alpha\right)\right)\cos\left(\theta\right) + P(\left(\theta\nu - \theta\right)\sin\left(\theta\right) + \left(\nu - 2\log\left(r\right) + 1\right)\cos\left(\theta\right))}{E\left(\alpha + \sin\left(\alpha\right)\right)}$$

$$u_\theta = \frac{Eb_2\left(\alpha + \sin\left(\alpha\right)\right)\sin\left(\theta\right) + P\left(\theta\nu\cos\left(\theta\right) - \theta\cos\left(\theta\right) + 2\log\left(r\right)\sin\left(\theta\right)\right)}{E\left(\alpha + \sin\left(\alpha\right)\right)}$$

```
Constants yet to be found: [b2]
```

As observed, one unknown constant, b_2, remains in the displacement expressions. Since it does not appear in the stress or strain formulas, it corresponds to a rigid body motion of the wedge. To eliminate this motion, we can impose a displacement boundary condition. Specifically, by setting $u_r = 0$ at $r = d$ and $\theta = 0$, the rigid body movement is removed, and the constant b_2 can be determined accordingly.

The snippet is as follows:

```
#sln_cs = {**sln_SBCs, **sln_equi, **sln_equI}
sln_ur = sp.solve([urg.subs(sln_cs).subs({r:d0, θ:0})], cs)
display(sln_ur)

# Update unknown constants:
sln_cs = {**sln_SBCs, **sln_equi, **sln_equI, **sln_ur}
sln_cs = {key: value.subs(sln_cs).simplify() for key, \
                                          value in sln_cs.items()}
print(f" Current len of sln_cs={len(sln_cs)}; Total cs={len(cs)}")
```

```
# The final displacements:
ur = sp.simplify(urg.subs(sln_cs))
#display(Math(f"\\text{{Final displacements: }}"))
#display(Math(f"\\text{{$u_r$ = }}\\small {latex(ur)}"))

uθ = sp.simplify(sp.logcombine(uθg.subs(sln_cs).simplify()))
#display(Math(f"\\text{{$u_θ$ = }}\\small {latex(uθ)}"))

ur_  = P*(θ*(ν-1)*sin(θ)+2*cos(θ)*log(d0)-2*cos(θ)*log(r))/(E*(α+sin(α)))
uθ_  = P*(θ*(ν-1)*cos(θ) + (ν-2*log(d0)+1)*sin(θ) + 2*log(r)*sin(θ))\
                                                          /(E*(α+sin(α)))
ur_halfSpace = ur_.subs(α,pi).simplify().factor()
uθ_halfSpace = uθ_.subs(α,pi).simplify()
```

$$\left\{b_2 : \frac{P\nu}{E\alpha + E\sin(\alpha)} - \frac{2P\log(d_0)}{E\alpha + E\sin(\alpha)} + \frac{P}{E\alpha + E\sin(\alpha)}\right\}$$

```
Current len of sln_cs=7; Total cs=7
```

It is found that the constant b_2 has a rather complicated expression.

Having now determined all the constants, we can rewrite the final displacement expressions for the wedge under a vertical point force as follows, which will serve as a reference for subsequent analyses:

$$\begin{aligned} u_r &= \frac{P\left(\theta\left(\nu-1\right)\sin\theta + 2\cos\theta\log(d_0) - 2\cos\theta\log(r)\right)}{E\left(\alpha+\sin\alpha\right)} \\ u_\theta &= \frac{P\left(\theta\left(\nu-1\right)\cos\theta + \left(\nu - 2\log\left(d_0\right)+1\right)\sin\theta + 2\log\left(r\right)\sin\theta\right)}{E\left(\alpha+\sin\alpha\right)} \end{aligned} \tag{9.6}$$

When setting $\alpha = \pi$, the wedge becomes a half-space. We can then compare the displacement solution with that obtained in Chapter 8. The following snippet confirms this:

```
ur_strssMthd = P/(sp.pi*E)*( θ*(v-1)*sp.sin(θ) + 2*sp.log(d0)*sp.cos(θ)\
                   - 2*sp.log(r)*sp.cos(θ) )

uθ_strssMthd = P/(sp.pi*E)*( θ*(v-1)*sp.cos(θ) + (v + 1)*sp.sin(θ) - \
                              2*sp.log(d0)*sp.sin(θ) + 2*sp.log(r)*sp.sin(θ))
# check the differences:
((ur_halfSpace-ur_strssMthd).simplify(),
 (uθ_halfSpace-uθ_strssMthd).simplify())                  # should be (0, 0)
```

(0, 0)

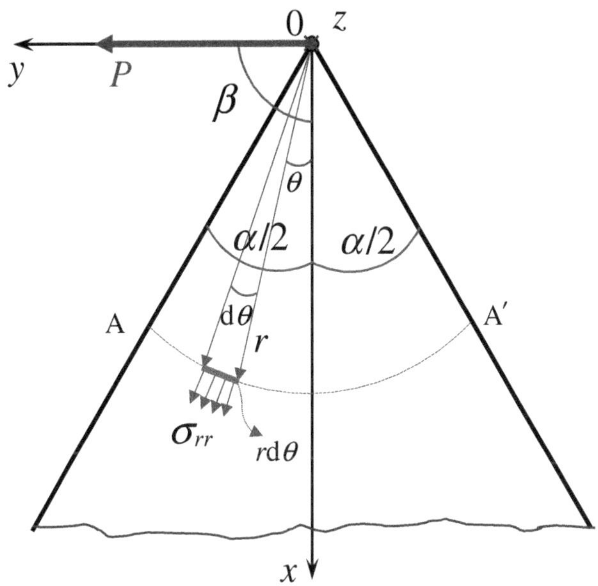

Figure 9.2. A wedge with angle α subjected to a horizontal point force P at its tip.

9.2 Wedges under a horizontal point load

The setup of this problem is the same as in the previous section. A polar coordinate system is used, with $\theta = 0$ aligned along the vertical left-right mirror-symmetric line, so that the wedge remains geometrically symmetric. The only difference is that the point force P acts in the horizontal direction, as shown in Fig. 9.2. Consequently, most of the Python functions developed in the previous section can be directly used to solve this problem.

9.2.1 *General expressions for displacements*

Since the force is no longer in the y-direction, the symmetry properties of the displacement components are reversed: the radial displacement u_r is anti-symmetric, while the tangential displacement u_θ is symmetric with respect to θ. They can be expressed in the following form:

$$
\begin{aligned}
u_r(r,\theta) &= a_0 \log(r) \sin\theta + a_1\theta\cos\theta + a_2\sin\theta + a_3\theta \\
u_\theta(r,\theta) &= b_0 \log(r) \cos\theta + b_1\theta\sin\theta + b_2\cos\theta
\end{aligned}
\tag{9.7}
$$

where a_i and b_i are arbitrary unknown constants. The reasoning for including these terms is as follows:

1. Since the stress σ_{rr} is the derivative of the radial displacement u_r, and $\log(r)$ is the antiderivative of $1/r$, the term $\log(r)$ should be included in the displacement expressions.
2. Since the problem is antisymmetric with respect to θ, u_r should also be antisymmetric with respect to θ. When the sign of θ changes, u_r changes sign as well. Therefore, basis terms that are odd functions of θ are included.
3. Due to the same antisymmetry property, the tangential displacement component u_θ must be symmetric with respect to θ. Thus, basis terms that are even functions of θ are used in the second equation in Eq. (9.1).

The code snippet for generating the displacement expressions is as follows:

```
# Define variables for displacement method for axial symmetric problems:
E, ν = symbols('E, ν', positive=True)         # given material constants
P = symbols('P', real=True)                  # given vertical point force
r = sp.symbols('r', positive=True)             # variable for radial axis
θ = sp.symbols('θ', real=True)              # variable for tangential axis
α = sp.symbols('α')                                           # wedge angle

ur_rbs_ = [log(r)*sin(θ), θ*cos(θ), sin(θ), θ]
ur_rbs = sp.Matrix(ur_rbs_)

uθ_rbs_ = [log(r)*cos(θ), θ*sin(θ), cos(θ)]
uθ_rbs = sp.Matrix(uθ_rbs_)

# unknown constants for displacements:
ai = sp.symbols(f'a:{len(ur_rbs)}')                     # unknown constants
display(Math(f" \\text{{Unknown constants for $u$}}={latex(ai)}"))
ai = Matrix([ai[i] for i in range(len(ur_rbs))]) #form coeffs in vector

bi = sp.symbols(f'b:{len(uθ_rbs)}')                     # unknown constants
display(Math(f" \\text{{Unknown constants for $v$}}={latex(bi)}"))
bi = Matrix([bi[i] for i in range(len(uθ_rbs))])          # form a vector

cs = ai.col_join(bi)                   # form a single vector of constants

urg = (ai.T@Matrix(ur_rbs))[0]          # Form the general displacements
uθg = (bi.T@Matrix(uθ_rbs))[0]
display(Math(f"\\text{{Assumed displacements, $u_r(r,θ)$}}=\
                                                       {latex(urg)}"))
display(Math(f"\\hspace{{4.4cm}} \\text{{ $u_θ(r,θ)$}}={latex(uθg)}"))
```

Unknown constants for $u = (a_0,\ a_1,\ a_2,\ a_3)$

Unknown constants for $v = (b_0,\ b_1,\ b_2)$

$$\text{Assumed displacements, } u_r(r,\theta) = a_0 \log(r)\sin(\theta) + a_1\theta\cos(\theta) + a_2\sin(\theta) + a_3\theta$$
$$u_\theta(r,\theta) = b_0 \log(r)\cos(\theta) + b_1\theta\sin(\theta) + b_2\cos(\theta)$$

We have constructed these displacements, which contain a total of seven unknown constants. These constants must now be determined using equilibrium and stress boundary conditions, similar to the previous case. All of these steps require the corresponding stress expressions.

9.2.2 *Expressions for stresses*

Next, we compute the strain expressions from the displacements, and then obtain the stresses using the previously defined function `strainStressFrDisp()`:

```
εvg, σvg = strainStressFrDisp(urg, uθg, r, θ, E,ν)

display(Math(f" \\text{{General expression for stresses:}} "))
display(Math(f" \\text{{$σ_{{v}}$=}} {latex(σvg)}"))
display(Math(f" \\text{{$ε_{{v}}$=}} {latex(εvg)}"))
```

General expression for stresses:

$$\sigma_v = \begin{bmatrix} -\frac{E(a_0\sin(\theta)+\nu(a_3\theta+(a_1\theta+b_1\theta)\cos(\theta)+(a_0\log(r)+a_2-b_0\log(r)+b_1-b_2)\sin(\theta)))}{r(\nu^2-1)} \\ \frac{E(-a_3\theta+(-a_1\theta-b_1\theta)\cos(\theta)+(-a_0\nu-a_0\log(r)-a_2+b_0\log(r)-b_1+b_2)\sin(\theta))}{r(\nu^2-1)} \\ \frac{E(a_3+(-a_1\theta-b_1\theta)\sin(\theta)+(a_0\log(r)+a_1+a_2-b_0\log(r)+b_0-b_2)\cos(\theta))}{2r(\nu+1)} \end{bmatrix}$$

$$\varepsilon_v = \begin{bmatrix} \frac{a_0\sin(\theta)}{r} \\ \frac{b_1\theta\cos(\theta)+(-b_0\log(r)+b_1-b_2)\sin(\theta)}{r} + \frac{a_1\theta\cos(\theta)+a_3\theta+(a_0\log(r)+a_2)\sin(\theta)}{r} \\ \frac{b_0\cos(\theta)}{2r} - \frac{b_1\theta\sin(\theta)+(b_0\log(r)+b_2)\cos(\theta)}{2r} + \frac{-a_1\theta\sin(\theta)+a_3+(a_0\log(r)+a_1+a_2)\cos(\theta)}{2r} \end{bmatrix}$$

These general stresses contain the unknown constants. The procedure for determining these constants is the same as in the previous section, and many of the code snippets can be reused with minor modifications.

9.2.3 *Enforcing differential equilibrium equations*

The same Python function `applyEquilibrium()` can be used to enforce the differential equilibrium equations:

```
br = 0; bθ = 0                                     # Consider zero body forces
sln_cs, sln_equi = applyEquilirium(σvg, r, θ, br, bθ, bases, ijkl, cs)
```

```
Total number of expressions = 22
Number of nonzero expressions = 8
```

$$\left\{a_0 : b_0,\ a_1 : -b_1,\ a_2 : \frac{b_0(\nu+1)}{\nu-3} + \frac{b_1(\nu+1)}{\nu-3} + b_2,\ a_3 : 0\right\}$$

```
Current len of sln_cs=4; Total cs=7
Constants yet to be found: [b1, b2, b0]
```

9.2.4 *Enforcing stress boundary conditions*

The SBCs are the same as those given in Eq. (9.4), and the same Python function `applySBCs()` can be used:

```
σv = σvg.subs(sln_cs)
sln_cs, sln_SBCs = applySBCs(σv, r, θ, α/2, -α/2,0,0,0,0, sln_cs, cs)
```

```
Total number of expressions = 4
Number of nonzero expressions = 4
```

$$\left\{b_0 : -\frac{2b_1}{\nu-1}\right\}$$

```
Current len of sln_cs=5; Total cs=7
```

$$\left\{a_0 : -\frac{2b_1}{\nu-1},\ a_1 : -b_1,\ a_2 : \frac{b_1\nu + b_1 + b_2\nu - b_2}{\nu-1},\ a_3 : 0,\ b_0 : -\frac{2b_1}{\nu-1}\right\}$$

```
Constants yet to be found: [b1, b2]
```

9.2.5 *Enforcing integral equilibrium conditions*

For this, we make the same circular cut A-A$'$ as shown in Fig. 9.1, along which we require

$$\begin{aligned}\sum F_x = 0 : \int_{-\alpha/2}^{\alpha/2} \sigma_{rr} r d\theta \cos\theta = 0\\ \sum F_y = 0 : \int_{-\alpha/2}^{\alpha/2} \sigma_{rr} r d\theta \sin\theta + P = 0\end{aligned} \tag{9.8}$$

Here, we first compute the forces by projecting the stress onto the x-direction, which should be zero since the force now acts in the horizontal direction, as shown in the first equation in Eq. (9.8). The total stress projection in the y-direction must balance the point force P, as given in the second equation in Eq. (9.8). These two equations can be used to determine some of the remaining constants, and `applyIntegralEqui()` will accomplish this task:

```
σv = σvg.subs(sln_cs)                                    # update stresses
Px = 0; Py = P
sln_cs, sln_equI = applyIntegralEqui(σv, r,θ, α/2,-α/2,sln_cs,cs,Px,Py)

sln_cs_= gr.split_dict_n(sln_cs, 2)
display(sln_cs_[0]); display(sln_cs_[1])

σv = sp.simplify(σv.subs(sln_cs))                        # update stresses
display(Math(f" \\text{{ Current stresses $σ_{{v}}$=}} \
                                    {latex(sp.simplify(σv).T)}"))
```

```
 Total number of expressions = 2
 Number of nonzero expressions = 1
 Current len of sln_cs=6; Total cs=7
Constants yet to be found: [b2]
```

$$\left\{a_0 : \frac{2P}{E\left(-\alpha + \sin\left(\alpha\right)\right)},\ a_1 : \frac{P\left(1-\nu\right)}{E\left(\alpha - \sin\left(\alpha\right)\right)},\right.$$

$$\left. a_2 : \frac{-Eb_2\alpha + Eb_2\sin\left(\alpha\right) - P\nu - P}{E\left(-\alpha + \sin\left(\alpha\right)\right)}\right\}$$

$$\left\{a_3 : 0,\ b_0 : \frac{2P}{E\left(-\alpha + \sin\left(\alpha\right)\right)},\ b_1 : \frac{P\left(\nu - 1\right)}{E\left(\alpha - \sin\left(\alpha\right)\right)}\right\}$$

Current stresses σ_v= $\left[\frac{2P\sin\left(\theta\right)}{r\left(-\alpha+\sin\left(\alpha\right)\right)} \quad 0 \quad 0\right]$

We have successfully determined the unknown constants required for computing the stresses and have updated the stress expressions accordingly. It is found that the only nonzero stress is σ_{rr}, which corresponds to the antisymmetric part of the stress given in Eq. (7.9) obtained using the stress method. The following snippet confirms this:

```
σrr_ = 2*P*sin(θ)/(r*(-α + sin(α)))
σ_rr_strssMthd = 2*P*sp.sin(θ)/(r*(-α + sp.sin(α)))
(σrr_ - σ_rr_strssMthd).simplify()                          # should be 0
```

0

9.2.6 *Validation of solution*

It is good practice to validate the obtained solution by verifying that all conditions are satisfied. This can be done using the `checkConditions()` function developed in the previous section:

```
br = 0; bθ = 0
Px = 0; Py = P
checkConditions(σv, r, θ, br, bθ, α/2, -α/2, Px, Py)
```

Are the equilibrium equations satisfied? True

Are the SBCs satisfied? True

Are the FBCs satisfied? True

It is confirmed that the stress solution obtained is correct. We can now update the strains and displacements:

```
εv = sp.simplify(εvg.subs(sln_cs))            # Update the strain vector
display(Math(f" \\text{{ Current strains $ε_{{v}}$=}} \
                                   {latex(sp.simplify(εv).T)}"))
#print(εv)
εrr_ = 2*P*sin(θ)/(E*r*(-α + sin(α)))          # re-define for later use
εθθ_ = 2*P*ν*sin(θ)/(E*r*(α - sin(α)))
```

$$\text{Current strains } \varepsilon_v = \begin{bmatrix} \frac{2P\sin(\theta)}{Er(-\alpha+\sin(\alpha))} & \frac{2P\nu\sin(\theta)}{Er(\alpha-\sin(\alpha))} & 0 \end{bmatrix}$$

```
# Update the displacements:
ur = urg.subs(sln_cs).expand().simplify().collect(P).collect(E)\
                        .collect(sin(θ)).collect(cos(θ))
display(Math(f"\\text{{Current displacements: }}"))
display(Math(f"\\text{{$u_r$ = }}\\small {latex(ur)}"))

uθ = uθg.subs(sln_cs).expand().simplify().collect(P).collect(E)
display(Math(f"\\text{{$u_θ$ = }}\\small {latex(uθ)}"))

# Check how many unknown constants remain:
list_keys = list(sln_cs.keys())
cs_Eql = list(set(list(cs))-set(list_keys))
print(f"Constants yet to be found: {cs_Eql}")
```

Current displacements:

$$u_r = \frac{E(b_2\alpha - b_2 \sin(\alpha))\sin(\theta) + P((-\theta\nu + \theta)\cos(\theta) + (\nu - 2\log(r) + 1)\sin(\theta))}{E(\alpha - \sin(\alpha))}$$

$$u_\theta = \frac{Eb_2(\alpha - \sin(\alpha))\cos(\theta) + P(\theta\nu\sin(\theta) - \theta\sin(\theta) - 2\log(r)\cos(\theta))}{E(\alpha - \sin(\alpha))}$$

```
Constants yet to be found: [b2]
```

As observed, there remains one unknown constant, b_2, in the displacement expressions, which does not appear in the stress expressions. This indicates that it represents a rigid body movement, similar to the previous problem. We can therefore set $u_r = 0$ at $r = d$ and $\theta = 0$, and use this condition to determine b_2:

```
sln_cs = {**sln_SBCs, **sln_equi, **sln_equI}
sln_ur = sp.solve([uθg.subs(sln_cs).subs({θ:0})], cs)
display(sln_ur)

# Update unknown constants:
sln_cs = {**sln_SBCs, **sln_equi, **sln_equI, **sln_ur}
sln_cs = {key: value.subs(sln_cs).simplify() for key, \
                                        value in sln_cs.items()}
print(f" Current len of sln_cs={len(sln_cs)}; Total cs={len(cs)}")

# The final displacements:
ur = sp.simplify(urg.subs(sln_cs))
#display(Math(f"\\text{{Final displacements: }}"))
#display(Math(f"\\text{{$u_r$ = }}\\small {latex(ur)}"))

uθ = sp.simplify(sp.logcombine(uθg.subs(sln_cs).simplify()))
#display(Math(f"\\text{{$u_θ$ = }}\\small {latex(uθ)}"))

ur_ = P*(θ*ν*cos(θ) - θ*cos(θ) - ν*sin(θ) - sin(θ))/(E*(-α + sin(α)))
uθ_ = P*θ*(ν - 1)*sin(θ)/(E*(α - sin(α)))

ur_halfSpace = ur_.subs(α,pi).simplify().factor()
uθ_halfSpace = uθ_.subs(α,pi).simplify()
#display((ur_halfSpace, uθ_halfSpace))
```

$$\left\{b_2 : -\frac{2P\log(r)}{-E\alpha + E\sin(\alpha)}\right\}$$

```
Current len of sln_cs=7; Total cs=7
```

We have determined all the constants. The final displacements for a wedge under a horizontal point force can now be found as follows for future reference:

$$u_r = -\frac{P\left[(1-\nu)\theta\cos\theta + (1+\nu)\sin\theta\right]}{E\left(-\alpha+\sin\alpha\right)}$$
$$u_\theta = -\frac{P\theta(1-\nu)\sin\theta}{E\left(\alpha-\sin\alpha\right)} \tag{9.9}$$

When setting $\alpha = \pi$, the wedge becomes a half-space. The displacements become

$$u_r = \frac{P\left[(1-\nu)\theta\cos\theta + (1+\nu)\sin\theta\right]}{\pi E}$$
$$u_\theta = -\frac{P(1-\nu)\theta\sin\theta}{\pi E} \tag{9.10}$$

Note that if the applied point force acts in an arbitrary direction, the solution can be obtained by simple superposition of the results derived in this chapter. Readers are encouraged to try this approach, which is widely applied to solid mechanics problems.

9.3 Review questions

1. Consider a wedge subjected to a point force P in the direction of β, as shown in Fig. 7.1. Use the displacement method to find the full set of solutions, including displacements, stresses, and strains. Compare the solutions with those obtained in Sections 9.1 and 9.2.
2. Use the displacement method to solve Question 7.1.
3. Use the displacement method to solve Question 7.2.

9.4 Remarks

This chapter discusses the displacement method for solving wedges subjected to point loads. We conclude this chapter with a few remarks:

1. When a wedge is subjected to a point load in an arbitrary direction, the problem is neither symmetric nor antisymmetric, even though the geometry is symmetric.
2. Such an asymmetric problem can be decomposed into two problems: one symmetric and the other antisymmetric.

3. The displacement methods proposed in this chapter are effectively applied to both problems, as the displacements can be assumed conveniently for wedges. The final solution is the sum of the solutions to these two problems.
4. Sets of Python codes are provided for all these tasks mostly automatically.
5. The decomposition technique demonstrated in this chapter is useful for many other types of problems, as long as the geometry of the solid is symmetric.

The following chapter presents a displacement method for arches subjected to surface pressures.

Chapter 10

Displacement Method for an Arch Bridge

```
# Place cursor in this cell, and press Ctrl+Enter to import dependencies.
import sys                              # for accessing the computer system
sys.path.append('../grbin/')              # Add in the path to your system

from commonImports import *         # Import dependencies from '../grbin/'
import grcodes as gr                     # Import the module of the author
importlib.reload(gr)                  # When grcodes is modified, reload it

from continuum_mechanics import vector
from continuum_mechanics.solids import sym_grad, strain_stress
init_printing(use_unicode=True)          # For latex-like quality printing
np.set_printoptions(precision=4,suppress=True)    # Digits in print-outs
```

Building upon the previous chapters, this chapter presents a general displacement method through a case study: an exact closed-form solution for an arch bridge subjected to pressure on its top surface. Due to the complexity of the problem, the analytical expressions obtained during the derivation are typically lengthy. Special effort has been made to simplify the key formulas into concise forms for easier reading and practical use. The derivation is primarily carried out using SymPy snippets, making this chapter code-intensive. For problems of this complexity, obtaining a solution without SymPy (or a comparable computational tool) would be extremely difficult, if not impossible.

The aim of this chapter is to provide a challenging example that utilizes Python codes to solve complex solid mechanics problems not previously addressed, highlighting the capabilities of modern computational tools and innovative solution strategies.

10.1 Setting of the problem

Consider a problem similar to that studied in Section 3.4: a thin, partially circular curved arch of uniform thickness, with an inner radius a and an outer radius b. It is subjected to a pressure p on the outer surface, as shown in Fig. 10.1. All other settings are identical. This defines a general 2D solid in the polar coordinate system, typical of widely used arch bridges. To simplify the problem, the body force (the weight of the bridge) is assumed to be zero, as it is usually small compared to the applied load. Alternatively, the body force can be approximately included by converting the weight into a distributed load.

We propose a displacement method to obtain the full set of solutions, including displacements, stresses, and strains. The strategy, concept, and procedure are similar to those discussed in the previous chapters on displacement methods, although the complexity of this problem is significantly higher.

10.2 Construction of displacements

In any displacement method, the key is to assume the displacements appropriately. We shall make use of the mirror symmetry property of the problem:

Symmetry with respect to θ: It is evident that the displacement component in the radial direction, u_r, is symmetric with respect to the axis

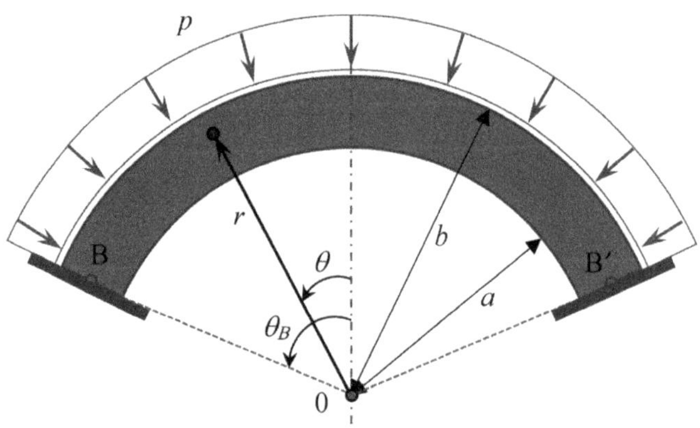

Figure 10.1. A thin, partially circular arch of uniform thickness loaded by pressure on its top surface.

$\theta = 0$ and therefore must be an even function of θ. Conversely, the tangential component, u_θ, must be antisymmetric and hence an odd function of θ. Accordingly, the displacements can be assumed as

$$\begin{aligned} u_r(r,\theta) &= \frac{a_0\cos(\theta)}{r^2} + \frac{a_1\cos(\theta)}{r} + a_{10}r\cos(\theta) + a_{12}r^2\cos(\theta) \\ &+ \frac{a_2}{r} + a_3 r + a_4\cos(\theta) + a_7 + a_{11}r\log(r)\cos(\theta) \\ &+ a_5\theta\sin(\theta) + a_6 r\theta\sin(\theta) + a_8 r\log(r) + a_9\log(r)\cos(\theta) \\ u_\theta(r,\theta) &= \frac{b_0\sin(\theta)}{r^2} + \frac{b_1\sin(\theta)}{r} + b_2\sin(\theta) + b_3\theta\cos(\theta) \\ &+ b_5 r\theta + b_6\log(r)\sin(\theta) b_4 r\theta\cos(\theta) + b_7 r\sin(\theta) \\ &+ b_8 r\log(r)\sin(\theta) + b_9 r^2\sin(\theta) \end{aligned} \tag{10.1}$$

where a_i and b_i are unknown constants, totaling 34, corresponding to 16 basis functions that are all elementary functions. The choice of these bases largely follows the guidelines below, which are similar to those used in the displacement system [1], but with special considerations for the polar coordinate system:

1. Use the lowest-order bases possible, assuming that the displacements lie in a space spanned by these basis functions. If this assumption is not valid, the order must be increased and additional bases included.
2. The chosen bases should satisfy the symmetry conditions. This drastically reduces the number of required bases while maintaining the desired order. For this problem, u_r must be an even function of θ, and u_θ must be an odd function of θ. The choices in Eq. (10.1) satisfy these requirements. Note that the lowest constant basis, 1, associated with the unknown constant a_7, is included only in u_r and not in u_θ, due to the same symmetry conditions.
3. The selection of sine and cosine functions is based on the sinusoidal nature of the problem. The choice of a single angle θ (rather than the double angle 2θ used in the previous section) arises from the presence of only one plane of mirror symmetry in this problem. The inclusion of logarithmic basis functions is due to the stresses in this problem (see Section 3.4), which contain the term $1/r$, whose antiderivative is $\log r$. This arises because stresses are related to the derivatives of displacements.
4. For the coordinate r, terms with both negative and positive powers are used, which is typical in a polar coordinate system.

The displacements assumed in Eq. (10.1) satisfy the major requirements for this problem and include a large number of arbitrary constants, allowing them to fulfill all detailed conditions.

Note that the validity of this assumption depends on the problem. If the displacements do not lie in the space spanned by the assumed basis functions, the displacement method will fail. Conversely, if the displacements do lie in this space, a solution for the displacements can be found. The strains and stresses can then be straightforwardly obtained, as they require only differentiation, which can always be performed for any linear combination of elementary basis functions [1,2].

Since the assumption of the displacements is essentially based on educated guessing, experience and a thorough understanding of the problem are crucial. Choosing an appropriate set of bases for a given problem is a challenging but critical step.

The code snippet for generating the expressions for the displacements is as follows:

```
# Define variables for displacement method for asymmetric problems:
E, ν = symbols('E, ν', positive=True)          # given material constants
r = sp.symbols('r', positive=True) # symbolic variable for radial axis
θ = sp.symbols('θ', real=True)                          # for tangential axis
R = sp.Matrix([r, θ])                                      # position vector
br = 0; bθ = 0                                              # zero body force

# Assumed basis functions for displacements (do not change the order):
urABases=[cos(θ)/r**3, cos(θ)/r**2, cos(θ)/r, 1/r, r, cos(θ),θ*sin(θ),\
    r*θ*sin(θ), 1, r*log(r), cos(θ)*log(r), r*cos(θ), r*log(r)*cos(θ),\
              r**2, r**2*log(r)*cos(θ), r**2*cos(θ), r**3, r**3*cos(θ)]
urABases = sp.Matrix(urABases)

uθABases=[sin(θ)/r**3, sin(θ)/r**2, sin(θ)/r, sin(θ), θ*cos(θ),\
  r*θ*cos(θ), θ*r, sin(θ)*log(r), r*sin(θ),r*log(r)*θ,r*log(r)*sin(θ),\
             r**2*sin(θ), r**2, r**2*log(r)*sin(θ), r**3,  r**2*sin(θ)]
uθABases = sp.Matrix(uθABases)

_, order_n = urABases[0].as_coeff_exponent(r)
_, order_p = urABases[-1].as_coeff_exponent(r)
order = order_p - order_n + 1              # compute the order difference
print(f"The highest inverse order of the monomial = {order}")

# unknown constants for displacements:
ai = sp.symbols(f'a:{len(urABases)}')                   # unknown constants
display(Math(f" \\text{{Unknown constants:}}"))
display(Math(f" \\text{{for $u_r$}}={latex(ai)}"))
ai = Matrix([ai[i] for i in range(len(urABases))])        # form a vector
```

```
bi = sp.symbols(f'b:{len(uθABases)}')                    # unknown constants
display(Math(f" \\text{{for $u_θ$}}={latex(bi)}"))
bi = Matrix([bi[i] for i in range(len(uθABases))])

cs = ai.col_join(bi)                    # form a single vector of constants
display(Math(f" \\text{{Total number of unknowns}}={latex(len(cs))}"))

urA = (ai.T@Matrix(urABases))[0]                    # Assumed displacements
uθA = (bi.T@Matrix(uθABases))[0]
display(Math(f"\\text{{Assumed displacements}}:"))
display(Math(f" \\text{{ $u_r$}}=\\small {latex(sum(urA.args[:8]))}"))
display(Math(f" \\hspace{{1cm}}+\\small  {latex(sum(urA.args[8:13]))}"))
display(Math(f" \\hspace{{1cm}}+\\small  {latex(sum(urA.args[13:]))}"))
display(Math(f" \\text{{ $u_θ$}}=\\small {latex(sum(uθA.args[:7]))}"))
display(Math(f" \\hspace{{1cm}}+\\small  {latex(sum(uθA.args[7:12]))}"))
display(Math(f" \\hspace{{1cm}}+\\small  {latex(sum(uθA.args[12:]))}"))
```

```
The highest inverse order of the monomial = 7
```

Unknown constants:

for $u_r = (a_0, a_1, a_2, a_3, a_4, a_5, a_6, a_7, a_8, a_9, a_{10}, a_{11}, a_{12}, a_{13}, a_{14}, a_{15}, a_{16}, a_{17})$

for $u_\theta = (b_0, b_1, b_2, b_3, b_4, b_5, b_6, b_7, b_8, b_9, b_{10}, b_{11}, b_{12}, b_{13}, b_{14}, b_{15})$

Total number of unknowns $= 34$

Assumed displacements:

$$u_r = \frac{a_0 \cos(\theta)}{r^3} + \frac{a_1 \cos(\theta)}{r^2} + a_{13} r^2 + a_{16} r^3 + \frac{a_3}{r} + a_4 r + a_5 \cos(\theta) + a_8$$

$$+ a_{10} \log(r) \cos(\theta) + a_{11} r \cos(\theta) + a_{15} r^2 \cos(\theta) + a_{17} r^3 \cos(\theta)$$

$$+ \frac{a_2 \cos(\theta)}{r} + a_{12} r \log(r) \cos(\theta) + a_{14} r^2 \log(r) \cos(\theta) + a_6 \theta \sin(\theta)$$

$$+ a_7 r \theta \sin(\theta) + a_9 r \log(r)$$

$$u_\theta = \frac{b_0 \sin(\theta)}{r^3} + \frac{b_1 \sin(\theta)}{r^2} + b_{11} r^2 \sin(\theta) + b_{12} r^2 + b_{14} r^3 + b_{15} r^2 \sin(\theta)$$

$$+ b_3 \sin(\theta) + \frac{b_2 \sin(\theta)}{r} + b_4 \theta \cos(\theta) + b_6 r \theta + b_7 \log(r) \sin(\theta)$$

$$+ b_8 r \sin(\theta) + b_{10} r \log(r) \sin(\theta) + b_{13} r^2 \log(r) \sin(\theta) + b_5 r \theta \cos(\theta)$$

$$+ b_9 r \theta \log(r)$$

The constructed displacements contain a total of 34 unknown constants. These constants are determined by enforcing the differential equilibrium equations, integral equilibrium equations, stress boundary conditions (SBCs), and displacement boundary conditions (DBCs). Computing these constants requires stress expressions, which can be obtained from the assumed displacements through differentiation using the constitutive equations.

10.3 Computing stresses

First, the strain expressions are computed from the displacements. The stresses are then obtained using the constitutive equations. This procedure can be performed using the functions `gr.displacement2StrainPolar()` and `gr.strain2stress2D()` developed in the previous chapter:

```
# The formulas for computing strains using the displacements:
# εrr = urA.diff(r)
# εθθ = urA/r + uθA.diff(θ)/r
# γrθ = urA.diff(θ)/r + uθA.diff(r) - uθA/r
# εvA_= sp.Matrix([εrr, εθθ, γrθ/2])                    # forms a vector

εvA = sp.Matrix(gr.displacement2StrainPolar([urA, uθA], R))

# Compute stress vector using the strain vector:
σvA = sp.Matrix(gr.strain2stress2D(E, ν, εvA, Ptype='stress'))
# executed in 2m 47.0s, finished 06:02:38 2025-10-25
```

The symbolic computations take more than three minutes on the author's laptop. The expressions for stresses and strains are very lengthy because of the large number of basis functions used in the displacement field. The following snippet shows the expression of only one component of the current stress:

```
## To take a look at the assumed strain and stresses, but lengthy:
#display(Math(f" \\text{{General solution for stresses:}} "))
σ_rrv = sp.simplify(σvA[0,0]) # pick only one
display(Math(f" \\text{{$σ_{{rrv}}$=}}\\scriptsize {latex(σ_rrv)}"))
#display(Math(f" \\text{{$σ_{{v}}$=}} {latex(sp.simplify(σvA))}"))
#display(Math(f" \\text{{$ε_{{v}}$=}} {latex(sp.simplify(εvA))}"))
#executed in 1m 7.20s, finished 06:24:41 2025-10-25
```

$$\sigma_{rrv} = \frac{\begin{aligned}&E(3a_0\cos(\theta) + 2a_1 r\cos(\theta) - a_{10}r^3\cos(\theta) - r^4(a_{11}\cos(\theta)\\ &+ a_{12}\log(r)\cos(\theta) + a_{12}\cos(\theta) + 2a_{13}r + 2a_{14}r\log(r)\cos(\theta)\\ &+ a_{14}r\cos(\theta) + 2a_{15}r\cos(\theta) + 3a_{16}r^2 + 3a_{17}r^2\cos(\theta) + a_4\\ &+ a_7\theta\sin(\theta) + a_9\log(r) + a_9) + r^2(a_2\cos(\theta) + a_3) - (a_0\cos(\theta)\\ &+ a_1 r\cos(\theta) + b_0\cos(\theta) + b_1 r\cos(\theta) + b_2 r^2\cos(\theta)\\ &+ r^3(b_{10}r\log(r)\cos(\theta) + b_{11}r^2\cos(\theta) + b_{13}r^2\log(r)\cos(\theta)\\ &+ b_{15}r^2\cos(\theta) + b_3\cos(\theta) - b_4\theta\sin(\theta) + b_4\cos(\theta) - b_5 r\theta\sin(\theta)\\ &+ b_5 r\cos(\theta) + b_6 r + b_7\log(r)\cos(\theta) + b_8 r\cos(\theta) + b_9 r\log(r))\\ &+ r^3(a_{10}\log(r)\cos(\theta) + a_{11}r\cos(\theta) + a_{12}r\log(r)\cos(\theta) + a_{13}r^2\\ &+ a_{14}r^2\log(r)\cos(\theta) + a_{15}r^2\cos(\theta) + a_{16}r^3 + a_{17}r^3\cos(\theta)\\ &+ a_4 r + a_5\cos(\theta) + a_6\theta\sin(\theta) + a_7 r\theta\sin(\theta) + a_8 + a_9 r\log(r))\\ &+ r^2(a_2\cos(\theta) + a_3)))\end{aligned}}{r^4(^2-1)}$$

Readers may uncomment the lines in the preceding code cell to view the complete general solutions for stresses and strains.

10.4 Enforcing differential equilibrium equations

The computed stresses are then used to formulate the equilibrium equations, as illustrated in the following snippet:

```
# Define equilibrium equation that needs to be satisfied:
br = 0; bθ = 0                                # Consider zero body forces
equi1=(σvA[0].diff(r)+σvA[2].diff(θ)/r+(σvA[0]-σvA[1])/r+br)\
                                                  .simplify().factor()
equi2=(σvA[2].diff(r)+σvA[1].diff(θ)/r+2*σvA[2]/r+bθ)\
                                                  .simplify().factor()
#display(equi1, equi2)                            # about a page long
#executed in 1m 55.0s, finished 10:38:17 2025-02-27
```

The expressions for the equilibrium equations are about one page long. Typical terms in the numerator of an equilibrium equation are

$$-\frac{\begin{aligned}&E(a_0\nu\cos(\theta) + \cdots + 3a_1 r^2\cos(\theta) + a_2 r^2\theta\nu\sin(\theta)\\ &+ \cdots - 3a_4 r^2\log(r)\cos(\theta) + \cdots)\end{aligned}}{2r^4(\nu-1)(\nu+1)} \tag{10.2}$$

To ensure that the equilibrium equations vanish, we use `sp.simplify()` and/or `sp.factor()` to convert the expressions into rational form. The equations are enforced by setting their numerators to zero. As shown above, the numerator is a linear combination of the combined basis functions. We group these terms according to the basis functions in the following steps for each of the two equilibrium equations:

Step 1: Group all terms containing the bases $\sin(\theta)$ and $\cos(\theta)$ (including their zeroth-order terms). The code snippet is as follows:

```
exprs = []
exprs.append(sp.numer(equi1).expand().coeff(sin(θ)).expand())
exprs.append(sp.numer(equi2).expand().coeff(cos(θ)).expand())
exprs.append(sp.numer(equi1).expand().coeff(cos(θ)).expand())
exprs.append(sp.numer(equi2).expand().coeff(sin(θ)).expand())
exprs.append(sp.numer(equi1).expand().coeff(cos(θ),0).coeff(sin(θ),0)\
                                                        .expand())
exprs.append(sp.numer(equi2).expand().coeff(cos(θ),0).coeff(sin(θ),0)\
                                                        .expand())
#display(exprs)                                         # 3/4 page long
```

These grouped expressions are given in a Python list. Typical terms in a grouped expression are combined monomials of r and $\log(r)$:

$$\begin{aligned}&[-Ea_2r^2\theta\nu + 3Ea_2r^2\theta + \cdots - Ea_4r^2\nu\log(r)\\&\quad + 3Ea_4r^2\log(r) - Ea_5r^3\nu + \cdots]\end{aligned} \tag{10.3}$$

Step 2: Group the terms containing various combinations of the bases θ and $\log(r)$:

For each grouped expression, further organize the terms into a list containing the bases of θ and $\log(r)$ (including their zeroth-order terms). These lists are then concatenated, resulting in a comprehensive list of expressions that are purely polynomials in r. The code snippet is as follows:

```
exprsθ    = [e.coeff(θ,1).coeff(log(r),0) for e in exprs]
exprsLog  = [e.coeff(log(r),1).coeff(θ,0) for e in exprs]
exprs00   = [e.coeff(θ,0).coeff(log(r),0) for e in exprs]
exprs11   = [e.coeff(θ,1).coeff(log(r),1) for e in exprs]
all_exprs = exprsθ + exprsLog + exprs00 + exprs11          # concatenate
#display(all_exprs)                                     # a half page long
```

Some expressions in this long list may be zero and are removed. The remaining expressions are then used to solve for the unknown constants.

A typical term in the cleaned-up list is a pure polynomial in r:

$$[\cdots + Eb_4r^3\nu + \cdots + Eb_{13}r^5 + \cdots - 6Eb_0 + 3Eb_1r\nu + \cdots - 6Ea_{13}r^5 - 16Ea_{16}r^6 + \cdots] \tag{10.4}$$

All these steps are implemented using **gr.constantUpdate()**, developed in the previous chapter. Note that the highest-order monomials in the numerator correspond to the largest order differences of r in the bases used to construct the displacements. The coefficient of each monomial (including the zeroth-order term) must vanish. This produces a large set of conditions, which are then used to determine the unknown constants. The corresponding code snippet is as follows:

```
sln_cs = {}                 # initialize solution of unknown constants, cs
sln_equi = gr.constantUpdate(all_exprs, r, order+1, cs, prnt=False)

# Update the solution for the unknown constants:
sln_cs = {**sln_equi}
sln_cs = {key: value.subs(sln_cs) for key, value in sln_cs.items()}
print(f" Current len of sln_cs={len(sln_cs)}; Total cs={len(cs)}")
sln_cs_ = gr.split_dict_n(sln_cs, 4)
display(Math(f"\\text{{Constants found: }}"))
display(Math(f"\\text{{ }} \\small {latex(sln_cs_[0])}"))
display(Math(f"\\text{{ }} \\scriptsize {latex(sln_cs_[1])}"))
display(Math(f"\\text{{ }} \\small {latex(sln_cs_[2])}"))
display(Math(f"\\text{{ }} \\small {latex(sln_cs_[3])}"))

# Check how many constants are found and how many unknowns remain:
list_keys = list(sln_cs.keys())
cs_Eql = list(set(list(cs))-set(list_keys))
print(f"Constants yet to be found: {cs_Eql}")
```

```
Total number of expressions = 192
Number of nonzero expressions = 32
Current len of sln_cs=25; Total cs=34
```

Constants found:

$$\{a_0 : 0, a_1 : b_1, a_{10} : -b_7, a_{11} : 0, a_{12} : 0, a_{13} : 0\}$$

$$\left\{a_{14} : 0, a_{15} : \frac{b_{11}(1-3\nu)}{\nu+5} + \frac{b_{15}(1-3\nu)}{\nu+5}, a_{16} : 0, a_{17} : 0, a_2 : 0, a_5 : -b_3 + \frac{b_4(\nu+1)}{\nu-3} + \frac{b_7(-\nu-1)}{\nu-3}\right\}$$

$$\left\{a_6 : b_4, a_7 : 0, a_8 : 0, a_9 : b_6\left(\frac{1}{2}-\frac{\nu}{2}\right), b_0 : 0, b_{10} : 0\right\}$$

$$\{b_{12} : 0, b_{13} : 0, b_{14} : 0, b_2 : 0, b_5 : 0, b_8 : 0, b_9 : 0\}$$

```
Constants yet to be found: [b11, b4, b7, a4, b1, b15, b3, b6, a3]
```

10.5 General expressions for displacements

As seen, 192 expressions were generated, of which 32 are nonzero. These 32 expressions are used to determine 27 constants, some of which are expressed in terms of the remaining constants. Many of these constants are found to be zero, leaving only 9 constants to be determined. Although this process is complex, it is fully automated in code, eliminating the need for manual calculations. We can now update the displacements.

The following updates the displacements to produce general formulas that satisfy the equilibrium equations and contain only the 9 remaining unknown constants:

```
# Compute the general expression for the displacements:
urg = urA.subs(sln_cs)
uθg = uθA.subs(sln_cs)

display(Math(f"\\text{{General displacements obeying equilibrium}}:"))
display(Math(f"\\text{{ $u_r$}}=\\scriptsize {latex(sum(urg.args[:5]))}"))
display(Math(f"\\hspace{{1cm}}+\\scriptsize {latex(sum(urg.args[5:]))}"))
display(Math(f"\\text{{ $u_θ$}}=\\scriptsize {latex(sum(uθg.args))}"))
```

General displacements obeying equilibrium:

$$u_r = \frac{a_3}{r} + a_4 r + \frac{b_1 \cos(\theta)}{r^2} + b_4\theta\sin(\theta)$$

$$+ \left(-b_3 + \frac{b_4(\nu+1)}{\nu-3} + \frac{b_7(-\nu-1)}{\nu-3}\right)\cos(\theta) b_6 r\left(\frac{1}{2} - \frac{\nu}{2}\right)\log(r)$$

$$+ -b_7\log(r)\cos(\theta) + r^2\left(\frac{b_{11}(1-3\nu)}{\nu+5} + \frac{b_{15}(1-3\nu)}{\nu+5}\right)\cos(\theta)$$

$$u_\theta = \frac{b_1\sin(\theta)}{r^2} + b_{11}r^2\sin(\theta) + b_{15}r^2\sin(\theta) + b_3\sin(\theta) + b_4\theta\cos(\theta)$$

$$+ b_6 r\theta + b_7\log(r)\sin(\theta)$$

We have obtained the general formulas of the displacements. These formulas are rewritten as follows for future reference:

$$u_r(r,\theta) = \frac{a_3}{r} + a_4 r + \frac{b_1\cos\theta}{r^2} + b_4\theta\sin\theta - \left(b_3 - \frac{\nu+1}{\nu-3}(b_4 - b_7)\right)\cos\theta$$

$$+ r\frac{b_6}{2}(1-\nu)\log r - b_7\log r\cos\theta + r^2\frac{1-3\nu}{\nu+5}(b_{11}+b_{15})\cos\theta$$

$$u_\theta(r,\theta) = \frac{b_1 \sin\theta}{r^2} + b_{11}r^2\sin\theta + b_{15}r^2\sin\theta + b_3\sin\theta$$
$$+ b_4\theta\cos\theta + b_6 r\theta + b_7\log r\sin\theta \tag{10.5}$$

These general displacements satisfy the differential equilibrium equations independent of these undetermined constants. Consequently, they can be applied to a wide range of problems using the displacement method in the polar coordinate system.

The displacements and stresses still contain 9 undetermined constants, which must be determined by enforcing conditions specific to each problem.

10.6 General expressions for stresses

The following snippet updates the stresses:

```
σvg_0 = sp.simplify(σvA.subs(sln_cs))                    # update stresses
```

```
#display(Math(f"\\text{{ General stresses: }}"))
display(Math(f"\\text{{$σ_{{rrv}}$=}}\\scriptsize {latex(σvg_0[0,0])}"))
#display(Math(f"\\text{{$σ_{{v}}$=}}\\small {latex(σvg_0)}")) # lengthy
```

$$\sigma_{rrv} = \frac{\begin{gathered}E(-r^3(\nu-3)(-4r(b_{11}+b_{15})(3\nu-1)\cos(\theta)\\ +(\nu+5)(2a_4-b_6(\nu-1)\log(r)-b_6(\nu-1)))\\ -\nu(r^2(-2r^2(b_{11}+b_{15})(\nu-3)(3\nu-1)\cos(\theta)\\ +(\nu-3)(\nu+5)(2a_4r+2b_4\theta\sin(\theta)-b_6r(\nu-1)\log(r)\\ -2b_7\log(r)\cos(\theta))+2(\nu+5)(-b_3(\nu-3)+b_4(\nu+1)\\ -b_7(\nu+1))\cos(\theta))+2(\nu-3)(\nu+5)(a_3r+2b_1\cos(\theta)\\ +r^2(b_{11}r^2\cos(\theta)+b_{15}r^2\cos(\theta)+b_3\cos(\theta)-b_4\theta\sin(\theta)\\ +b_4\cos(\theta)+b_6r+b_7\log(r)\cos(\theta))))\\ +2(\nu-3)(\nu+5)(a_3r+2b_1\cos(\theta)+b_7r^2\cos(\theta)))\end{gathered}}{2r^3(\nu-3)(\nu+5)(\nu^2-1)}$$

The resulting set of general stresses satisfies the equilibrium equations. These expressions are considerably simpler, as many of the unknown constants have been determined to be zero. The following snippet verifies that the current stresses satisfy the equilibrium equations:

```
def checkEquilibriumPolar(σvg, R, br, bθ):
    '''Check the equilibrium equations for a given stress vector and
    body forces br, bθ in the polar coordinate system.'''
    r, θ = R
    equi1=(σvg[0].diff(r)+σvg[2].diff(θ)/r+(σvg[0]-σvg[1])/r+br)
    equi2=(σvg[2].diff(r)+σvg[1].diff(θ)/r+2*σvg[2]/r+bθ)

    return (equi1.simplify(), equi2.simplify())

# check whether the equilibrium equation are satisfied:
checkEquilibriumPolar(σvg_0, [r, θ], br, bθ)          # should be (0, 0)
```

(0, 0)

It is confirmed that these stresses satisfy the equilibrium equations. As a good practice, the conditions should always be checked in a complicated derivation process. For this purpose, a Python function has been created.

The displacements are redefined in Python as follows, so that the general displacements can be used directly without rerunning the previous code. The stresses can be computed from the redefined displacements, which should match the stresses obtained earlier by updating the constants:

```
# Define variables for displacement method, as a new starting point:
E, ν = symbols('E, ν', positive=True)           # given material constants
r = sp.symbols('r', positive=True)  # symbolic variable for radial axis
θ = sp.symbols('θ', real=True)                       # for tangential axis
R = sp.Matrix([r, θ])                                    # position vector
br = 0; bθ = 0                                           # zero body force
a, b = sp.symbols('a,b',positive=True)              # variables of radii
p = sp.symbols('p', real=True)                               # the pressure
```

```
# Re-define the unknown constants, and then the general displacements:

#a2, a3, b2, b5, b0, b9, b3, b6= sp.symbols("a2, a3, b2,b5,b0,b9,b3,b6")
b6, a4, b3, a3, b7, b1, b15, b4, b11 = sp.symbols\
                              ("b6, a4, b3, a3, b7, b1, b15, b4, b11")

urg_= a3/r + a4*r + b1*cos(θ)/r**2 + b4*θ*sin(θ) + b6*r*(Rational(1,2)\
    - ν/2)*log(r)-b7*log(r)*cos(θ)+r**2*(b11*(1-3*ν)/(ν+5)+b15*(1-3*ν)\
    /(ν+5))*cos(θ) + (-b3 + b4*(ν+1)/(ν-3) + b7*(-ν-1)/(ν-3))*cos(θ)
#display((urg_-urg).simplify())
```

```
uθg_= b1*sin(θ)/r**2 + b11*r**2*sin(θ) + b15*r**2*sin(θ) + b3*sin(θ) \
    + b4*θ*cos(θ) + b6*r*θ + b7*log(r)*sin(θ)
#display((uθg_-uθg).simplify())

R = sp.Matrix([r, θ])                                    # position vector
u_g = [urg_, uθg_]                 # general displacements with 5 constants
εvg = gr.displacement2StrainPolar(u_g, R)                # general strains
#display(εg)

# Compute the general stresses with with 8 constants and check:
σvg = sp.simplify(Matrix(gr.strain2stress2D(E, ν, εvg,Ptype='stress')))
#display(Math(f"\\text{{General stresses $σ_{{v}}$=}} {latex(σvg)}"))

# Check against the stress obtained using the assumed displacements:
display(Math(f" \\text{{Agrees with the stresses obtained earlier? }}\
                                        {latex((σvg_0-σvg).simplify().T)}"))
```

Agrees with the stresses obtained earlier? $\begin{bmatrix}0 & 0 & 0\end{bmatrix}$

These remaining unknown constants must be determined by applying the stress boundary conditions (SBCs), integral equilibrium or force boundary conditions (FBCs), and finally the displacement boundary conditions (DBCs). We begin with the SBCs.

10.7 Enforcing stress boundary conditions

For the problem shown in Fig. 10.1, the stress boundary conditions on the inner and outer curved surfaces of the arch are

$$\begin{aligned} \sigma_{rr} = \sigma_{r\theta} = 0 \quad &\text{at} \quad r = a \\ \sigma_{rr} = \sigma_{r\theta} = 0 \quad &\text{at} \quad r = b \end{aligned} \tag{10.6}$$

The following snippet enforces these SBCs to determine some of the unknown constants. First, we compute the expressions for the stresses and then form the sub-expressions, which are the coefficients of the basis functions of r and θ. The process is similar to that used for enforcing the equilibrium equations, but somewhat simpler due to the smaller number of unknowns. The snippet is as follows:

```
σrrg, σθθg, σrθg = σvg                              # unpack the stress vector

exprs = []
expr = sp.simplify(σrrg.expand().subs(r,a).coeff(cos(θ),0))
exprs.append(expr)
exprs.append(σrrg.expand().subs(r,a).coeff(cos(θ)).simplify())
exprs.append(σrθg.expand().subs(r,a).coeff(sin(θ)).simplify())

expr = σrrg.expand().subs(r,b).coeff(cos(θ),0).simplify() + p
exprs.append(expr)
exprs.append(σrrg.expand().subs(r,b).coeff(cos(θ)).simplify())
exprs.append(σrθg.expand().subs(r,b).coeff(sin(θ)).simplify())

#display(exprs[0]);display(exprs[1]);display(exprs[2]))
```

We now use these expressions to determine some of the unknown constants. Due to the complexity of the problem, the solutions for these constants are still lengthy, so we introduce intermediate constants to simplify the expressions. The snippet is as follows:

```
# Solve SBCs for some of the unknown constants:

cs = [b6, a4, b3, a3, b7, b1, b15, b4, b11]
sln_cs = {}
sln_SBC = sp.solve(exprs, cs)      # solution of some unknown constants

# Update unknown constants:
sln_cs = {**sln_SBC}
sln_cs = {key: sp.expand_log(value.subs(sln_cs).expand().collect(E)\
  .collect(p).collect(a3).collect(b4)) for key, value in sln_cs.items()}
#display(sln_cs)                               # lengthy and needs simplified

# Simplify formulas for the constants using intermedia constants Ai:
A1, A2, A3, A4, A5, A6, A7 = sp.symbols('A1, A2, A3, A4, A5, A6, A7')
A1_ = sln_cs[a4].coeff(a3)
A2_ = sln_cs[a4].coeff(p)
A3_ = sln_cs[b1].coeff(b4)
A4_ = sln_cs[b15].coeff(b4)
A5_ = sln_cs[b6].coeff(a3)
A6_ = sln_cs[b6].coeff(p)
A7_ = sln_cs[b7].coeff(b4)

dic_A  = {A1_:A1, A2_:A2, A3_:A3, A4_:A4, A5_:A5, A6_:A6, A7_:A7}
Idic_A = {k: v for v, k in dic_A.items()}
Idic_A = {k: v.simplify() for k, v in Idic_A.items()}
```

```
# update unknown constants:
sln_cs = {key: value.subs(dic_A) for key, value in sln_cs.items()}
display(Math(f"\\text{{Constants found:}}=\\scriptsize {latex(sln_cs)}"))
Idic_A_2 = gr.split_dict_n(Idic_A, 2)
display(Math(f"\\text{{}}\\scriptsize {latex(Idic_A_2[0])}"))
display(Math(f"\\text{{}}\\scriptsize {latex(Idic_A_2[1])}"))

# compute the current displacements:
urg = urg_.subs(sln_cs).expand().collect(p).collect(cos(θ))\
            .collect(b4).collect(r*log(r)).collect(A6).collect(A5*a3)
uθg = uθg_.subs(sln_cs).simplify().collect(p).collect(sin(θ))
display(Math(f"\\text{{Current $u_{{r}}$=}}\\scriptsize \
     {latex(sum(urg.args[:2]))}\\text{{+}} {latex(sum(urg.args[3:]))}"))
display(Math(f"\\text{{+}}\\scriptsize {latex(urg.args[2])}"))
display(Math(f"\\text{{Current $u_{{θ}}$=}}\\scriptsize {latex(uθg)}"))

# compute the current stress after SBCs
σv = sp.simplify(σvg.subs(sln_cs))
#display(Math(f"\\text{{General stresses $σ_{{v}}$=}} {latex(σv)}"))
                                                              # Lengthy
# Check SBCs and equilibrium:
#display((σv[0].subs(r,a).simplify(), σv[0].subs(r,b).simplify()))
display(checkEquilibriumPolar(σv.subs(Idic_A), [r, θ], br, bθ))
```

Constants found: $=\{a_4 : A_1a_3 + A_2p, b_1 : A_3b_4, b_{15} : A_4b_4 - b_{11}, b_6 : A_5a_3 + A_6p, b_7 : A_7b_4\}$

$$\left\{A_1 : \frac{a^2 - b^2 + \log\left(a^{a^2(1-\nu)}b^{b^2(\nu-1)}\right)}{a^2b^2\log\left(a^{\nu+1}b^{-\nu-1}\right)}, A_2 : \frac{\log\left(a^{\nu-1}\right) - 1}{E\log\left(\frac{a}{b}\right)}, A_3 : \frac{a^2b^2(\nu+1)}{4\left(a^2+b^2\right)}\right\}$$

$$\left\{A_4 : \frac{\nu - 5}{4(a^2+b^2)}, A_5 : \frac{2(-a^2+b^2)}{a^2b^2\log\left(a^{\nu+1}b^{-\nu-1}\right)}, A_6 : \frac{2}{E\log\left(\frac{a}{b}\right)}, A_7 : \frac{\nu}{2} - \frac{1}{2}\right\}$$

Current $u_r = \frac{a_3}{r} + p\left(A_2r + A_6r\left(\frac{1}{2} - \frac{\nu}{2}\right)\log(r)\right) + A_1a_3r + A_5a_3r$ $\left(\frac{1}{2} - \frac{\nu}{2}\right)\log(r) + b_4\theta\sin(\theta) + \left(-b_3 + b_4\left(\frac{A_3}{r^2} - \frac{3A_4r^2\nu}{\nu+5} + \frac{A_4r^2}{\nu+5} - \frac{A_7\nu}{\nu-3} - A_7\log(r) - \frac{A_7}{\nu-3} + \frac{\nu}{\nu-3} + \frac{1}{\nu-3}\right)\right)\cos(\theta)$

Current $u_\theta = A_5a_3r\theta + A_6pr\theta + b_4\theta\cos(\theta) + \left(\frac{A_3b_4}{r^2} + A_4b_4r^2 + A_7b_4\log(r) + b_3\right)\sin(\theta)$

$(0,\ 0)$

The unknown constants are now reduced to three: a_3, b_3, and b_4. We next enforce the integral equilibrium conditions, also referred to as force boundary conditions (FBCs).

10.8 Enforcing integral equilibrium conditions

Consider a free body diagram of the arch covering $0 \leq \theta \leq \theta_B$, where θ_B is chosen arbitrarily, as shown in Fig. 10.2. The free body diagram is taken from the left half of the arch due to symmetry. It is important to note that the diagram is arbitrary, and θ_B does not need to correspond to the left end of the arch. This implies that the FBCs must be satisfied for any such free body diagram, which is a strong requirement.

The FBCs require the following integral equilibrium conditions to hold:

$$
\begin{aligned}
&\sum M_z = 0: \int_a^b \sigma_{\theta\theta}\big|_{\theta=\theta_B} r dr + \int_a^b \sigma_{\theta\theta}\big|_{\theta=0} r dr = 0 \\
&\sum F_x = 0: F_{\theta 0} - F_{\theta_B} \cos\theta_B - F_{rB} \sin\theta_B + \int_0^{\theta_B} p \sin\theta r d\theta = 0 \\
&\sum F_y = 0: F_{r0} + F_{rB} \cos\theta_B - F_{\theta_B} \sin\theta_B - \int_0^{\theta_B} p \cos\theta r d\theta = 0
\end{aligned}
\tag{10.7}
$$

In the preceding integral evaluation, we used the following observations:

1. The arch is circular, so the applied pressure points toward the origin. Its contribution to the moment M_z is therefore zero. Consequently, we

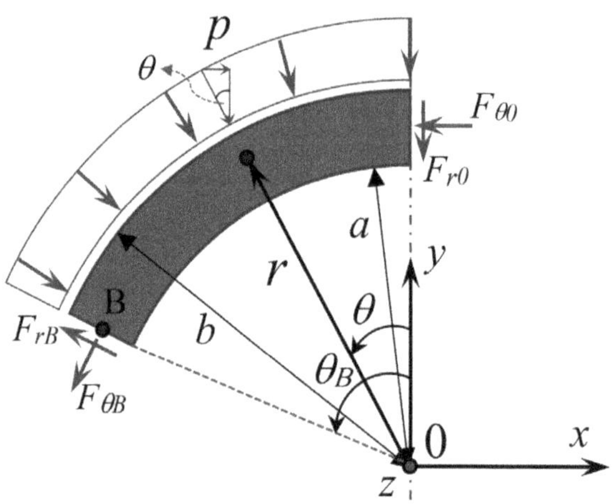

Figure 10.2. A free body diagram of the arch covering $0 \leq \theta \leq \theta_B$, with θ_B taken arbitrarily and not necessarily at the left end of the arch.

only need to consider the contribution of the normal stress $\sigma_{\theta\theta}$ on the cross-section.

2. Shear forces F_{r0} at the cross-sections $\theta = 0$ and $\theta = \theta_B$ do not contribute to M_z. Normal forces F_r at these cross-sections, however, do contribute to M_z.
3. The contribution of the pressure to the force F_x is obtained by projecting it onto the x-axis at any cross-section within $0 \leq \theta \leq \theta_B$, resulting in the corresponding integral.
4. Similarly, the contribution of the pressure to the force F_y is obtained by projecting it onto the y-axis at any cross-section within $0 \leq \theta \leq \theta_B$, leading to the corresponding integral.

10.8.1 *Enforcing the moment condition*

The following snippet enforces the conditions given in Eq. (10.7). First, the moment equilibrium $\sum M_z$:

```
# Compute the moment using current stresses with known constants:
σrr, σθθ, σrθ = σv
Int = sp.integrate
θB = sp.symbols('θ_B', real=True)
Mz = (Int(σθθ.subs(θ,θB)*r, (r,a,b)) + Int(σθθ.subs(θ,0)*r, (r,a,b)))
#display(Mz.expand().collect(cos(θB)).collect(b4).collect(a3))
```

The expression for M_z is two pages long and is therefore not displayed. Readers can view it by uncommenting the corresponding code line. Since M_z must be satisfied for any arbitrary θ_B, it is grouped into two parts: Group 1 consists of all terms containing the common factor $\cos(\theta_B)$, and Group 2 contains all remaining terms. Each of these two grouped expressions must vanish independently.

It is observed that the coefficients of $\cos(\theta_B)$ all involve the unknown constant b_4. Importantly, the coefficients of b_4 automatically become zero (these are the `termsM1` in the snippet below). The following code snippet confirms this:

```
termsM1=Mz.expand().coeff(cos(θB)) # group-1, terms with factor cos(θB)
termsM2=Mz.expand().coeff(cos(θB),0)    # group-2 of all the rest terms
termsM1_ = termsM1.expand().coeff(b4)           # further grouping by b4
display(termsM1_.simplify().subs(Idic_A).simplify())      # should be 0
```

0

Therefore, we need only to enforce `termsM2`, which can be used to solve for a_3:

```
termsM2_ = termsM2.subs(Idic_A).simplify().collect(p).collect(a3)
#display(termsM2_)  # Lengthy
sln_FBC = sp.solve([termsM2_], cs)
sln_FBC = {key: value.simplify() for key, value in sln_FBC.items()}
display(sln_FBC)

# simplify it and redefine for later use:
a3_numer, a3_denom = sln_FBC[a3].as_numer_denom()
a3_ = a3_numer.factor()/a3_denom.coeff(E).factor()/E
#display(Math(f"\\text{{$a_3$=}} {latex(a3_)}"))
```

$$\left\{ a_3 : \frac{a^2b^2p\left(a^2\nu + a^2 - b^2\nu - b^2 + \log\left(a^{2b^2(\nu+1)}b^{-2b^2(\nu+1)}\right)\right)}{E\left(a^4 - 4a^2b^2\log(a)^2 - 4a^2b^2\log(b)^2 - 2a^2b^2 + b^4 + \log\left(b^{\log\left(a^{8a^2b^2}\right)}\right)\right)} \right\}$$

We found a_3 by enforcing the moment equilibrium condition. It is rewritten manually in the following easy-to-read form via factorization:

$$a_3 = \frac{-a^2b^2p(\nu+1)\left(a^2 + 2b^2\log(a) - 2b^2\log(b) - b^2\right)}{E\left(a^2 - 2ab\log(a) + 2ab\log(b) - b^2\right)\left(a^2 + 2ab\log(a) - 2ab\log(b) - b^2\right)} \tag{10.8}$$

It depends on the pressure p. It is zero if $p = 0$.

10.8.2 *Property of a_3*

Since a_3 is responsible for enforcing this FBC, let us examine its behavior for a given arch with arbitrary parameters a, b, E, and ν. This can be done using the techniques described in Ref. [1].

The expression for a_3 is quite complicated, so we first set $a = 1$. This means that b will represent the ratio b/a with $b > 1$, which does not result in any loss of generality. With this setting, a_3 becomes a 1D function of b, with the other parameters fixed at a specific set of values. We can then examine its value. The snippet is as follows:

```
dic_={a:1, E:1, p:1}
a3_f = a3_.coeff((ν+1)).subs(dic_) # define a3_ as a function of b
display(Math(f"\\text{{Simplified (without coefficient) \
                                   $a_3(b)$=}}\small {latex(a3_f)}"))
display(Math(f"\\text{{Sampled values of $a_3$ = }} \
       {latex(a3_f.subs(b,1.01))}b^2; {latex(a3_f.subs(b,10.))}b^2"))
display(Math(f"\\text{{ Limit of $a_3$ when $b→1$ = }} \
                                   {latex(sp.limit(a3_f, b, 1))}"))
display(Math(f"\\text{{ Limit of $a_3$ when $b→∞$ = }} \
                                   {latex(sp.limit(a3_f, b, oo))}"))
```

Simplified (without coefficient)

$$a_3(b) = \frac{b^2\left(-2b^2\log(b) - b^2 + 1\right)}{\left(-b^2 - 2b\log(b) + 1\right)\left(-b^2 + 2b\log(b) + 1\right)}$$

Sampled values of $a_3 = -3090964.07912972b^2; -7.28514946846739b^2$

Limit of a_3 when $b \to 1 = -\infty$

Limit of a_3 when $b \to \infty = -\infty$

These results imply the following properties of a_3:

$$\begin{aligned} &1: \quad a_3 < 0 \quad \text{for all} \quad b > a > 0 \\ &2: \quad \text{as} \quad b \to a, a_3 \to \quad \infty \\ &3: \quad \text{as} \quad b \to \infty, a_3 \to -\infty \end{aligned} \tag{10.9}$$

This can be proven by writing the simplified a_3 in terms of three sub-functions:

$$a_3 = \frac{b^2 f_1}{f_2 f_3} \tag{10.10}$$

where

$$\begin{aligned} f_1 &= -2b^2\log b - b^2 + 1 \\ f_2 &= -b^2 - 2b\log b + 1 \\ f_3 &= -b^2 + 2b\log b + 1 \end{aligned} \tag{10.11}$$

The property of f_1 is found using the following Python function (using x as a general independent variable).

```
def exam_f(f, x, a, b=oo):
    '''exam a given function f(x) in domain (a, b)'''
    print(f" f(a)={sp.limit(f, x, a)}, f(b)={sp.limit(f, x, b)}")
    df = f.diff(x)
    print(f" df(a)={sp.limit(df, x, a)}, df(b)={sp.limit(df, x, b)}")
    sln_x = sp.solve(df, x)#, domain=sp.Interval(a, b))
    display(Math(f" \\frac{{df}}{{dx}}={latex(df)}, \
                    \\text{{ its stationary point }}= {latex(sln_x)}"))
    d2f = f.diff(x,2)
    print(f" d2f(a)={sp.limit(d2f, x, a)}, d2f(b)={sp.limit(d2f,x,b)}")
    d2sln_x = sp.solve(d2f, x)#, domain=sp.Interval(a, b))
    display(Math(f" \\frac{{d^2f}}{{dx^2}}={latex(d2f)}, \
                  \\text{{ its stationary point }}= {latex(d2sln_x)}"))
```

Let's compute the derivatives of these sub-functions:

```
x = sp.symbols('x', positive=True)          # define independent variable
condition = x > 1                                         # with a condition

f1 = -2*x**2*log(x) - x**2 + 1                                 # define f1
f2 = -x**2 - 2*x*log(x) + 1                                    # f2
f3 = -x**2 + 2*x*log(x) + 1                                    # f3
exam_f(f1, x, 1, oo)
#exam_f(f2, x, 1, oo)
#exam_f(f3, x, 1, oo)
```

```
f(a)=0, f(b)=-oo
df(a)=-4, df(b)=-oo
```

$$\frac{df}{dx} = -4x\log(x) - 4x, \text{ its stationary point } = \left[e^{-1}\right]$$

```
d2f(a)=-8, d2f(b)=-oo
```

$$\frac{d^2f}{dx^2} = -4(\log(x) + 2), \text{ its stationary point } = \left[e^{-2}\right]$$

The results show that the function $f_1(\infty) = -\infty$, its first derivative is also negative, and it does not have any roots within $(1, \infty)$. In addition, its second derivative is negative and has no roots within $(1, \infty)$, indicating that the function is concave downward [1]. Therefore, $f_1(b)$ is a negative, monotonically decreasing function, approaching $-\infty$ as $b \to \infty$.

The same behavior is observed for $f_2(b)$ and $f_3(b)$. Readers may perform a similar examination using the `exam_f()` function.

Now, using Eq. (10.10), we can prove Eq. (10.9):

1. a_3 is negative and monotonically decreasing over $(1, \infty)$ because f_1, f_2, and f_3 are all negative and monotonically decreasing.
2. As $b \to a$, $a_3 \to -\infty$ because $a_3 \to \frac{0}{0 \times 0} \to -\infty$, with f_1, f_2, and f_3 approaching zero at the same rate with respect to b.
3. As $b \to \infty$, $a_3 \to -\infty$ because $a_3 \to \frac{\infty}{\infty \times \infty} \to 0$ from the negative side, while f_1, f_2, and f_3 approach ∞ at the same rate with respect to b.

All these functions, f_1, f_2, f_3, and a_3, are plotted using the following snippet:

```
# Convert SymPy expressions to Python functions:

func1 = sp.lambdify(x, f1,'numpy')
func2 = sp.lambdify(x, f2,'numpy')
func3 = sp.lambdify(x, f3,'numpy')
func4 = sp.lambdify(x, x**2*f1/f2/f3,'numpy')  # a3
labels=[r'$f_{1}$', r'$f_{1}$', r'$f_{1}$', r'$a_{3}$']

x_axis  = np.linspace(1.25, 8, 400)              # x values for the plot
x_data = [x_axis,  x_axis, x_axis,  x_axis]
y_data = [func1(x_data[0]), func2(x_data[1]), func3(x_data[0]),
          func4(x_data[1])]
#plt.ion()
#gr.plot_fs(x_data, y_data, labels, '$b$', 'functions', \
#                                     'imagesSMp/a3func',ioff=False)
#                                  # Use ioff=False to display the plot
#plt.show()
```

10.8.3 *Monotonic behavior of special functions*

The monotonic characteristics of the special functions f_1, f_2, f_3, and a_3 are illustrated in Fig. 10.3.

Figure 10.3 demonstrates how f_1, f_2, and f_3 combine to form a_3 in order to satisfy the boundary conditions.

10.9 Verifying force equilibrium conditions

It turns out that the second and third equations in Eq. (10.7) are already satisfied. The following code snippets confirm this:

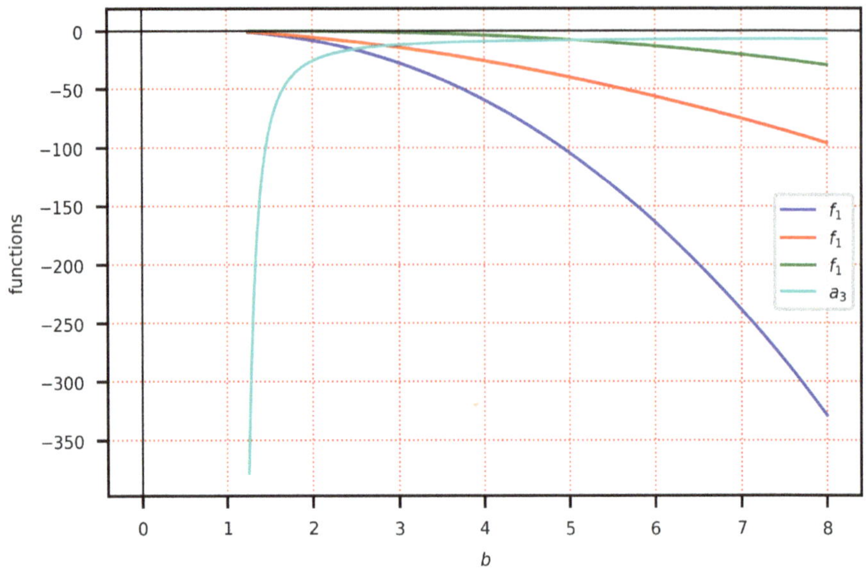

Figure 10.3. Monotonic behavior of special functions from the arch problem. These functions consist of monomials and logarithmic bases.

10.9.1 *Internal forces on the cross-sections*

We first compute all the forces on the cross-sections at $\theta = 0$ and $\theta = \theta_B$ via integration over these cross-sections:

```
Fr0 = Int(σrθ.subs(θ,0), (r,a,b)).subs(Idic_A).simplify()
display(Math(f"\\text{{at $θ=0: F_{{r0}}$=}} {latex(Fr0)}"))

Fθ0 = Int(σθθ.subs(θ,0), (r,a,b)).subs(Idic_A).simplify().collect(b4)\
                                                        .collect(p*b)
Fθ0 = gr.splitFraction(Fθ0)
display(Math(f"\hspace{{1.7cm}}\\text{{$F_{{θ0}}$=}}\\scriptsize \
                                                    {latex(Fθ0)}"))
FrB = Int(σrθ.subs(θ,θB), (r,a,b)).subs(Idic_A).simplify()
display(Math(f"\\text{{at $θ=θ_B: F_{{rB}}$=}} {latex(FrB)}"))

FθB = Int(σθθ.subs(θ,θB), (r,a,b)).subs(Idic_A).simplify()
FθB = sp.expand_log(FθB).collect(cos(θB)).collect(E).collect(p*b)
FθB = gr.splitFraction(FθB).collect(a**2+b**2).collect(b4)
display(Math(f"\hspace{{1.9cm}}\\text{{$F_{{θB}}$=}}\\scriptsize \
                                                    {latex(FθB)}"))
```

at $\theta = 0 : F_{r0} = 0$

$$F_{\theta 0} = \frac{bp\left(-2a^2 - 2b^2\right)}{2a^2 + 2b^2} + \frac{b_4\left(Ea^2 - Eb^2 + \log\left(a^{-E(a^2+b^2)}b^{E(a^2+b^2)}\right)\right)}{2a^2 + 2b^2}$$

$$\text{at } \theta = \theta_B : F_{rB} = \frac{Eb_4\left(a^2 - b^2 + \log\left(a^{-a^2-b^2}b^{a^2+b^2}\right)\right)\sin(\theta_B)}{2(a^2+b^2)}$$

$$F_{\theta B} = \frac{Eb_4\left(a^2 - b^2 + (a^2+b^2)\left(-\log(a) + \log(b)\right)\right)\cos(\theta_B)}{2a^2+2b^2} + \frac{bp\left(-2a^2 - 2b^2\right)}{2a^2+2b^2}$$

10.9.2 *Sum of internal forces*

It is observed that the internal forces consist of two parts: one associated with the pressure p and the other with the constant b_4. The following snippet computes the sum of all these forces required by the last two equations in Eq. (10.7):

```
Fx=(Fθ0-FrB*sin(θB)-FθB*cos(θB)+p*Int(sin(θB-θ)*b, (θ,0,θB))).simplify()
Fy=(Fr0+FrB*cos(θB)-FθB*sin(θB)-p*Int(cos(θB-θ)*b, (θ,0,θB))).simplify()
display(Math(f"\\text{{$\sum F_{{x}}$=}} {latex(Fx)}"))
display(Math(f"\\text{{$\sum F_{{y}}$=}} {latex(Fy)}"))
```

$$\sum F_x = 0$$

$$\sum F_y = 0$$

It is observed that these two equilibrium equations are automatically satisfied. This occurs because the current stresses already satisfy both the differential equilibrium equations and the stress boundary conditions, which involve the externally applied pressure p. The following snippet confirms this:

```
# Checking:
display((σv[0].subs(Idic_A).subs(r,a).simplify(),
         σv[0].subs(Idic_A).subs(r,a).simplify()))
display(checkEquilibriumPolar(σv.subs(Idic_A), [r, θ], br, bθ))  # (0,0)
```

$(0,\ 0)$
$(0,\ 0)$

It is now clear that all the equilibrium conditions (both differential and integral) as well as the SBCs have been satisfied. However, two unknown constants, b_3 and b_4, remain to be determined using the displacement boundary conditions (DBCs) at the two ends of the arch.

10.10 Enforcing displacement conditions

Let us assume that both displacement components are zero at the midpoints of the cross-section at the two ends of the arch, which are labeled as points B and B′ in Fig. 10.1:

$$\begin{aligned} u_r &= 0 \quad \text{at} \quad \theta = \pm\theta_B, \quad r = \frac{a+b}{2} \\ u_\theta &= 0 \quad \text{at} \quad \theta = \pm\theta_B, \quad r = \frac{a+b}{2} \end{aligned} \tag{10.12}$$

The following snippet defines these displacement boundary conditions. Due to the complexity of the equations, the DBCs are expected to be quite lengthy. To facilitate the solution process, we introduce intermediate constants D_i, which simplify the expressions and allow `sp.solve()` to obtain the solution within a reasonable time. The code snippet is as follows:

```
# Simplify the expressions for DBCs:

D0, D1, D2, D3, D4, D5 = sp.symbols('D0, D1, D2, D3, D4, D5')
urg = urg_.subs(sln_cs)
expr_ur_θB=urg.subs({θ:θB, r:(a+b)/2}).expand().collect(b3).collect(b4)
expr_uθ_θB=uθg.subs({θ:θB, r:(a+b)/2}).expand().collect(b3).collect(b4)

D0_  = expr_ur_θB.coeff(b3, 0).coeff(b4, 0)       # simplify the results
D1_  = expr_ur_θB.coeff(b4, 0).coeff(b3, 1)
D2_  = expr_ur_θB.coeff(b4, 1).coeff(b3, 0)

D3_  = expr_uθ_θB.coeff(b3, 0).coeff(b4, 0)
D4_  = expr_uθ_θB.coeff(b4, 0).coeff(b3, 1)
D5_  = expr_uθ_θB.coeff(b4, 1).coeff(b3, 0)

dic_D  = {D0_:D0, D1_:D1, D2_:D2, D3_:D3, D4_:D4, D5_:D5}
Idic_D = {k: v.simplify() for v, k in dic_D.items()}
Idic_D_2 = gr.split_dict_n(Idic_D, 2)

expr_ur_θB_= expr_ur_θB.subs(dic_D)
expr_uθ_θB_= expr_uθ_θB.subs(dic_D)
display(Math(f"\\text{{Simplified expr, $u_{{r}}$=0 at arch end: }}\
                                              {latex(expr_ur_θB_)}=0"))
display(Math(f"\\text{{Simplified expr, $u_{{θ}}$=0 at arch end: }}\
                                              {latex(expr_uθ_θB_)}=0"))
```

Simplified expr, $u_r = 0$ at arch end: $D_0 + D_1 b_3 + D_2 b_4 = 0$

Simplified expr, $u_\theta = 0$ at arch end: $D_3 + D_4 b_3 + D_5 b_4 = 0$

We have obtained very simple linear algebraic equations for the two unknowns, b_3 and b_4, which can be solved easily. However, the expressions for some of the D_i are quite lengthy. Let us perform some bookkeeping to make these expressions more concise in terms of the original parameters that define the problem, such as the radii of the arch and the material properties.

The following Python function can assist with this task. Since it is intended only for expression simplification, readers may skip reading it and simply use it:

```
def splitFraction(expr):
    '''Separate numerator of expr with the common denominator'''
    numerator, denominator = expr.as_numer_denom()
    numerator_terms = numerator.as_ordered_terms()
    separated_terms = [term / denominator for term in numerator_terms]
    return sum(separated_terms)
```

```
# Write Di in terms of the problem parameters and simplifications:

tmp = sp.expand_log(Idic_D[D0].subs(Idic_A)).simplify()
tmp1 = sp.expand_log(splitFraction(tmp).args[1])
tmp2 = sp.expand_log(splitFraction(tmp).args[0]).collect((1-ν))\
    .collect((-a**2+b**2+(1-ν)*(-a**2*log(a)+b**2*log(b)))).collect(a)\
    .collect(b).collect((ν-1)).collect((-(1-ν)*log(a)-1)).collect(a3*E)
Idic_D[D0] = tmp1 + tmp2

D0diff = sp.expand_log((Idic_D[D0] - D0_.subs(Idic_A))).simplify()==0

tmp = Idic_D[D2].subs(Idic_A).simplify()
tmp = splitFraction(tmp).collect(a**2).collect(b**2).collect((3*ν-1))\
         .collect(cos(θB)).subs((log(a+b)-log(2)),(log((a+b)/2)))\
                                         .subs(a**2+2*a*b+b**2, (a+b)**2)
Idic_D[D2] = tmp
D2diff = sp.expand_log((Idic_D[D2] - D2_.subs(Idic_A))).simplify()==0

tmp=Idic_D[D3].subs(Idic_A).simplify().collect(a**2-b**2).collect(E*a3)
Idic_D[D3] = tmp
D3diff = sp.expand_log((Idic_D[D3] - D3_.subs(Idic_A))).simplify()==0

tmp = Idic_D[D5].subs(Idic_A).simplify()
tmp = splitFraction(tmp).collect(sin(θB)).subs((log(a+b)-log(2)),\
        (log((a+b)/2))).collect((ν+5)).subs(a**2+2*a*b+b**2, (a+b)**2)
tmp1 = tmp.args[1]
tmp2 = splitFraction(tmp.args[0]).collect(16)
Idic_D[D5] = (tmp1 + tmp2).collect(sin(θB))
D5diff = sp.expand_log((Idic_D[D5] - D5_.subs(Idic_A))).simplify()==0

print("Rewritten D_i correct?", D0diff, D2diff, D3diff, D5diff)
#display(Idic_D)                 # one may take look, but still lengthy
```

```
Rewritten D_i correct? True True True True
```

All the expressions for D_i are simplified and stored in the dictionary `Idic_D`. We can now solve the foregoing two equations for b_3 and b_4:

```
# Enforce the displacement boundary conditions at the arch ends:
sln_DBC = sp.solve([expr_ur_θB_, expr_uθ_θB_], [b3, b4])
sln_DBC = {k: v.simplify() for k, v in sln_DBC.items()}
display(sln_DBC)
```

$$\left\{b_3 : \frac{-D_0D_5 + D_2D_3}{D_1D_5 - D_2D_4},\ b_4 : \frac{D_0D_4 - D_1D_3}{D_1D_5 - D_2D_4}\right\}$$

We have successfully found all these constants.

10.11 Final displacements

Note that the expressions for the displacement component u_θ are simple enough, but u_r is very lengthy. Let us simplify it:

```
# Simplify the displacement expressions:

d0, d1, d2, d3, d4, d5, d6, d7=sp.symbols('d0, d1, d2, d3,d4,d5,d6,d7')
d_bases=sp.Matrix([cos(θ), θ*sin(θ), 1/r, r, cos(θ)*log(r), \
                   cos(θ)/r**2, r**2*cos(θ), r*log(r)])
disp_d = sp.Matrix([d0, d1, d2, d3, d4, d5, d6, d7])
expr = urg_.subs(sln_cs).expand()

d0_ = expr.coeff(cos(θ)).coeff(r, 0)
d1_ = expr.coeff(θ*sin(θ)).coeff(r, 0)
d2_ = expr.coeff(r,-1)
d3_ = expr.coeff(r, 1).coeff(log(r),0)
d4_ = expr.coeff(cos(θ)).coeff(log(r))
d5_ = expr.coeff(r,-2).coeff(cos(θ))
d6_ = expr.coeff(r, 2).coeff(cos(θ))
d7_ = expr.coeff(r, 1).coeff(log(r))

dic_d= {d0_:d0, d1_:d1, d2_:d2, d3_:d3, d4_:d4, d5_:d5, d6_:d6, d7_:d7}
Idic_d = {k: v.simplify() for v, k in dic_d.items()}
ur_final = (disp_d.T@d_bases)[0]
uθ_final = uθg_.subs(sln_cs).simplify()
display(Math(f"\\text{{Simplification correct?}}= \
             {latex((ur_final.subs(Idic_d)-urg).simplify()==0)}"))
#display(Math(f" \\text{{Final displacements: }} "))
#display(Math(f"\\text{{$u_{{θ}}$=}}\\small {latex(uθ_final)}"))

ur_final_=d0*cos(θ) + d1*θ*sin(θ) + d2/r + d3*r + d4*log(r)*cos(θ) \
          + d5*cos(θ)/r**2 + d6*r**2*cos(θ) + d7*r*log(r)
uθ_final_=A3*b4*sin(θ)/r**2 + A4*b4*r**2*sin(θ) + A5*a3*r*θ + A6*p*r*θ\
          + A7*b4*log(r)*sin(θ) + b3*sin(θ) + b4*θ*cos(θ)
```

Simplification correct? = True

The final displacements are rewritten as follows for future reference:

$$
\begin{aligned}
u_r &= d_0 \cos\theta + d_1\theta \sin\theta + \frac{d_2}{r} + d_3 r + d_4 \log r \cos\theta + \frac{d_5 \cos\theta}{r^2} \\
&\quad + d_6 r^2 \cos\theta + d_7 r \log r \\
u_\theta &= \frac{A_3 b_4 \sin\theta}{r^2} + A_4 b_4 r^2 \sin\theta + A_5 a_3 r\theta + A_6 p r\theta + A_7 b_4 \log r \sin\theta \\
&\quad + b_3 \sin\theta + b_4 \theta \cos\theta
\end{aligned}
\tag{10.13}
$$

All these constants in the formulas are given in the following dictionary:

$$
\begin{aligned}
&\left\{d_0 : \frac{-A_7 b_4 \nu - A_7 b_4 - b_3 (\nu - 3) + b_4 \nu + b_4}{\nu - 3},\right. \\
&\left. d_1 : b_4,\ d_2 : a_3,\ d_3 : A_1 a_3 + A_2 p\right\} \\
&\left\{d_4 : -A_7 b_4,\ d_5 : A_3 b_4,\ d_6 : \frac{A_4 b_4 \cdot (1 - 3\nu)}{\nu + 5},\right. \\
&\left. d_7 : -\frac{A_5 a_3 \nu}{2} + \frac{A_5 a_3}{2} - \frac{A_6 p \nu}{2} + \frac{A_6 p}{2}\right\}
\end{aligned}
\tag{10.14}
$$

The related constants are found earlier in Python dictionaries. All those are displayed using the following snippet:

```
#Idic_d_2 = gr.split_dict_n(Idic_d, 2)
#display(Idic_d_2[0]); display(Idic_d_2[1])

# Related constants:
display(Math(f"\\text{{}}\\scriptsize {latex(Idic_A_2[0])}"))
display(Math(f"\\text{{}}\\scriptsize {latex(Idic_A_2[1])}"))
display(Math(f"\\text{{}}\\scriptsize {latex(a3_)}"))
display(Math(f"\\text{{}}\\scriptsize {latex(sln_DBC)}"))

Idic_D_6 = gr.split_dict_n(Idic_D, len(Idic_D))
display(Math(f"\\text{{Constants $D_{{i}}$: }} "))
display(Math(f"\\text{{}}\\scriptsize {latex(Idic_D_6[0])}"))
display(Math(f"\\text{{}}\\scriptsize {latex(Idic_D_6[1])}"))
display(Math(f"\\text{{}}\\scriptsize {latex(Idic_D_6[2])}"))
display(Math(f"\\text{{}}\\scriptsize {latex(Idic_D_6[3])}"))
display(Math(f"\\text{{}}\\scriptsize {latex(Idic_D_6[4])}"))
display(Math(f"\\text{{}}\\scriptsize {latex(Idic_D_6[5])}"))
#sln_All = {**Idic_A, **sln_DBC.subs(Idic_D), **Idic_D, **{a3:a3_}}
```

$$\left\{A_1 : \frac{a^2 - b^2 + \log\left(a^{a^2(1-\nu)}b^{b^2(\nu-1)}\right)}{a^2b^2\log\left(a^{\nu+1}b^{-\nu-1}\right)},\ A_2 : \frac{\log\left(a^{\nu-1}\right) - 1}{E\log\left(\frac{a}{b}\right)},\right.$$
$$\left. A_3 : \frac{a^2b^2(\nu+1)}{4(a^2+b^2)}\right\}$$

$$\left\{A_4 : \frac{-\nu-5}{4(a^2+b^2)},\ A_5 : \frac{2(-a^2+b^2)}{a^2b^2\log\left(a^{\nu+1}b^{-\nu-1}\right)},\ A_6 : \frac{2}{E\log\left(\frac{a}{b}\right)},\ A_7 : \frac{\nu}{2} - \frac{1}{2}\right\}$$

$$\frac{a^2b^2p(\nu+1)\left(a^2 + 2b^2\log(a) - 2b^2\log(b) - b^2\right)}{E\left(a^2 - 2ab\log(a) + 2ab\log(b) - b^2\right)\left(a^2 + 2ab\log(a) - 2ab\log(b) - b^2\right)}$$

$$\left\{b_3 : \frac{-D_0D_5 + D_2D_3}{D_1D_5 - D_2D_4},\ b_4 : \frac{D_0D_4 - D_1D_3}{D_1D_5 - D_2D_4}\right\}$$

Constants D_i:

$$\left\{D_0 : \frac{2a_3}{a+b} + \frac{\begin{array}{l} Ea_3\left((-a-b)\left(-a^2+b^2+(\nu-1)\left(a^2\log(a) - b^2\log(b)\right)\right)\right. \\ \left.+(a+b)\left(a^2-b^2\right)(\nu-1)\log\left(\frac{a}{2}+\frac{b}{2}\right)\right) \\ +a^2b^2p(\nu+1)\left(-(a+b)(\nu-1)\log\left(\frac{a}{2}+\frac{b}{2}\right)\right. \\ \left.+(a+b)\left((\nu-1)\log(a) - 1\right)\right) \end{array}}{2Ea^2b^2(\nu+1)\left(\log(a) - \log(b)\right)}\right\}$$

$$\{D_1 : -\cos(\theta_B)\}$$

$$\left\{D_2 : \left(\frac{a^2b^2(\nu+1)}{(a+b)^2(a^2+b^2)} + (3\nu-1)\left(\frac{ab}{8(a^2+b^2)} + \frac{1}{16}\right)\right)\right.$$
$$\left.\cos(\theta_B) + \frac{\begin{array}{l}(\nu-3)\left(2\theta_B\sin(\theta_B) - (\nu-1)\log\left(\frac{a}{2}+\frac{b}{2}\right)\cos(\theta_B)\right) \\ +(-\nu(\nu-1)+\nu+3)\cos(\theta_B)\end{array}}{2(\nu-3)}\right\}$$

$$\left\{D_3 : -\frac{\theta_B\left(Ea_3(a+b)\left(a^2-b^2\right) - a^2b^2p(a+b)(\nu+1)\right)}{Ea^2b^2(\nu+1)\left(\log(a) - \log(b)\right)}\right\}$$

$$\{D_4 : \sin(\theta_B)\}$$

$$\left\{D_5 : \theta_B\cos(\theta_B) + \left(\frac{a^2b^2(\nu+1)}{(a+b)^2(a^2+b^2)} - \frac{(a+b)^2(\nu+5)}{16(a^2+b^2)}\right.\right.$$
$$\left.\left.+\frac{(\nu-1)\log\left(\frac{a}{2}+\frac{b}{2}\right)}{2}\right)\sin(\theta_B)\right\}$$

10.12 Final stresses

10.12.1 *Stress simplification*

Let's now simplify the stresses by introducing intermediate constants:

```
# Genearl stresses:  σv = sp.simplify(σvg.subs(sln_cs))
# Simplify the stress expressions, for σrr:
σrr, σθθ, σrθ = σv
σrr_tmp = sp.expand_log(σrr.subs(Idic_A).subs({a3:a3_}).simplify())\
            .collect(cos(θ)).collect(p).collect(E).collect(b4)
σrr_ = gr.splitFraction(σrr_tmp)
#display(Math(f"\\text{{Moment $σ_{{rr}}$=}}\\small {latex(σrr_)}"))
σdiff=(σrr_-sp.expand_log(σrr.subs(Idic_A).subs({a3:a3_}))).simplify()
display(Math(f" \\text{{Substitution correct? }} {latex(σdiff==0)}"))

s0, s1, s2, s3 = sp.symbols('s0, s1, s2, s3')
s_bases=sp.Matrix([1/r**2, 1, log(r)])
stress_s = sp.Matrix([s1, s2, s3])

σrr_numer, σrr_denom = σrr_.as_numer_denom()
σrr_numer_E = σrr_numer.coeff(E*b4).factor()
σrr_E = σrr_numer.coeff(E*b4).factor()/σrr_denom.factor() * E*b4

σrr_numer_p = σrr_numer.coeff(p).factor()
σrr_p = σrr_numer.coeff(p).factor()/σrr_denom.factor()
σrrp_numer, σrrp_denom = σrr_p.as_numer_denom()
s0_ = σrrp_denom.coeff(r**2)

expr = ((σrrp_numer/s0)/r**2).expand()
s1_ = expr.coeff(r,-2).coeff(log(r),0)
s2_ = expr.coeff(r, 0).coeff(log(r),0)
s3_ = expr.coeff(log(r),1)

dic_s= {s0_:s0, s1_:s1, s2_:s2, s3_:s3}

Idic_s = {k: v.simplify() for v, k in dic_s.items()}
σrr_final = p*(stress_s.T@s_bases)[0] + σrr_E
display(Math(f"\\text{{Simplification correct? }}= \
   {latex((σrr_final.subs(Idic_s).subs(s0,s0_)-σrr_).simplify()==0)}"))
display(Math(f" \\text{{Final stress: $σ_{{rr}}$=}}\\scriptsize \
                                                   {latex(σrr_final)}"))
```

Substitution correct? True

Simplification correct? $=$ True

Final stress: $\sigma_{rr} = -\frac{Eb_4\left(-a+r\right)\left(a+r\right)\left(-b+r\right)\left(b+r\right)\cos\left(\theta\right)}{2r^3(a^2+b^2)} + p\left(s_2 + s_3\log\left(r\right) + \frac{s_1}{r^2}\right)$

```
# Simplify the stress expressions, for σθθ:

σθθ_tmp = sp.expand_log(σθθ.subs(Idic_A).subs({a3:a3_}).simplify())\
                .collect(cos(θ)).collect(p).collect(E).collect(b4)
σθθ_ = gr.splitFraction(σθθ_tmp)
#display(Math(f"\\text{{Stress $σ_{{θθ}}$=}}\\small {latex(σθθ_)}"))
σdiff=(σθθ_-sp.expand_log(σθθ.subs(Idic_A).subs({a3:a3_}))).simplify()
display(Math(f" \\text{{Substitution correct? }} {latex(σdiff==0)}"))

s4, s5, s6 = sp.symbols('s4, s5, s6 ')
s_bases=sp.Matrix([1/r**2, 1, log(r)])

σθθ_numer, σθθ_denom = σθθ_.as_numer_denom()
σθθ_numer_E = σθθ_numer.coeff(E*b4).factor()
σθθ_E = σθθ_numer.coeff(E*b4).factor()/σθθ_denom.factor() * E*b4

σθθ_numer_p = σθθ_numer.coeff(p).factor()
σθθ_p = σθθ_numer.coeff(p).factor()/σθθ_denom.factor()
σθθp_numer, σθθp_denom = σθθ_p.as_numer_denom()

expr = ((σθθp_numer/s0)/r**2).expand()
s4_ = expr.coeff(r,-2).coeff(log(r),0)      # this is the same as -s1_
s5_ = expr.coeff(r, 0).coeff(log(r),0)
s6_ = expr.coeff(log(r),1)                   # this is the same as s3_
#display(s4_,s5_,s6_)
#                      s4, s5, s6
stress_s = sp.Matrix([-s1, s5, s3])

dic_s= {s0_:s0, s1_:s1, s2_:s2, s3_:s3, s5_:s5}
Idic_s = {k: v.simplify() for v, k in dic_s.items()}
σθθ_final = p*(stress_s.T@s_bases)[0] + σθθ_E
display(Math(f"\\text{{Simplification correct?}}= \
   {latex((σθθ_final.subs(Idic_s).subs(s0,s0_)-σθθ_).simplify()==0)}"))
display(Math(f" \\text{{Final stress: $σ_{{θθ}}$=}}\\scriptsize \
                                              {latex(σθθ_final)}"))
```

Substitution correct? True

Simplification correct? = True

Final stress: $\sigma_{\theta\theta} = -\frac{Eb_4\left(-a^2b^2 - a^2r^2 - b^2r^2 + 3r^4\right)\cos(\theta)}{2r^3(a^2+b^2)} + p\left(s_3\log(r) + s_5 - \frac{s_1}{r^2}\right)$

```
# Expression for σrθ requires no simplification:

σrθ_final = sp.expand_log(σrθ.subs(Idic_A).subs({a3:a3_}).simplify())\
                                             .collect(E).collect(b4)
display(Math(f"\\text{{Final stress $σ_{{rθ}}$=}}\\scriptsize \
                                              {latex(σrθ_final)}"))
```

$$\text{Final stress } \sigma_{r\theta} = \frac{Eb_4\left(-a^2b^2 + a^2r^2 + b^2r^2 - r^4\right)\sin(\theta)}{2r^3(a^2+b^2)}$$

10.12.2 *Simplified final stresses*

The general expression for the stresses are rewritten in the following form:

$$\begin{aligned}
\sigma_{rr} &= \frac{Eb_4}{2(a^2+b^2)r^3}\left(r^2 - a^2\right)\left(b^2 - r^2\right)\cos\theta + p\left(s_2 + s_3\log r + \frac{s_1}{r^2}\right) \\
\sigma_{\theta\theta} &= \frac{Eb_4}{2(a^2+b^2)r^3}\left[a^2(b^2+r^2) + r^2(b^2 - 3r^2)\right]\cos\theta + p\left(s_3\log r + s_5 - \frac{s_1}{r^2}\right) \\
\sigma_{r\theta} &= \frac{Eb_4}{2(a^2+b^2)r^3}\left(r^2 - a^2\right)\left(b^2 - r^2\right)\sin\theta
\end{aligned} \tag{10.15}$$

We have successfully obtained clean and concise formulas for the stresses for this quite complicated problem. The contributors to these stresses consist of two parts: one from the pressure p and the other from the constraints at the arch end boundary $\theta = \theta_B$, which is controlled by the constant b_4. Note that b_4 also depends implicitly on the pressure p.

The coefficients s_i in the foregoing equation are given as follows:

```
Idic_s_5 = gr.split_dict_n(Idic_s, len(Idic_s))

display(Math(f"\\text{{Constants $s_{{i}} $}}({latex(len(Idic_s))}):"))
display(Math(f"\\text{{}}\\small {latex(Idic_s_5[0])}"))
display(Math(f"\\text{{}}\\small {latex(Idic_s_5[1])}"))
display(Math(f"\\text{{}}\\small {latex(Idic_s_5[2])}"))
display(Math(f"\\text{{}}\\small {latex(Idic_s_5[3])}"))
display(Math(f"\\text{{}}\\small {latex(Idic_s_5[4])}"))
```

Constants $s_i(5)$:

$$\left\{s_0 : \left(a^2 - b^2 + \log\left(a^{-2ab}b^{2ab}\right)\right)\left(a^2 - b^2 + \log\left(a^{2ab}b^{-2ab}\right)\right)\right\}$$

$$\left\{ s_1 : \frac{a^2 b^2 \left(-a^2 + b^2 + \log\left(a^{-2b^2} b^{2b^2}\right)\right)}{s_0} \right\}$$

$$\left\{ s_2 : \frac{b^2 \left(4a^2 \log(a)^2 + a^2 - b^2 + \log\left(a^{2a^2} b^{-2b^2 - \log\left(a^{4a^2}\right)}\right)\right)}{s_0} \right\}$$

$$\left\{ s_3 : \frac{2b^2 \left(-a^2 + b^2 + \log\left(a^{-2a^2} b^{2a^2}\right)\right)}{s_0} \right\}$$

$$\left\{ s_5 : \frac{b^2 \left(4a^2 \log(a)^2 - a^2 + b^2 + \log\left(a^{-2a^2} b^{4a^2 - 2b^2 - \log\left(a^{4a^2}\right)}\right)\right)}{s_0} \right\}$$

10.13 Pure bending in pressured arches

Let us consider a special class of arches that undergo **pure bending** without any shear stress on their cross-sections, even though they are subjected to distributed pressure. This case is of particular interest because:

1. Such a condition cannot occur in regular straight beams subjected to distributed loads, where the internal shear force is directly related to the distributed pressure. If the pressure is nonzero, the shear force can never be zero.
2. When an arch experiences no shear stress, its design becomes simpler, and its performance can be expected to be superior.

An arch in pure bending exhibits the following characteristics:

1. The stresses in the arch are axisymmetric, meaning that they do not depend on the tangential coordinate θ and are therefore constant with respect to θ.
2. The stresses in such an arch depend only on the pressure applied to its surfaces (outer, inner, or both).
3. The shear stress and hence the shear force on any cross-section are zero everywhere in the arch.

10.13.1 *Condition for pure bending in pressured arches*

The condition for the existence of such an arch can be determined from Eq. (10.15) by setting $b_4 = 0$, after which the stresses become

$$\begin{aligned}\sigma_{rr} &= p\left(s_2 + s_3 \log{(r)} + \frac{s_1}{r^2}\right)\\ \sigma_{\theta\theta} &= p\left(s_3 \log{(r)} + s_5 - \frac{s_1}{r^2}\right)\\ \sigma_{r\theta} &= 0\end{aligned} \tag{10.16}$$

These stresses do not change with θ and hence are functions of only r. We redefine them explicitly for future use:

```
# Define the normal stresses of a pure bending arch by pressure p:
σrr_Mp = p*(s3*log(r) + s5 - s1/r**2)
σθθ_Mp = p*(s3*log(r) + s5 - s1/r**2)
```

10.13.2 *Formula for b_4*

The constant b_4 was found earlier when enforcing the displacement boundary conditions, Eq. (10.12). It is written in terms of intermediate constants D_i. Let us examine it in more detail:

```
# b4 found earlier when enforcing the DBCs:
display(Math(f"\\text{{$b_4$=}} {latex(sln_DBC[b4])},\\text{{ $D_4$=}}\
            {latex(Idic_D[D4])},\\text{{ $D_1$=}}{latex(Idic_D[D1])}"))
```

$$b_4 = \frac{D_0 D_4 - D_1 D_3}{D_1 D_5 - D_2 D_4}, \quad D_4 = \sin{(\theta_B)}, \quad D_1 = -\cos{(\theta_B)}$$

First, b_4 implicitly depends on the pressure p. This can be seen from its expression, but it is a little too lengthy to display. However, we can set $p = 0$ and compute b_4; it will be zero. The snippet is as follows:

```
b4_val = sln_DBC[b4].subs(Idic_D).subs({a3:a3_}).subs(p,0).simplify()
display(b4_val)                                          # should be 0
```

0

The **key idea** stems from the fact that b_4 also depends on the half-central angle of the arch θ_B. Therefore, we can design arches with the proper θ_B so that b_4 is zero. In this case, Eq. (10.15) becomes Eq. (10.16), and we have a pure bending arch. The condition is as follows:

$$D_0 D_4 = D_1 D_3 \quad \text{or} \quad D_4 = D_1 \frac{D_3}{D_0} \quad \text{or}$$
$$\tan\theta_B = -\frac{D_3}{D_0} \tag{10.17}$$

Therefore, the critical half-central angle for a pure moment in a pressured arch, denoted as θ_M, should satisfy the following equation:

$$\tan\theta_M = -\theta_M \times \underbrace{\frac{\text{coefficient of } D_3}{D_0}}_{k_{30}} = -\theta_M \times k_{30} \tag{10.18}$$

where k_{30} is the slope of these constants. In the foregoing derivation, we used the fact that the expression of D_3 has the coefficient θ_B:

```
display(Idic_D[D3])
```

$$-\frac{\theta_B \left(Ea_3 (a+b)\left(a^2-b^2\right) - a^2b^2p(a+b)(\nu+1)\right)}{Ea^2b^2(\nu+1)(\log(a)-\log(b))}$$

The following code snippet computes and simplifies k_{30}. In the process, we set $a = 1$, and hence $b > 1$ becomes the single parameter, which is the ratio of b/a:

```
k30 = (Idic_D[D3].coeff(θB)/Idic_D[D0]).subs(a3,a3_).simplify()
k30_numer, k30_denom = k30.as_numer_denom()
k30 = k30_numer.simplify().factor()/k30_denom.simplify().factor()
k30_numer, k30_denom = k30.subs(a,1).as_numer_denom()      # set a = 1

# Simplify the denominator:
k30_denom = k30_denom.simplify().factor().collect(b).collect(4*log(b))\
   .collect(2*log(b)).collect(4*log(b+1)).collect(2*log(b+1))\
   .collect(4*log(2)).collect(2*log(2)).collect(2).collect(4)\
                                                  .collect((ν-1))
k30 = k30_numer/k30_denom
#display(k30)                         # Lengthy, one may take a look at it
```

10.13.3 *Condition for* $b_4 = 0$

The existence of arches in a pure bending state depends on whether a value of θ_B can be found such that Eq. (10.18) is satisfied, resulting in $b_4 = 0$. This, in turn, depends on the value of k_{30}. Since the slope of $\tan(\theta_B)$ is 1 near $\theta_B = 0$, k_{30} must be smaller than -1 for $-\theta_B k_{30}$ to intersect with $\tan(\theta_B)$ within the range $0 < \theta_B < \pi/2$.

Since the expression for k_{30} is rather complicated and also depends on Poisson's ratio ν, an analytical proof that $k_{30} < -1$ is difficult. However, we can verify this numerically. The following code snippet demonstrates the procedure:

```
# check for some arches with various b and ν=0.3:
b_vals = [1.2, 1.5, 2., 3., 5., 10., 20.]
k30_vals = [k30.subs({b:bi,ν:0.3}).evalf(4) for bi in b_vals]
display(Math(f"\\text{{Values of }} θ_B=\\small {latex(k30_vals)}"))
```

Values of $\theta_B = [-1.001,\ -1.007,\ -1.02,\ -1.047,\ -1.08,\ -1.103,\ -1.105]$

It is seen that the values of k_{30} for the cases computed satisfy the condition $k_{30} < -1$. We have also computed cases for different Poisson's ratios and concluded that, as long as $0 < \nu < 0.5$, $k_{30} < -1$. For a negative Poisson's ratio, $k_{30} > -1$, and hence such an arch is not possible, as shown in the following plot:

```
x = np.linspace(-np.pi, np.pi, 1000)
θB_vals = np.array([30.072])*np.pi/180

f0 = np.tan(x)
k1 = -1.1032                        # k30 value for Poisson's ratio ν = 0.3
k2 = -0.84867                       # k30 value for Poisson's ratio ν = -0.3
f1 = -k1*x; f2 = -k2*x

plt.plot(x, f0, label='tan(x)', linewidth=1.5)
plt.plot(x, f1, label=r'$-k_{30ν } x$', linewidth=1.5)
plt.plot(x, f2, label=r'$-k_{30-ν } x$', linewidth=1.5)
plt.scatter(θB_vals, [np.tan(θB_vals[0])], s=10, c='r')

plt.ylim(-2, 2.0)
plt.xlabel(r'$θ$'); plt.ylabel('Functions')
plt.gca().set_aspect('equal', adjustable='box')
plt.grid(True); plt.legend()
plt.savefig('imagesSMp/tan_k30.png', dpi=500, bbox_inches='tight');
#plt.show()
```

The root of the nonlinear Eq. (10.18) is illustrated in Fig. 10.4. It lies at the intersection of the curve $\tan(\theta_B)$ and the straight line $-k_{30}\theta_B$. It is observed that when Poisson's ratio is positive, within the range $0 < \nu < 0.5$, there is one intersection point in $0 < \theta_B < \pi$. When Poisson's ratio is negative (for certain special materials), no intersection point exists within $0 < \theta_B < \pi$, and therefore, pure bending in a pressured arch is not possible for such materials.

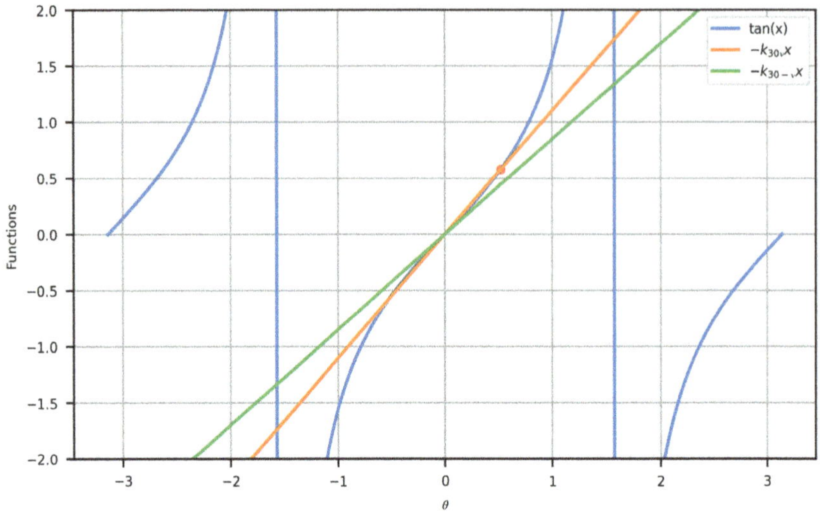

Figure 10.4. Intersection point (red dot) of the function $\tan(\theta_B)$ and the linear function $-k_{30}\theta_B$ in the domain $0 < \theta_B < \pi$.

10.13.4 *Examples of pressured arches in pure bending*

Next, let us compute θ_M for arches made of materials with $0 < \nu < 0.5$, such that Eq. (10.18) is satisfied. Although Eq. (10.18) appears simple, it is nonlinear, and we can only solve for θ_M numerically. To do this, we assign fixed numerical values to all parameters involved in k_{30}. We also need to provide an initial guess for θ_M. Clearly, $\theta_M = 0$ is a trivial solution to Eq. (10.18); however, for an arbitrary circular arch with a half-central angle $0 < \theta_B < \pi$, we should choose an initial value within this range. The following snippet finds the solutions for θ_M:

```
θM = sp.symbols('θ_M')          # half-central-angle for pure Moment arches

# set b/a values:
b_vals = [1.2, 1.5, 2., 3., 4., 5., 10., 20.] # [n for n in range(2, 6)]
# set the initial θM = np.pi/5 (one may try others):
k30_vals = [k30.subs({b:bi,ν:0.3}).evalf(5) for bi in b_vals]
sln_θM = [(sp.nsolve(sp.tan(θM)+θM*k30.subs({b:bi,ν:0.3}),\
                     θM, np.pi/5)*180/np.pi).evalf(13) for bi in b_vals]
dic_θM = dict(zip(b_vals, sln_θM))
dic_θM_2 = gr.split_dict_n(dic_θM, 2)         # split it to 2 for display
display(Math(f"\\text{{Arches with half-central-angle $θ_{{M}}$ \
                                    that undergoes pure bending:}}"))
display(Math(f"\\text{{}} \scriptsize {latex(dic_θM_2[0])}"))
display(Math(f"\\text{{}} \scriptsize {latex(dic_θM_2[1])}"))
display(Math(f"\\text{{}}D_{{30}}=\scriptsize {latex(k30_vals)}"))
```

Arches with half-central angle θ_M that undergoes pure bending:

$\{1.2 : 3.594468676019,\ 1.5 : 8.131397926469,\ 2.0 : 13.86127872484,\ 3.0 : 20.87231577965\}$

$\{4.0 : 24.65718229016,\ 5.0 : 26.83718162859,\ 10.0 : 30.07216340451,\ 20.0 : 30.2421307876\}$

$D_{30} = [-1.0013,\ -1.0068,\ -1.02,\ -1.0467,\ -1.0667,\ -1.0802,\ -1.1032,\ -1.1045]$

We found 10 different arches with paired b/a and half-central angle θ_M. Such an arch undergoes pure bending, the stresses in it are axisymmetric, do not change with θ, depend only on the applied pressure p, and the shear stress is zero, as given in Eq. (10.16). The following snippet checks the value for b_4 for such an arch:

```
b_1p8 = 1.8
θM_1p8 = sp.nsolve(sp.tan(θM)+θM*k30.subs({b:b_1p8, ν:0.3}),θM,np.pi/5)
display(Math(f"\\text{{Half central angle of the arch of $b/a$=1.8 is \
                    }} {latex(θM_1p8*180/np.pi)} \\text{{ degree}}"))
b4_val = sln_DBC[b4].subs(Idic_D).subs(a3,a3_)\
                    .subs({a:1, b:b_1p8, ν:0.3, θB:θM_1p8}).evalf()
display(Math(f"\\text{{The corresponding $b_4$ is }} {latex(b4_val)}"))
```

Half central angle of the arch of b/a=1.8 is 11.8055462304079 degree

The corresponding b_4 is 0

Note that for arch beams with $0 < \theta_B < \theta_M$, shear stresses will exist on the beam cross-section. However, these stresses are expected to be small because the $\tan(\theta_B)$ curve is usually very close to the linear function $-k_{30}\theta_B$ within the domain $0 < \theta_B < \theta_M$, as shown in Fig. 10.4.

Readers may attempt to determine θ_M for a negative Poisson's ratio and will find that θ_M can only be zero in such cases. Therefore, pure bending in a pressured arch with a negative Poisson's ratio is not possible.

10.13.5 *Displacements in pure bending arches*

Similar to the cases discussed in Chapter 3, the displacements may not be axisymmetric, even though the stresses are. The displacements can be obtained directly from Eq. (10.13) by setting $b_4 = 0$ in Eq. (10.14):

$$\begin{aligned} u_r &= d_0 \cos\theta + \frac{d_2}{r} + d_3 r + d_7 r \log r \\ u_\theta &= b_3 \sin\theta \end{aligned} \tag{10.19}$$

Note that these terms changing with θ represent rigid body movements. This is because when the arch is under pure bending, $b_4 = 0$, which leads to $d_0 = -b_3$. The following snippet confirms this:

```
display(Math(f"\\text{{$d_0$ is }} {latex(Idic_d[d0].subs(b4,0))}"))
```

d_0 is $-b_3$

When the strains are computed using the displacements given in Eq. (10.19) along with the strain–displacement relations, the terms $d_0 \cos\theta$ and $b_3 \sin\theta$ cancel each other out and do not contribute to the strains.

10.14 Forces on the cross-sections

With the stress components determined, we can compute the internal forces on the cross-section of the arch through integration. The expressions are as follows:

$$
\begin{aligned}
M_z &= \int_a^b \sigma_{\theta\theta}(r - \frac{a+b}{2})dr, \quad \text{moment w.r.t middle curve} \\
F_\theta &= \int_a^b \sigma_{\theta\theta} dr, \quad \text{normal force} \\
F_r &= \int_a^b \sigma_{r\theta} dr, \quad \text{shear force}
\end{aligned}
\tag{10.20}
$$

Here, M_z, F_θ, and F_r are, respectively, the internal moment, normal force, and shear force along the arch axis. Note that the moment is measured with respect to the centroidal axis of the arch cross-section. The following snippet computes all these internal forces. To produce easy-to-read formulas, simplification measures are applied with proper checks in place:

```
## Compute internal forces on the cross-section:

σrr, σθθ, σrθ = σv                          # unpack the stress components
Mz = sp.integrate(σθθ*(r-(a+b)/2), (r, a, b)).simplify()
#display(Mz)  # 1.5 pages long

# simplify the expression for moment:
Mz_tmp = sp.expand_log(Mz.subs(Idic_A).subs({a3:a3_}).simplify())\
                .collect(cos(θ)).collect(2*b*p).collect(E).collect(b4)\
                   .subs((a**3+a**2*b+a*b**2+b**3), (a+b)*(a**2+b**2))\
            .collect((a+b)*(a**2+b**2)).subs((-a**3-a**2*b+a*b**2+b**3)\
                            , -(a-b)*(a+b)**2).collect((a**2+b**2))
Mz_ = gr.splitFraction(Mz_tmp).collect(4).collect((a+b))
display(Math(f"\\text{{Moment $M_z$=}}\\scriptsize {latex(Mz_)}"))
display((Mz_- sp.expand_log(Mz.subs(Idic_A).subs({a3:a3_}))).simplify())
```

```
# simplify the expression for normal force:
Fθ = sp.integrate(σθθ, (r, a, b)).simplify()
Fθ_tmp = sp.expand_log(Fθ.subs(Idic_A).subs({a3:a3_}).simplify())\
                .collect(cos(θ)).collect(2*b*p).collect(E).collect(b4)\
                   .subs((a**3+a**2*b+a*b**2+b**3), (a+b)*(a**2+b**2))\
          .collect((a+b)*(a**2+b**2)).subs((-a**3-a**2*b+a*b**2+b**3),\
                                       -(a-b)*(a+b)**2).collect((a**2+b**2))
Fθ_ = gr.splitFraction(Fθ_tmp).collect(2).collect((a+b))
display(Math(f"\\text{{Normal force $F_θ$=}}\\scriptsize {latex(Fθ_)}"))
display((Fθ_-sp.expand_log(Fθ.subs(Idic_A).subs({a3:a3_}))).simplify())

# simplify the expression for shear force:
Fr = sp.integrate(σrθ, (r, a, b)).simplify()
Fr_tmp = sp.expand_log(Fr.subs(Idic_A).subs({a3:a3_}).simplify())\
    .collect(cos(θ)).collect(2*b*p).collect(E).collect(b4)\
    .subs((a**3+a**2*b+a*b**2+b**3), (a+b)*(a**2+b**2))\
    .collect((a+b)*(a**2+b**2)).subs((-a**3-a**2*b+a*b**2+b**3),\
                                       -(a-b)*(a+b)**2).collect((a**2+b**2))
Fr_ = gr.splitFraction(Fr_tmp).collect(2).collect((a+b))
display(Math(f"\\text{{Shear force $F_r$=}}\\scriptsize {latex(Fr_)}"))
display((Fr_-sp.expand_log(Fr.subs(Idic_A).subs({a3:a3_}))).simplify())
```

$$\text{Moment } M_z = \frac{Eb_4\left((-a+b)(a+b)^2 + (a+b) \times (a^2+b^2)\left(\log(a) - \log(b)\right)\right)\cos(\theta)}{4(a^2+b^2)} + \frac{bp(a+b)}{2}$$

0

$$\text{Normal force } F_\theta = \frac{Eb_4\left(a^2 - b^2 + (a^2+b^2)\left(-\log(a) + \log(b)\right)\right)\cos(\theta)}{2(a^2+b^2)} - bp$$

0

$$\text{Shear force } F_r = \frac{Eb_4\left(a^2 - b^2 + (a^2+b^2)\left(-\log(a) + \log(b)\right)\right)\sin(\theta)}{2(a^2+b^2)}$$

0

We rewrite these formulas as follows for future reference:

$$\begin{aligned}
M_z &= \frac{bp(a+b)}{2} - \frac{a+b}{2}\frac{Eb_4a^2}{2(a^2+b^2)}\underbrace{\left[1 - \frac{b^2}{a^2} + \left(1 + \frac{b^2}{a^2}\right)\log\frac{b}{a}\right]}_{\alpha}\cos\theta \\
F_\theta &= -bp + \frac{Eb_4a^2}{2(a^2+b^2)}\underbrace{\left[1 - \frac{b^2}{a^2} + \left(1 + \frac{b^2}{a^2}\right)\log\frac{b}{a}\right]}_{\alpha}\cos\theta \qquad (10.21) \\
F_r &= \frac{Eb_4a^2}{2(a^2+b^2)}\underbrace{\left[1 - \frac{b^2}{a^2} + \left(1 + \frac{b^2}{a^2}\right)\log\frac{b}{a}\right]}_{\alpha}\sin\theta
\end{aligned}$$

where

$$b_4 = \frac{D_0 D_4 - D_1 D_3}{D_1 D_5 - D_2 D_4} \tag{10.22}$$

Note that α is the function f_3 given in Eq. (10.11) studied earlier and is always negative for all $b > a > 0$.

For an arch undergoing only pure bending when pressured on the top surface, we have $\theta_B = \theta_M$, and the internal forces have the following concise formula:

$$\begin{aligned} M_z &= \frac{bp(a+b)}{2} \\ F_\theta &= -bp \\ F_r &= 0 \end{aligned} \tag{10.23}$$

As mentioned earlier, the moment and normal force are constants that depend only on p and do not vary with θ. The shear force is zero throughout the arch.

10.15 Final formulas for strains

Once the stresses are obtained, the strains can be computed using the constitutive equation, which involves only algebraic operations. Alternatively, the strains can be calculated from the displacements using `gr.displacement2StrainPolar()`, which involves differentiation. The snippet is as follows:

```
# Compute strains using the displacements:
εrr, εθθ, εrθ = gr.displacement2StrainPolar([ur_final, uθ_final], R)
display(Math(f"\\text{{Final strains:}} "))
display(Math(f"\\text{{$ε_{{rr}}$=}}\\small {latex(εrr)}"))

εθθ_final = εθθ.expand().collect(r**3).collect(sin(θ)).collect(cos(θ))
display(Math(f"\\text{{$ε_{{θθ}}$=}}\scriptsize \
                                  {latex(sum(εθθ_final.args[:6]))}"))
display(Math(f"\\text{{ + }}\scriptsize \
                                  {latex(sum(εθθ_final.args[6:]))}"))
εrθ_final = εrθ.simplify().collect(θ*cos(θ)).collect(sin(θ))\
                                  .collect(r**2*log(r)).collect(r**2)
display(Math(f"\\text{{$ε_{{rθ}}$=}}\scriptsize {latex(εrθ_final)}"))
```

Final strains:

$$\varepsilon_{rr} = -\frac{d_2}{r^2} + d_3 + \frac{d_4 \cos(\theta)}{r} - \frac{2d_5 \cos(\theta)}{r^3} + 2d_6 r \cos(\theta) + d_7 \log(r) + d_7$$

$$\varepsilon_{\theta\theta} = A_5 a_3 + A_6 p + \frac{d_2}{r^2} + d_3 + d_7 \log(r)$$
$$+ \frac{(-b_4\theta + d_1\theta)\sin(\theta) + (A_7 b_4 \log(r) + b_3 + b_4 + d_0 + d_4 \log(r))\cos(\theta)}{r}$$
$$+ \left(r(A_4 b_4 + d_6) + \frac{A_3 b_4 + d_5}{r^3}\right)\cos(\theta)$$

$$\varepsilon_{r\theta} = \frac{\begin{array}{c} r^2\theta(-b_4 + d_1)\cos(\theta) + \left(-3A_3 b_4 - d_5 + r^4(A_4 b_4 - d_6)\right. \\ \left. + r^2(A_7 b_4 - b_3 - d_0 + d_1 + (-A_7 b_4 - d_4)\log(r))\right)\sin(\theta) \end{array}}{2r^3}$$

These strains involve constants that have already been determined. We are now ready to present examples of computing displacements and stresses in a given arch with specific geometric and material parameters.

10.16 Numerical examples: A particular arch

Consider an idealized arch with the following normalized parameters:

$$\begin{aligned} &a = 1; \quad b = 2; \quad \theta_B = \pi/4 \quad \text{geometry} \\ &E = 1; \quad \nu = 0.3 \quad \text{material property} \\ &p = 1 \quad \text{loading} \end{aligned} \tag{10.24}$$

Readers may use actual values for their arch. It is recommended to use normalized parameters (see Section 2.8 of Ref. [2]) so that the data are close to 1. This ensures high-quality numerical solutions and minimizes loss of precision during computations.

10.16.1 *Data preparation*

We can now redefine all related constants explicitly in Python, allowing this section to be studied independently of the previous code. Most importantly, interested readers can copy the snippets for the formulas of various constants for their own applications. The following snippet produces these formulas. Readers can simply uncomment the lines to generate the formulas in Python code, resulting in Python dictionaries containing all constants:

```
## Snippet to produce the formulas of constants in Python code:
#print('Idic_A=', Idic_A);   print(); print('Idic_D=', Idic_D); print()
#print('Idic_a3=', Idic_a3); print(); print('Idic_d=', Idic_d); print()
#print('sln_DBC=', sln_DBC); print(); print('Idic_s=', Idic_s)
```

```
## close-form formulas for all these constants in problem parameters:

Idic_A= {A1: (a**2-b**2+log(a**(a**2*(1-ν))*b**(b**2*(ν-1))))/\
                                    (a**2*b**2*log(a**(ν+1)*b**(-ν-1))), \
         A2: (log(a**(ν-1))-1)/(E*log(a/b)), \
         A3: a**2*b**2*(ν+1)/(4*(a**2+b**2)), \
         A4: (-ν-5)/(4*(a**2+b**2)), A5: 2*(-a**2+b**2)/\
                                    (a**2*b**2*log(a**(ν+1)*b**(-ν-1))), \
         A6: 2/(E*log(a/b)), A7: ν/2-1/2}

Idic_D= {D0: 2*a3/(a+b)+(E*a3*((-a-b)*(-a**2+b**2+(ν-1)*(a**2*log(a)\
         -b**2*log(b)))+(a+b)*(a**2-b**2)*(ν-1)*log(a/2+b/2))+a**2*b**2\
         *p*(ν+1)*(-(a+b)*(ν-1)*log(a/2+b/2)+(a+b)*((ν-1)*log(a)-1)))/\
         (2*E*a**2*b**2*(ν+1)*(log(a)-log(b))),\
         D1: -cos(θB), \
         D2: (a**2*b**2*(ν+1)/((a+b)**2*(a**2+b**2))+(3*ν-1)\
         *(a*b/(8*(a**2+b**2))+1/16))*cos(θB)+((ν-3)*(2*θB\
         *sin(θB)-(ν-1)*log(a/2+b/2)*cos(θB))+(-ν*(ν-1)+ν+3)\
         *cos(θB))/(2*(ν-3)), D3: -θB*(E*a3*(a+b)*(a**2-b**2)\
         -a**2*b**2*p*(a+b)*(ν+1))/(E*a**2*b**2*(ν+1)*(log(a)-log(b))),\
         D4: sin(θB), D5: θB*cos(θB)+(a**2*b**2*(ν+1)/((a+b)**2\
         *(a**2+b**2))-(a+b)**2*(ν+5)/(16*(a**2+b**2))+(ν-1)\
                                     *log(a/2+b/2)/2)*sin(θB)}

Idic_a3= {a3: a**2*b**2*p*(a**2*ν+a**2-b**2*ν-b**2+log(a**(2*b**2\
         *(ν+1))/b**(2*b**2*(ν+1))))/(E*(a**4-4*a**2*b**2*log(a)**2\
         -4*a**2*b**2*log(b)**2-2*a**2*b**2+b**4\
                               +log(b**log(a**(8*a**2*b**2))))))}

Idic_d= {d0: (-A7*b4*ν-A7*b4-b3*(ν-3)+b4*ν+b4)/(ν-3), \
         d1: b4, d2: a3, d3: A1*a3+A2*p, d4: -A7*b4, d5: A3*b4, \
         d6: A4*b4*(1-3*ν)/(ν+5), d7:-A5*a3*ν/2+A5*a3/2-A6*p*ν/2+A6*p/2}

sln_DBC= {b3: (-D0*D5+D2*D3)/(D1*D5-D2*D4), \
          b4: (D0*D4-D1*D3)/(D1*D5-D2*D4)}

s0 =(a**2-b**2+log(b**(2*a*b)/a**(2*a*b)))*\
    (a**2-b**2+log(a**(2*a*b)/b**(2*a*b)))
Idic_s= {s1: a**2*b**2*(-a**2+b**2+log(b**(2*b**2)/a**(2*b**2)))/s0, \
         s2: b**2*(4*a**2*log(a)**2+a**2-b**2+log(a**(2*a**2)*\
             b**(-2*b**2-log(a**(4*a**2)))))/s0, \
         s3: 2*b**2*(-a**2+b**2+log(b**(2*a**2)/a**(2*a**2)))/s0, \
         s5: b**2*(4*a**2*log(a)**2-a**2+b**2+log(b**(4*a**2-2*b**2-\
             log(a**(4*a**2)))/a**(2*a**2)))/s0}
```

The following snippet computes the numerical values of all the necessary constants for a given arch.

Before performing the numerical computations, one may use `mpmath` to set the global precision to the desired number of decimal places if experiencing significant loss of precision. It is usually not necessary if the data are normalized. This snippet is as follows:

```
## To set global precision to desired decimal places:
#import mpmath
#N_precision = 50    # any desired integer; default is ~15-17 in Python
#mpmath.mp.dps = N_precision
```

Let us now carry out the numerical analysis for the specific arch bridge defined in this example:

```
a_ = 1; b_ = 2; p_ = 1     # we use "_" to signify this is a given value
E_ = 1; ν_ = 0.3
θB_ = np.pi/4

# use dictionary to pair the variables with the specific numerical data:
dic_data = {E:E_, ν:ν_, a:a_, b:b_, p:p_, θB: θB_}

# propagating the numerical data to all the dictionaries:
Idic_a3_= {k: v.subs(dic_data).evalf() for k, v in Idic_a3.items()}
Idic_A_ = {k: v.subs(dic_data).evalf() for k, v in Idic_A.items()}
Idic_D_ = {k: v.subs(dic_data).evalf() for k, v in Idic_D.items()}
sln_DBC_= {k: v.subs(Idic_D_).evalf()  for k, v in sln_DBC.items()}
Idic_d_ = {k: v.subs(dic_data).evalf() for k, v in Idic_d.items()}
Idic_s_ = {k: v.subs(dic_data).evalf() for k, v in Idic_s.items()}

# form a single dictionary and propagate data:
dicAll_ = {**Idic_A_, **sln_DBC_,**Idic_a3_,**Idic_d_,**Idic_s_}
dicAll_ = {k: v.subs(dic_data).evalf() for k, v in dicAll_.items()}
dicAll_ = {k: v.subs(dicAll_).evalf()  for k, v in dicAll_.items()}
dicAll_ = {k: v.subs(dicAll_).evalf()  for k, v in dicAll_.items()}

#display(dicAll_) # to check: all constants must be in numerical values
```

10.16.2 *Plots of displacement distribution*

We can now compute and plot the displacement distribution using this set of specific values. Here, we redefine all the related formulas explicitly in Python. We also provide the snippet to produce the formulas. Readers can simply uncomment the following lines to produce the formulas for a Python code:

```
##Redefine the general function for the displacements:
## Snippet to produce the formulas in Python code:
#display(ur_final); display(uθ_final); print(ur_final); print(uθ_final)

ur_ = d0*cos(θ) + d1*θ*sin(θ) + d2/r + d3*r + d4*log(r)*cos(θ) \
                        + d5*cos(θ)/r**2 + d6*r**2*cos(θ) + d7*r*log(r)
uθ_ = A3*b4*sin(θ)/r**2 + A4*b4*r**2*sin(θ) + A5*a3*r*θ + A6*p*r*θ \
                        + A7*b4*log(r)*sin(θ) + b3*sin(θ) + b4*θ*cos(θ)
```

```
# Compute the displacements with numerical constant values:

ur_rt = ur_.subs(dicAll_).subs(dic_data).expand()
uθ_rt = uθ_.subs(dicAll_).subs(dic_data).expand()

display(Math(f"\\text{{Displacements with numerical constants for\
                                         this particular arch:}}"))
display(Math(f"\\text{{$u_r$ = }}\scriptsize \
                              {latex(sum(ur_rt.evalf(4).args[:6]))}"))
display(Math(f"\\hspace{{1cm}}\\text{{}}\scriptsize \
                              {latex(sum(ur_rt.evalf(4).args[6:]))}"))
display(Math(f"\\text{{$u_θ$ = }}\scriptsize {latex(uθ_rt.evalf(4))}"))
```

Displacements with numerical constants for this particular arch:

$$u_r = 18.71r\log(r) - 44.96r + 40.62\theta\sin(\theta) + 14.22\log(r)\cos(\theta)$$

$$+ 69.28\cos(\theta) - \frac{33.85}{r} - 0.2031r^2\cos(\theta) + \frac{10.56\cos(\theta)}{r^2}$$

$$u_\theta = -10.77r^2\sin(\theta) + 53.46r\theta + 40.62\theta\cos(\theta) - 14.22\log(r)\sin(\theta)$$

$$- 95.68\sin(\theta) + \frac{10.56\sin(\theta)}{r^2}$$

The distribution of the displacements along the arch can be plotted using `gr.plot_fs()`. The snippet is as follows:

```
# Convert SymPy expressions to Numpy functions:

func1 = sp.lambdify(θ, ur_rt.subs({r: a_}),'numpy')
func2 = sp.lambdify(θ, ur_rt.subs({r: b_}),'numpy')
func3 = sp.lambdify(θ, ur_rt.subs({r: (a_+b_)/2}),'numpy')
func4 = sp.lambdify(θ, uθ_rt.subs({r: a_}),'numpy')
func5 = sp.lambdify(θ, uθ_rt.subs({r: b_}),'numpy')
func6 = sp.lambdify(θ, uθ_rt.subs({r: (a_+b_)/2}),'numpy')
labels=[r'$u_{r}, r=a$', r'$u_{r}, r=b$', r'$u_{r}, r=(a+b)/2$',
        r'$u_{θ}, r=a$', r'$u_{θ}, r=b$', r'$u_{θ}, r=(a+b)/2$' ]
```

```
t_data  = np.linspace(-θB_, θB_, 400)          # θ values for the plot
x_data = [t_data,  t_data, t_data,  t_data, t_data, t_data]
y_data = [func1(x_data[0]), func2(x_data[1]), func3(x_data[0]),
          func4(x_data[1]), func5(x_data[2]), func6(x_data[2])]
plt.ion()
gr.plot_fs(x_data, y_data, labels, '$θ(rad)$', 'displacements', \
                                        'imagesSMp/Arch_pUt',ioff=True)
                                   # Use ioff=False to display the plot
#plt.show()
```

The distribution of displacements along the arch is shown in Fig. 10.5. It is observed that the displacement u_r is symmetric, while the displacement u_θ is antisymmetric with respect to $\theta = 0$. The displacement in the radial direction is much smaller than that in the tangential direction. The largest u_θ displacement occurs at the two ends on the arch surfaces.

The following snippet plots the distribution of displacements along the radial direction:

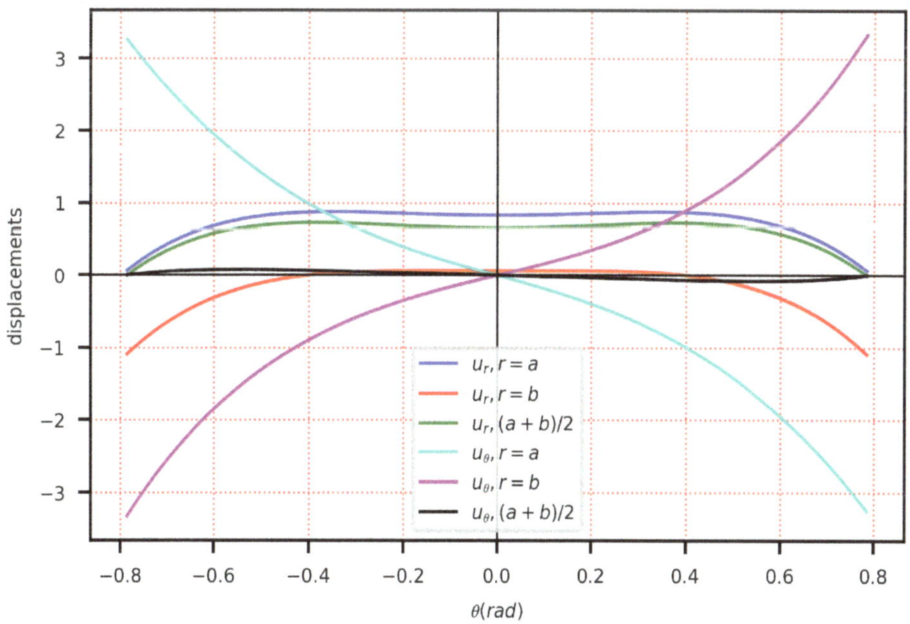

Figure 10.5. Distribution of the two displacement components u_r and u_θ along the arch.

```
# Convert SymPy expressions to Numpy functions:

func1 = sp.lambdify(r, ur_rt.subs({θ:   0.}),'numpy')
func2 = sp.lambdify(r, ur_rt.subs({θ:np.pi/8.}),'numpy')
func3 = sp.lambdify(r, ur_rt.subs({θ:np.pi/4.}),'numpy')
func4 = sp.lambdify(r, uθ_rt.subs({θ:np.pi/8.}),'numpy')
func5 = sp.lambdify(r, uθ_rt.subs({θ:np.pi/4.}),'numpy')
labels=[r'$u_{r}, θ=0$',r'$u_{r}, θ=\pi/8$', r'$u_{r}, θ=\pi/4$',
        r'$u_{θ}, θ=\pi/8$', r'$u_{θ}, θ=\pi/4$' ]

r_data  = np.linspace(1, 2, 400)                    # r values for the plot
x_data = [r_data,  r_data, r_data,  r_data, r_data  ]
y_data = [func1(x_data[0]), func2(x_data[1]), func3(x_data[0]),
          func4(x_data[1]), func5(x_data[2]) ]
plt.ion()
gr.plot_fs(x_data, y_data, labels, '$r$', 'displacements', \
                                 'imagesSMp/Arch_pUr',ioff=True)
                                # Use ioff=False to display the plot
#plt.show()
```

The two displacement components, u_r and u_θ, along the radial direction are plotted in Fig. 10.6. Observations are made at the top ($\theta = 0$), middle ($\theta = \pi/8$), and left end ($\theta = \pi/4$) cross-sections of the arch. The displacement u_θ reaches its largest values on the surfaces and at the ends of the arch.

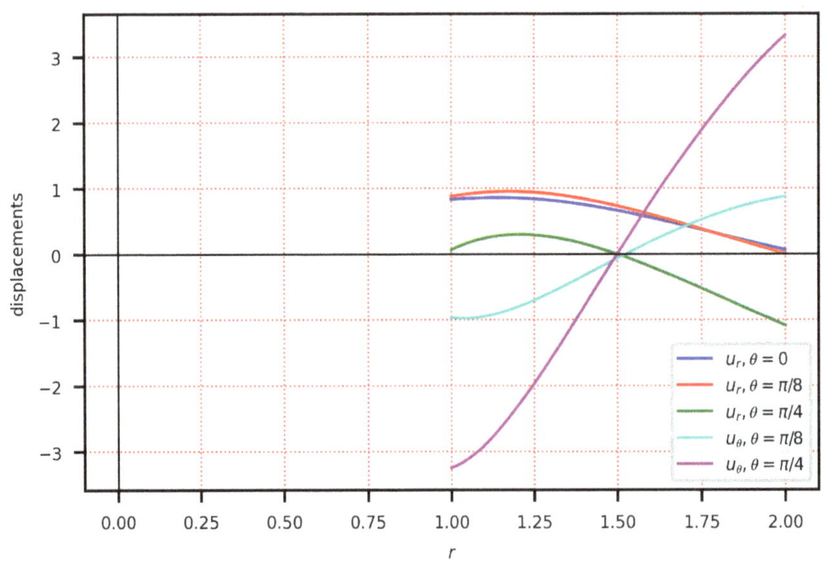

Figure 10.6. Distribution of the two displacement components u_r and u_θ along the radial direction.

10.16.3 *Plots of stress distribution*

We can now compute and plot the stress distribution. There are two options for computing the stresses. One approach uses the already computed displacements, applying `gr.displacement2StrainPolar()` to obtain strains, followed by `gr.strain2stress2D()` to obtain stresses. In this approach, we do not need the explicit stress formulas, which is a key feature offered by the displacement method. The following snippet illustrates this procedure:

```
εv_rt = sp.Matrix(gr.displacement2StrainPolar([ur_rt, uθ_rt], R))

# Compute stress vector using the strain vector:
σv_rt = sp.Matrix(gr.strain2stress2D(E_, ν_, εv_rt, Ptype='stress'))
σv_rt

# check whether the equilibrium equation are satisfied:
eqn1, eqn2 = checkEquilibriumPolar(σv_rt, [r, θ], br, bθ)
display(Math(f" \\text{{Satisfaction of eequilibrium:}}"))
display((eqn1.subs(dicAll_).expand().evalf(3), eqn2.evalf(3)))

# check SBCsf at r = a and b:
display(Math(f" \\text{{Satisfaction of SBCs:}}" ))
display((σv_rt[0].subs(r,a_).subs(Idic_A_).evalf(3),
         σv_rt[0].subs(r,b_).subs(Idic_A_).evalf(3)))
display((σv_rt[2].subs(r,a_).subs(Idic_A_).evalf(3),
         σv_rt[2].subs(r,b_).subs(Idic_A_).evalf(3)))

σrr_sp, σθθ_sp, σrθ_sp = σv_rt.expand()

display(Math(f" \\text{{Particular solution for stresses:}}"))
display(Math(f" \\text{{$σ_{{rr}}$ = }}\\small \
                              {latex(σrr_sp.evalf(4))}"))

display(Math(f" \\text{{$σ_{{θθ}}$ = }}\\small \
                              {latex(σθθ_sp.evalf(4))}"))

display(Math(f" \\text{{$σ_{{rθ}}$ = }}\\small \
                              {latex(σrθ_sp.evalf(4))}"))
```

Satisfaction of equilibrium:

$$\left(-8.88 \cdot 10^{-16} \cos(\theta) + \frac{2.49 \cdot 10^{-14} \cos(\theta)}{r^2}, \right.$$
$$\left. \frac{1.0 \left(2.84 \cdot 10^{-14} r^2 - 3.55 \cdot 10^{-15}\right) \sin(\theta)}{r^4} \right)$$

Satisfaction of SBCs:

$$\left(-7.11 \cdot 10^{-15} \cos(\theta) - 7.11 \cdot 10^{-15},\ -3.11 \cdot 10^{-15} \cos(\theta) - 1.0\right)$$
$$\left(5.33 \cdot 10^{-15} \sin(\theta),\ 3.55 \cdot 10^{-15} \sin(\theta)\right)$$

Particular solution for stresses:

$$\sigma_{rr} = -4.062r\cos(\theta) + 26.73\log(r) - 26.04 + \frac{20.31\cos(\theta)}{r} + \frac{26.04}{r^2} - \frac{16.25\cos(\theta)}{r^3}$$

$$\sigma_{\theta\theta} = -12.19r\cos(\theta) + 26.73\log(r) + 0.6929 + \frac{20.31\cos(\theta)}{r} - \frac{26.04}{r^2} + \frac{16.25\cos(\theta)}{r^3}$$

$$\sigma_{r\theta} = -4.062r\sin(\theta) + \frac{20.31\sin(\theta)}{r} - \frac{16.25\sin(\theta)}{r^3}$$

In the code above, we performed checks, confirming that the final stress solutions satisfy the equilibrium equations and the SBCs (to machine precision).

Alternatively, we can use the previously obtained general stress formulas to compute the stresses by substituting the constant values, as demonstrated in the following snippet:

```
# redefine explicit formulas for stresses:

#print('σrr_final_=',σrr_final);print()
#print('σθθ_final_=',σθθ_final);print();print('σrθ_final_=', σrθ_final)

σrr_final_=-E*b4*(-a + r)*(a + r)*(-b + r)*(b + r)*cos(θ)/\
                        (2*r**3*(a**2 + b**2)) + p*(s2 + s3*log(r) + s1/r**2)
σθθ_final_=-E*b4*(-a**2*b**2 - a**2*r**2 - b**2*r**2 + 3*r**4)*cos(θ)/\
                        (2*r**3*(a**2 + b**2)) + p*(s3*log(r) + s5 - s1/r**2)
σrθ_final_= E*b4*(-a**2*b**2 + a**2*r**2 + b**2*r**2 - r**4)*sin(θ)/\
                                                   (2*r**3*(a**2 + b**2))
σθθ_rt = σθθ_final_.subs(dicAll_).subs(dic_data).expand()
display(Math(f"\\text{{Stress solution for the given arch:}}"))
display(Math(f"\\text{{$σ_{{θθ}}$=}}\\small {latex(σθθ_rt.evalf(4))}"))
```

Stress solution for the given arch:

$$\sigma_{\theta\theta} = -12.19r\cos(\theta) + 26.73\log(r) + 0.6929 + \frac{20.31\cos(\theta)}{r} - \frac{26.04}{r^2} + \frac{16.25\cos(\theta)}{r^3}$$

As shown, the solution for the particular stress obtained by the two different methods is the same. This, in a way, confirms that our solutions are correct. Readers may confirm the same for the other two stress components.

The following snippet computes and plots the distribution of stresses along the arch:

```
# Convert SymPy expressions to Numpy functions:

func1 = sp.lambdify(θ, σrr_sp.subs({r: a_}),'numpy')
func2 = sp.lambdify(θ, σrr_sp.subs({r: b_}),'numpy')
func3 = sp.lambdify(θ, σrr_sp.subs({r: (a_+b_)/2}),'numpy')
func4 = sp.lambdify(θ, σθθ_sp.subs({r: a_}),'numpy')
func5 = sp.lambdify(θ, σθθ_sp.subs({r: b_}),'numpy')
func6 = sp.lambdify(θ, σθθ_sp.subs({r: (a_+b_)/2}),'numpy')
func7 = sp.lambdify(θ, σrθ_sp.subs({r: (a_+b_)/2}),'numpy')
labels=[r'$σ_{rr}, r=a$', r'$σ_{rr}, r=b$', r'$σ_{rr}, r=(a+b)/2$',
        r'$σ_{θθ}, r=a$', r'$σ_{θθ}, r=b$', r'$σ_{θθ}, r=(a+b)/2$',
        r'$σ_{rθ}, (a+b)/2$' ]

t_data  = np.linspace(-θB_, θB_, 400)           # θ values for the plot
x_data = [t_data,  t_data, t_data,  t_data, t_data, t_data, t_data]
y_data = [func1(x_data[0]), func2(x_data[1]), func3(x_data[0]),
          func4(x_data[1]), func5(x_data[2]), func6(x_data[2]),
          func7(x_data[2])]
plt.ion()
gr.plot_fs(x_data, y_data, labels, '$θ(rad)$', 'displacements', \
                                   'imagesSMp/Arch_pSt',ioff=True)
                                     # Use ioff=False to display the plot
#plt.show()
```

The distribution of stresses along the arch is plotted in Fig. 10.7. It is observed that the two normal stresses, σ_{rr} and $\sigma_{\theta\theta}$, are symmetric, while the shear stress $\sigma_{r\theta}$ is antisymmetric with respect to $\theta = 0$. The tangential normal stress $\sigma_{\theta\theta}$ is much larger than the other stress components. Its maximum values occur at the two ends on the arch surfaces.

Note that the shear stress $\sigma_{r\theta}$ is zero on both surfaces of the arch due to the stress boundary conditions. It reaches a maximum along the middle axis of the arch. In addition, it is zero at $\theta = 0$ due to symmetry and increases almost linearly toward the ends. As shown in Fig. 10.7, the shear stress is much smaller than the tangential normal stress $\sigma_{\theta\theta}$.

The following snippet plots the distribution of stresses along the radial direction:

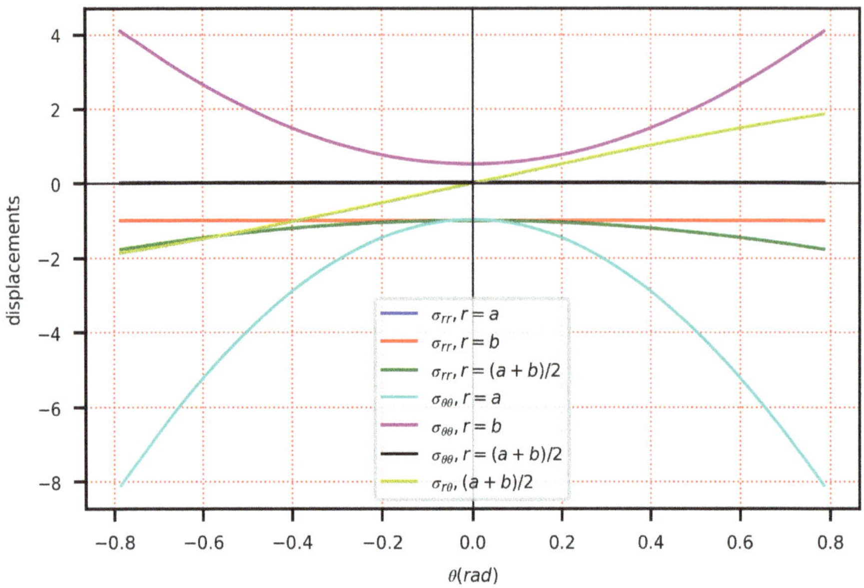

Figure 10.7. Distribution of the three stress components along the arch.

```
# Convert SymPy expressions to Python functions:
σrr_sp, σθθ_sp, σrθ_sp = σv_rt        # σv_rt is from the displacements

func1 = sp.lambdify(r, σrr_sp.subs({θ:   0.}),'numpy')
func2 = sp.lambdify(r, σrr_sp.subs({θ: np.pi/8.}),'numpy')
func3 = sp.lambdify(r, σrr_sp.subs({θ: np.pi/4.}),'numpy')
func4 = sp.lambdify(r, σθθ_sp.subs({θ:   0.}),'numpy')
func5 = sp.lambdify(r, σθθ_sp.subs({θ: np.pi/8.}),'numpy')
func6 = sp.lambdify(r, σθθ_sp.subs({θ: np.pi/4.}),'numpy')
func8 = sp.lambdify(r, σrθ_sp.subs({θ: np.pi/8.}),'numpy')
func9 = sp.lambdify(r, σrθ_sp.subs({θ: np.pi/4.}),'numpy')

labels=[r'$σ_{rr}, θ=0$', r'$σ_{rr}, θ=\pi/8$', r'$σ_{rr}, θ=\pi/4$',\
        r'$σ_{θθ}, θ=0$', r'$σ_{θθ}, θ=\pi/8$', r'$σ_{θθ}, θ=\pi/4$',\
        r'$σ_{rθ},θ=\pi/8$',r'$σ_{rθ},θ=\pi/4$']

r_data  = np.linspace(1, 2, 40)                    # r values for the plot
x_data = [r_data, r_data, r_data, r_data, r_data, r_data,r_data,r_data]
y_data = [func1(x_data[0]), func2(x_data[1]), func3(x_data[2]),
          func4(x_data[3]), func5(x_data[4]), func6(x_data[5]),
          func8(x_data[6]), func9(x_data[7])]
plt.ion()
gr.plot_fs(x_data, y_data, labels, '$r$', 'stresses', \
                                    'imagesSMp/Arch_pSr',ioff=True)
                                  # Use ioff=False to display the plot
#plt.show()
```

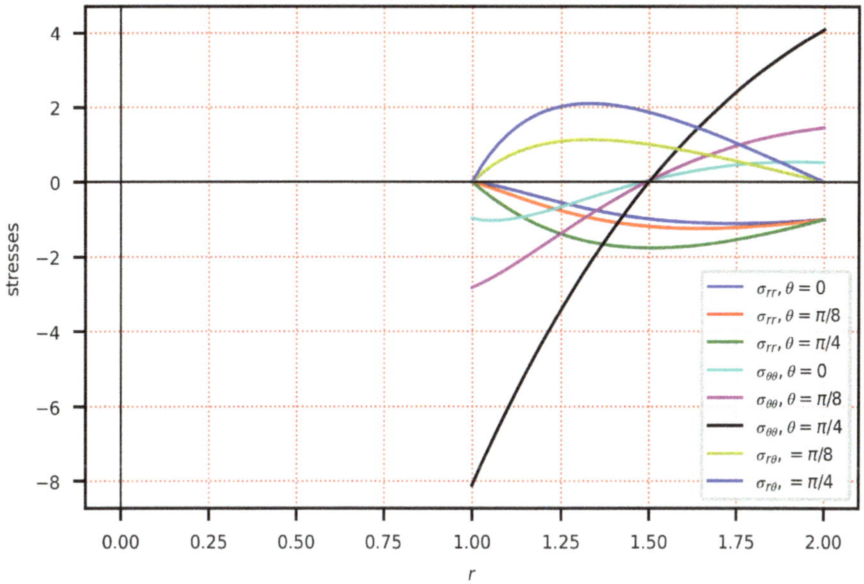

Figure 10.8. Distribution of the three stress components along the radial direction on vertical, middle, and end (45°) cross-sections of the arch.

The distributions of the three stress components, σ_{rr}, $\sigma_{\theta\theta}$, and $\sigma_{r\theta}$, along the radial direction are plotted in Fig. 10.8 for various θ values. The maximum and minimum stresses are the normal stress $\sigma_{\theta\theta}$, which occurs on the inner and outer surfaces of the arch at the $\theta = \pi/4$ cross-section, corresponding to the left end of the arch. Unlike in a straight beam, the normal stress is not linearly distributed along a cross-section. The value of $\sigma_{\theta\theta}$ on the inner surface of the semicircular arch is particularly high, as shown in Fig. 10.8. This is because the concave surface of the arch is prone to stress concentration, making it a likely initiation point if the material yields.

Note again that the shear stress $\sigma_{r\theta}$ is much smaller than the tangential normal stress $\sigma_{\theta\theta}$.

10.16.4 *Contour plots of displacements*

To plot the displacement contours, we first express the displacements in Cartesian coordinates using the relations (see Chapter 2):

$$r = \sqrt{x^2 + y^2}, \quad \theta = \arctan\left(\frac{-x}{y}\right) \tag{10.25}$$

where $\theta = 0$ is along the y-axis.

Next, we convert the expressions to Numpy functions and use `gr.contour_plot()` to plot the contours. The following snippet can be used to achieve this:

```
# Contour of displacement component u_r:
x, y = sp.symbols('x, y', real=True)
dic_rt = {r:sp.sqrt(x**2+y**2), θ:sp.atan(-x/y)}

ur_xy = ur_rt.subs(dic_rt)                     # Displacements in Cartesian
uθ_xy = uθ_rt.subs(dic_rt)

σrr_xy = σrr_sp.subs(dic_rt)                        # Stresses in Cartesian
σθθ_xy = σθθ_sp.subs(dic_rt)
σrθ_xy = σrθ_sp.subs(dic_rt)

display(Math(f" \\text{{Displacement in $x,y$ (as an example):}} "))
display(Math(f" \\text{{ $u_{{r}}(x, y)$}} =\\scriptsize \
                              {latex(sum(uθ_xy.evalf(4).args[:4]))}+···"))
```

Displacement in x, y (as an example):

$$u_r(x, y) = \frac{14.22x \log\left(\sqrt{x^2+y^2}\right)}{y\left(\frac{x^2}{y^2}+1.0\right)^{0.5}} + \frac{95.68x}{y\left(\frac{x^2}{y^2}+1.0\right)^{0.5}} - 53.46\left(x^2+y^2\right)^{0.5} \operatorname{atan}\left(\frac{x}{y}\right) - \frac{40.62 \operatorname{atan}\left(\frac{x}{y}\right)}{\left(\frac{x^2}{y^2}+1.0\right)^{0.5}} + \cdots$$

We can now plot the contour of the stresses using `gr.contour_plot()`, which also produces the minimum and maximum values of the stress. To plot the stresses on the shape of the arch, we create a mask. The snippet is as follows:

```
# Contour plot: Create a meshgrid for a rectangular domain in x-y plane

dX = 0.71*b_;     dY = 1.0*b_                          # plot 2dX × dY area
x_ = np.linspace(-dX, dX, 800)
y_ = np.linspace(dY*0.001, dY, 400)
X, Y = np.meshgrid(x_, y_)

# define mask to cover the region beyong the arch for plot:
mask = ((X/a_)**2 + (Y/a_)**2 < 1) | ((X/b_)**2 + (Y/b_)**2 > 1) \
      | (Y - X < 0) | (Y + X < 0)

np_ur_xy = lambdify((x,y), ur_xy, 'numpy')               # convert to numpy
Z = np_ur_xy(X, Y)                      # compute the values over the mesh grid
Z[mask] = np.nan   # mask the area beyond the arch, print nothing there
```

```
cntur = gr.contour_plot(X,Y, Z, x_lbl='x-axis',y_lbl='y-axis',
               title='Displacement $u_{{r}}$',lvls=80, ioff=True);

plt.savefig('imagesSMp/Arch_p_ur.png', dpi=500);          # save to file
#plt.show()
```

```
Minimum value in the domain: -1.08409
Maximum value in the domain: 0.951289
```

A contour plot of the radial displacement u_r within the arch subjected to uniform pressure on the top surface is shown in Fig. 10.9. The mirror symmetry of u_r with respect to the y-axis is clearly observed.

The largest radial displacements (in the negative direction) occur on the top surface and at the two ends. Additionally, significant radial displacements are observed near the bottom surface around $\theta = \pm\pi/8$.

The following snippet plots the tangential displacement component u_θ using the same procedure:

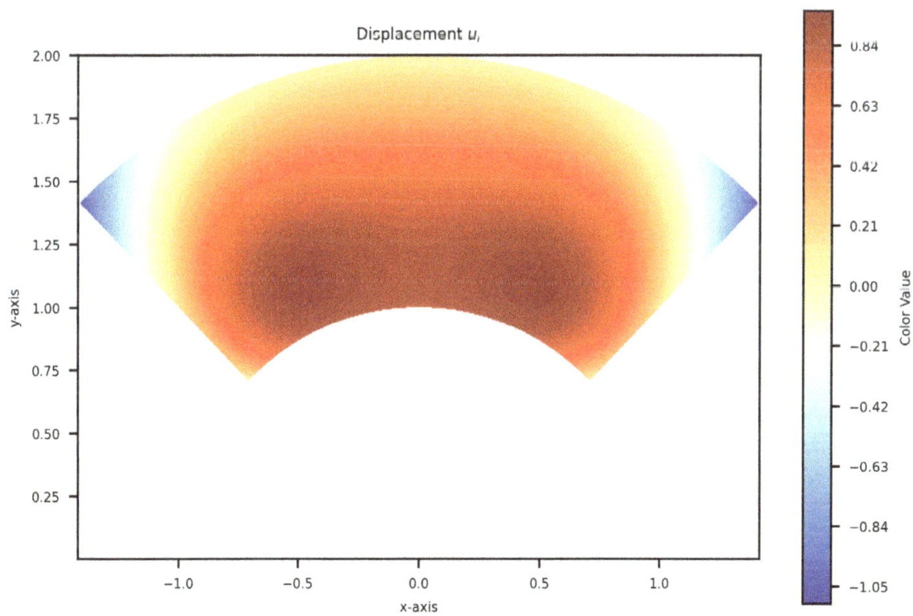

Figure 10.9. Contour plot of the radial displacement u_r in the arch subjected to uniform pressure on its top surface.

```
np_uθ_xy = lambdify((x,y), uθ_xy, 'numpy')              # convert to numpy
Z = np_uθ_xy(X, Y)                      # compute the values over the mesh grid
Z[mask] = np.nan   # mask the area beyond the arch, print nothing there

cntur = gr.contour_plot(X,Y, Z, x_lbl='x-axis',y_lbl='y-axis',
                title='Displacement $u_{{θ}}$',lvls=80, ioff=True);

plt.savefig('imagesSMp/Arch_p_ut.png', dpi=500);          # save to file
#plt.show()
```

```
Minimum value in the domain: -3.31883
Maximum value in the domain: 3.31883
```

A contour plot of the tangential displacement u_θ within the arch subjected to uniform pressure on the top surface is shown in Fig. 10.10. The antisymmetric distribution of u_θ with respect to the y-axis is observed. It is zero along the vertical cross-section and small in that region.

The largest (positive or negative) displacements occur at the four corners of the arch, as expected because these areas are the least constrained.

The following snippet plots the contour of the normal stress component $\sigma_{\theta\theta}$:

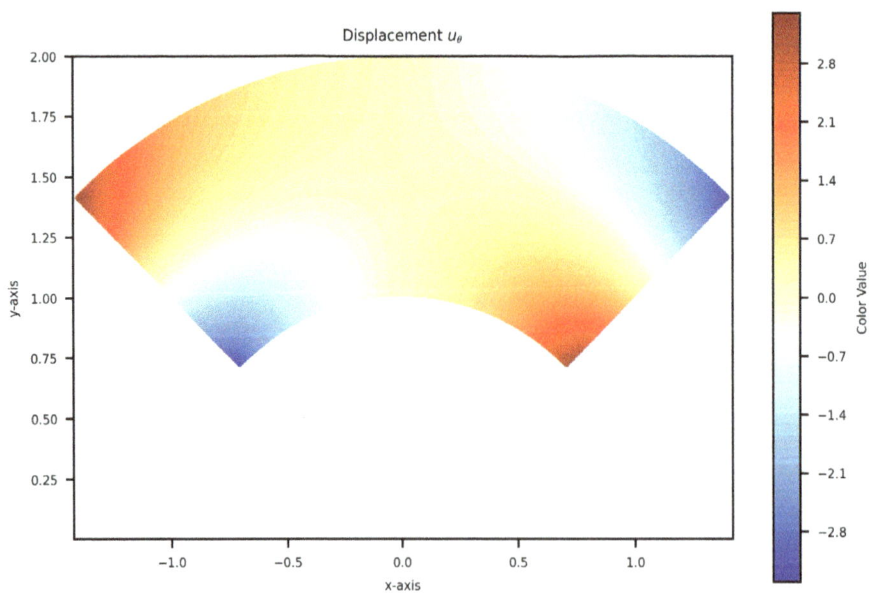

Figure 10.10. Contour plot of the tangential displacement u_θ in the arch subjected to uniform pressure on its top surface.

```
np_σθθ_xy = lambdify((x,y), σθθ_xy, 'numpy')          # convert to numpy
Z = np_σθθ_xy(X, Y)                                  # values over the mesh grid
Z[mask] = np.nan                                           # mask the area
cntur = gr.contour_plot(X,Y, Z, x_lbl='x-axis',y_lbl='y-axis',
                   title='Hoop stress $σ_{{θθ}}$',lvls=80, ioff=True);
plt.savefig('imagesSMp/Arch_p_Stt.png', dpi=500);         # save to file
#plt.show()
```

```
Minimum value in the domain: -7.96148
Maximum value in the domain: 4.08464
```

10.16.5 *Contour plots of stresses*

A contour plot of the normal stress $\sigma_{\theta\theta}$ within the arch subjected to uniform pressure on the top surface is shown in Fig. 10.11. The mirror symmetry of $\sigma_{\theta\theta}$ with respect to the y-axis is observed. Its values are relatively small near the vertical cross-section.

The largest (positive or negative) stresses occur at the four corners of the arch, particularly at the two bottom corners, reflecting strong constraints from the supporting boundaries at the arch ends. The design of the supports in an arch bridge is critical, as all the loading on the bridge surface is ultimately transferred to these two supporting ends.

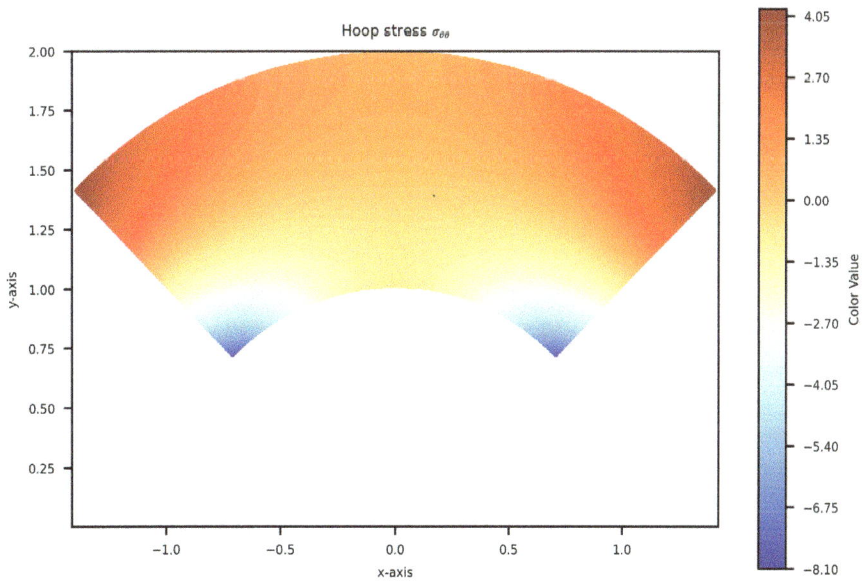

Figure 10.11. Contour plot of the normal stress $\sigma_{\theta\theta}$ in the arch subjected to uniform pressure on its top surface.

The following snippet plots the contour of the normal stress component σ_{rr}:

```
np_σrr_xy = lambdify((x,y), σrr_xy, 'numpy')          # convert to numpy
Z = np_σrr_xy(X, Y)                               # values over the mesh grid
Z[mask] = np.nan                                           # mask the area
cntur = gr.contour_plot(X,Y, Z, x_lbl='x-axis',y_lbl='y-axis',
                  title='Normal stress $σ_{{rr}}$',lvls=80, ioff=True);
plt.savefig('imagesSMp/Arch_p_Srr.png', dpi=500);         # save to file
#plt.show()
```

```
Minimum value in the domain: -1.7648
Maximum value in the domain: -1.51442e-05
```

A contour plot of the normal stress σ_{rr} within the arch subjected to uniform pressure on the top surface is shown in Fig. 10.12. The mirror symmetry of σ_{rr} with respect to the y-axis is observed. Its values are zero on the bottom surface and negative on the top surface, due to the imposed stress boundary conditions.

The largest (positive or negative) stresses occur at the supporting points at the ends of the arch, where the radial displacements are fully constrained.

The following snippet plots the contour of the shear stress component $\sigma_{r\theta}$:

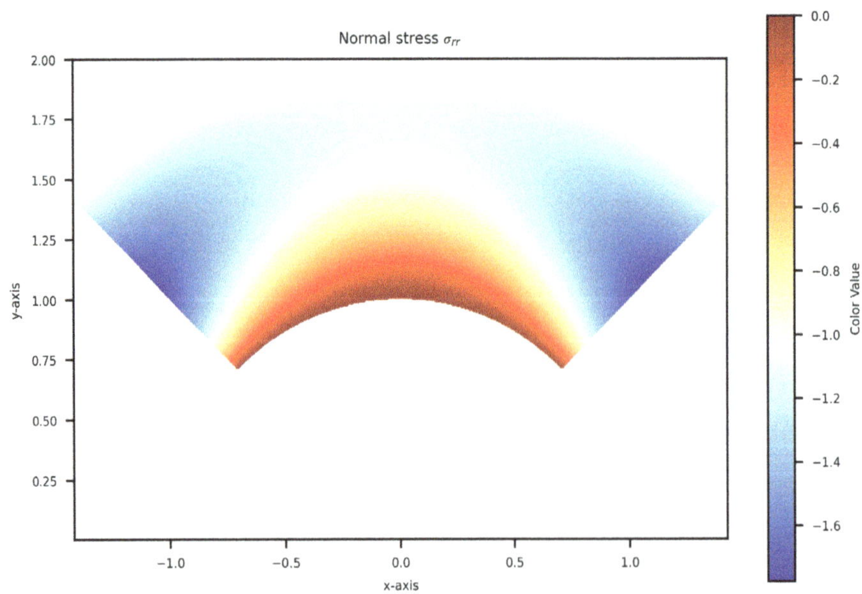

Figure 10.12. Contour plot of the normal stress σ_{rr} in the arch subjected to uniform pressure on its top surface.

```
np_σrθ_xy = lambdify((x,y), σrθ_xy, 'numpy')          # convert to numpy
Z = np_σrθ_xy(X, Y)                                  # values over the mesh grid
Z[mask] = np.nan                                            # mask the area
cntur = gr.contour_plot(X,Y, Z, x_lbl='x-axis',y_lbl='y-axis',
                    title='Hoop stress $σ_{{rθ}}$',lvls=80, ioff=True);
plt.savefig('imagesSMp/Arch_p_Srt.png', dpi=500);           # save to file
#plt.show()
```

```
Minimum value in the domain: -2.09287
Maximum value in the domain: 2.09287
```

A contour plot of the shear stress $\sigma_{r\theta}$ within the arch subjected to uniform pressure on the top surface is shown in Fig. 10.13. The antisymmetric distribution of $\sigma_{r\theta}$ with respect to the y-axis is observed. Its values are zero near the vertical cross-section and relatively small in that region. The shear stresses are zero on both surfaces, as imposed by the stress boundary conditions.

The largest (positive or negative) stresses occur at the supporting points at the ends of the arch, closer to the bottom surface. Note that for arches with $\theta_B = \theta_M$, the shear stresses are zero.

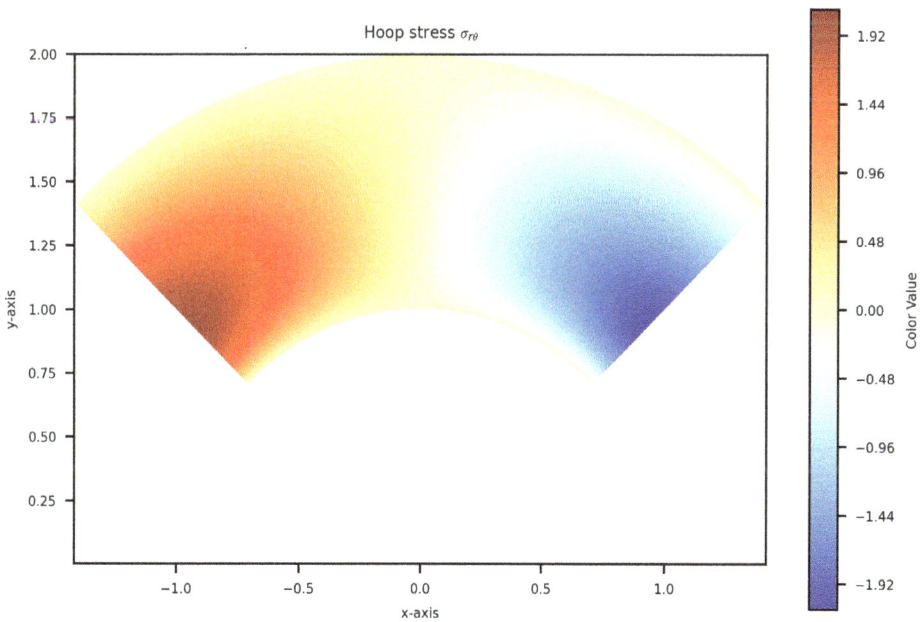

Figure 10.13. Contour plot of the shear stress $\sigma_{r\theta}$ in the arch subjected to uniform pressure on its top surface.

10.16.6 *Plots of forces on cross-sections*

The following snippets compute and plot the forces on the cross-sections of the arch. We first redefine the formulas explicitly:

```
#print(Mz_); print(); print(Fr_); print(); print(Fθ_)

a2b2 = a**2+b**2; logab = (log(a)-log(b))
Mz_=E*b4*((-a+b)*(a+b)**2+(a+b)*a2b2*logab)*cos(θ)/(4*a2b2)+b*p*(a+b)/2
Fr_=E*b4*(a**2-b**2-a2b2*logab)*sin(θ)/(2*a2b2)
Fθ_=E*b4*(a**2-b**2-a2b2*logab)*cos(θ)/(2*a2b2)-b*p

Mz_rt = Mz_.subs(dicAll_).subs(dic_data).evalf().expand()
Fr_rt = Fr_.subs(dicAll_).subs(dic_data).evalf().expand()
Fθ_rt = Fθ_.subs(dicAll_).subs(dic_data).evalf().expand()

display(Math(f"\\text{{Formulas of internal forces with numerical \
                                   constants for the example arch:}}"))
display(Math(f"\\text{{$M_z$=}} {latex(Mz_rt)}"))
display(Math(f"\\text{{$F_r$=}} {latex(Fr_rt)}"))
display(Math(f"\\text{{$F_θ$=}} {latex(Fθ_rt)}"))
```

Formulas of internal forces with numerical constants for the example arch:

$M_z = 3.0 - 2.83804125292923\cos(\theta)$

$F_r = 1.89202750195282\sin(\theta)$

$F_\theta = 1.89202750195282\cos(\theta) - 2.0$

```
# Convert SymPy expressions to Numpy functions:

func1 = sp.lambdify(θ, Mz_rt,'numpy')
func2 = sp.lambdify(θ, Fr_rt,'numpy')
func3 = sp.lambdify(θ, Fθ_rt,'numpy')
labels=[r'$M_{z}$', r'$F_{r}$', r'$F_{θ}$']

t_data  = np.linspace(-θB_, θB_, 400)          # θ values for the plot
x_data = [t_data,  t_data, t_data]
y_data = [func1(x_data[0]), func2(x_data[1]), func3(x_data[0])]
plt.ion()
gr.plot_fs(x_data, y_data, labels, '$θ(rad)$', 'Moment and Forces', \
                                         'imagesSMp/Arch_MzFs',ioff=True)
                                       # Use ioff=False to display the plot
#plt.show()
```

The distribution of internal forces along the arch is shown in Fig. 10.14. It can be observed that the moment M_z and the axial normal force F_θ are symmetric, while the shear force F_r is antisymmetric with respect to $\theta = 0$. All these forces are small near the vertical cross-section and reach their maximum values at the two supporting ends, which is typical. This is

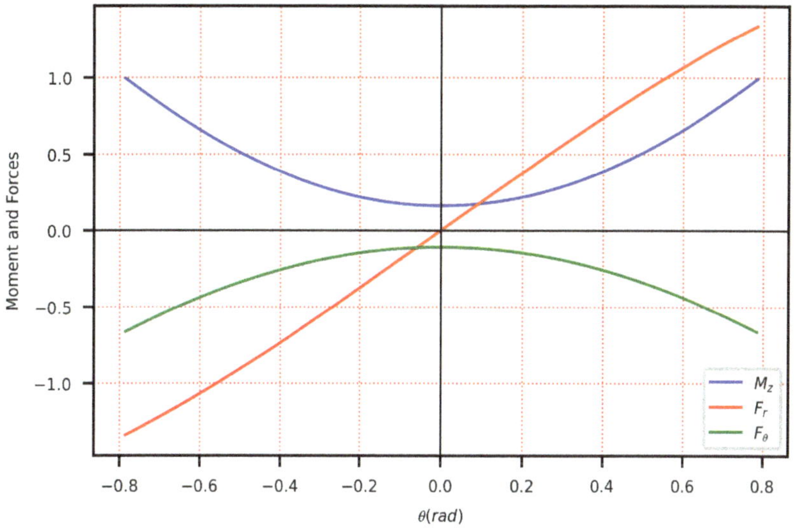

Figure 10.14. Distribution of the internal moment and forces along the arch.

the main reason why significant attention is required when designing the supports of an arch bridge.

10.16.7 *Checking FBCs*

Finally, we use the following snippet to check the FBCs by evaluating the forces on the boundary at $\theta = \theta_B$:

```
sumMz = (Int(σθθ_sp.subs(θ,θB)*r, (r,a_,b_)) \
      + Int(σθθ_sp.subs(θ,0 )*r, (r,a_,b_))).subs(θB, θB_).evalf()
display(Math(f"\\text{{$\sum M_{{z}}$=}} {latex(sumMz)}"))

Fr0 = Fr_.subs(θ,0); Fθ0 = Fθ_.subs(θ,0)
FrB = Fr_.subs(θ,θB_); FθB = Fθ_.subs(θ,θB_)

Fx=(Fθ0-FrB*sin(θB_)-FθB*cos(θB_)+p*Int(sin(θB_-θ)*b, (θ,0,θB_)))
Fy=(Fr0+FrB*cos(θB_)-FθB*sin(θB_)-p*Int(cos(θB_-θ)*b, (θ,0,θB_)))
display(Math(f"\\text{{$\sum F_{{x}}$=}} {latex(Fx)}"))
display(Math(f"\\text{{$\sum F_{{y}}$=}} {latex(Fy)}"))
```

$$\sum M_z = -2.24818281807499 \cdot 10^{-14}$$

$$\sum F_x = 0$$

$$\sum F_y = 0$$

These FBCs are all satisfied.

10.17 Arches subject to pure bending

The displacement method presented above is a general approach. It can be applied to many other similar problems, as long as the displacement field lies within the space spanned by the basis functions used in Eq. (10.1). This section demonstrates, as an example, how the method can be used to obtain the solution for a circular curved beam subjected to a pure bending moment on its cross-section, which was studied in Section 3.3 using the stress method.

10.17.1 *Determination of the constants*

In the case of pure moment bending, there are no displacement boundary conditions. Instead, we have integral equilibrium conditions or FBCs. Therefore, all formulas derived in the chapter prior to applying the FBCs can be used. The FBCs for the pure moment bending problem are illustrated in Fig. 10.15 (left), where the applied pressure is zero. The free body diagram (right) corresponds to the left half of the arch.

For this pure moment bending problem, the FBCs require the following integral equilibrium conditions to be satisfied:

$$
\begin{aligned}
&\sum F_r = 0 : \int_a^b \sigma_{r\theta}\big|_{\theta=\theta_B} dr = 0 \\
&\sum F_\theta = 0 : \int_a^b \sigma_{r\theta}\big|_{\theta=\theta_B} dr = 0 \\
&\sum M_z = 0 : \int_a^b \sigma_{\theta\theta}\big|_{\theta=\theta_B} \left(r - \frac{a+b}{2}\right) dr = -M
\end{aligned}
\tag{10.26}
$$

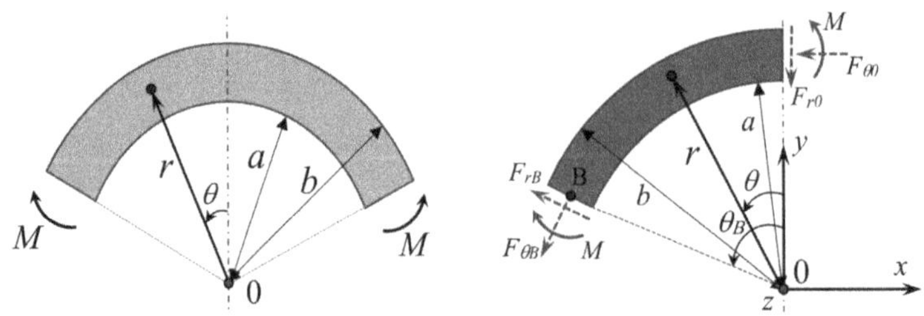

Figure 10.15. Left: An arch subjected to pure bending. Right: A free body diagram of the arch covering $\theta_B \leq \theta \leq 0$, where θ_B is taken arbitrarily and not necessarily at the left end of the arch.

This time, the moment is measured with respect to the middle axis of the arch's cross-section. Both the normal and shear forces are zero at every cross-section of the arch.

We first examine the shear force condition, which is the first equation in Eq. (10.26), using the general stresses previously obtained from the displacement. The corresponding snippet is as follows:

```
# Compute the moment using the general stresses obtained earlier:
σrr, σθθ, σrθ = σv.subs(p,0)                              # setting p=0

Fr = sp.integrate(σrθ.subs(θ,θB), (r, a, b)).simplify()
sln_Mb4 = sp.solve(Fr, b4)
display(Math(f"\\text{{Solution for $b_4$ }}= {latex(sln_Mb4)}"))
```

Solution for b_4 $= [0]$

We found that the constant b_4 must be zero. Let's now compute the normal force F_θ, using the second equation in Eq. (10.26).

```
Fθ = sp.integrate(σrθ.subs({θ:θB,b4:0}), (r, a, b)).simplify()
display(Fθ)                                                # should be 0
```

0

It is already satisfied when setting $b_4 = 0$. Finally, let us use the last equation in Eq. (10.26):

```
# Compute the moment on the cross-section:
Mz = sp.integrate(σθθ.subs({θ:θB,b4:0})*(r-(a+b)/2),(r,a,b)).simplify()
Mz_expr = Mz.collect(A5).collect(a**3).collect(b**3).collect((ν+1))
#display(Mz_expr)          # it shows that the expression has constant a3

# define the given moment on the cross-section:
M = sp.symbols('M')
sln_M = sp.solve(Mz_expr-M, a3)[0].collect(A5).collect(a**3)\
                                     .collect(b**3).collect((ν+1))
a3_M = sln_M.subs(Idic_A).simplify()           # solve it for constant a3
display(Math(f"\\text{{Solution for $a_3$}}=\\scriptsize {latex(a3_M)}"))
```

Solution for

$$a_3 = \frac{M \log\left(\left(a^{\nu+1} b^{-\nu-1}\right)^{\frac{4a^2b^2}{E}}\right)}{a^4 - 4a^2b^2\log(a)^2 - 4a^2b^2\log(b)^2 - 2a^2b^2 + b^4 + \log\left(b^{\log\left(a^{8a^2b^2}\right)}\right)}$$

Finally, we have determined a_3 for arches subjected to a pure bending load.

10.17.2 *Stress solutions*

We begin by computing the shear stress:

```
display(σrθ.subs({b4:0}))                                    # should be 0
```

0

It is zero, as expected for pure moment loading.

Next, we compute the two normal stresses:

```
σrrM=σrr.subs({b4:0}).subs(Idic_A).simplify().subs(a3,a3_M).simplify()
σθθM=σθθ.subs({b4:0}).subs(Idic_A).simplify().subs(a3,a3_M).simplify()
display(Math(f"\\text{{Normal stresses:}}"))
display(Math(f"\\text{{$σ_{{rr}}$}}=\small {latex(σrrM)}"))
display(Math(f"\\text{{$σ_{{θθ}}$}}=\small {latex(σθθM)}"))
```

Normal stresses:

$$\sigma_{rr} = \frac{M \log\left(\left(a^{a^2(-b^2+r^2)} b^{b^2(a^2-r^2)} r^{r^2(-a^2+b^2)}\right)^{\frac{4}{r^2}}\right)}{a^4 - 4a^2b^2\log(a)^2 - 4a^2b^2\log(b)^2 - 2a^2b^2 + b^4 + \log\left(b^{\log\left(a^{8a^2b^2}\right)}\right)}$$

$$\sigma_{\theta\theta} = \frac{4M\left(-a^2r^2 + b^2r^2 + \log\left(a^{a^2(b^2+r^2)} b^{-b^2(a^2+r^2)} r^{r^2(-a^2+b^2)}\right)\right)}{r^2\left(a^4 - 4a^2b^2\log(a)^2 - 4a^2b^2\log(b)^2 - 2a^2b^2 + b^4 + \log\left(b^{\log\left(a^{8a^2b^2}\right)}\right)\right)}$$

Finally, we compare the solutions with those obtained in Section 3.3:

```
# Reference solutions:
C_ab= (b**2/a**2-1)**2-4*b**2/a**2*(log(b/a))**2
σrr_ref=4*M*(-a**2*b**2*log(a)+a**2*b**2*log(b)+a**2*r**2*log(a)-\
    a**2*r**2*log(r)-b**2*r**2*log(b)+b**2*r**2*log(r))/(C_ab*a**4*r**2)

σθθ_ref=4*M*(r**2*(-a**2+b**2+log(a**(a**2)*r**(-a**2+b**2)/b**(b**2)))+\
                                      log((a/b)**(a**2*b**2)))/(C_ab*a**4*r**2)

print('Are the stresses the same? ')
display((sp.expand_log(σrrM - sp.expand_log(σrr_ref)).simplify()==0,
         sp.expand_log(σθθM - sp.expand_log(σθθ_ref)).simplify()==0))
```

```
Are the stresses the same?

(True, True)
```

We obtained the same stress solution. Readers may similarly verify the displacement solutions.

This example demonstrates the effectiveness of the displacement method in obtaining solutions for arches subjected to pure moment loads.

10.18 Review questions

1. Derive the formulas for displacements and stresses for the same arch given in Section 10.1 and Fig. 10.1, but with pressure applied over the inner surface of the arch. You may use the general displacement formulas given in Eq. (10.5).
2. Derive the displacement formulas for the arch subjected to bending, as shown in Fig. 10.15. You may use the general displacement formulas given in Eq. (10.5).
3. Consider a thin, semicircular arch of uniform thickness with an inner radius a and an outer radius b. It is subjected to a pair of horizontal forces P acting on the cross-sections at its two ends, as shown in the left figure in Fig. 10.16. Determine the displacements and stresses in the arch using the displacement method presented in this chapter. You may use the appropriate formulas already derived in this chapter.
4. Consider a thin, semicircular arch of uniform thickness with an inner radius a and an outer radius b. It is subjected to a pair of vertical forces P acting on the cross-sections at its two ends, as shown in the right figure in Fig. 10.16. Determine the displacements and stresses in the arch using the displacement method presented in this chapter. You may use the appropriate formulas already derived in this chapter.

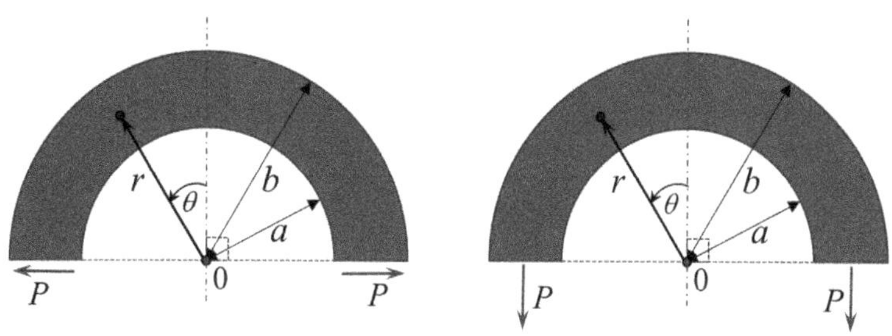

Figure 10.16. A thin, semicircular arch of uniform thickness loaded by a pair of forces acting on the cross-sections at its two ends. Left: horizontal forces; Right: vertical forces.

10.19 Remarks

This chapter examines a circular arch subjected to surface pressure using the displacement method. The following remarks can be highlighted:

1. A general displacement method is presented for the first time for this type of complex yet practical problem.
2. A set of exact closed-form solutions has been derived, with additional efforts made to simplify the key formulas into concise forms for easier reading and application.
3. A set of Python scripts is also provided for formula derivation, numerical computation, and results plotting, which can be applied to other similar problems.
4. Procedures and formulas are presented for the first time for designing arches in a pure bending state under pressure, where the shear stress is zero.

References

[1] G. R. Liu, *Solid Mechanics with Python: Cartesian Coordinates*, World Scientific, New Jersey, 2024.

[2] G. R. Liu, *Numbers and Functions: Theory, Formulation, and Python Codes*, World Scientific, New Jersey, 2024.

[3] G. R. Liu, *Calculus: A Practical Course with Python*, World Scientific, New Jersey, 2025.

Index

N

O

P

R

S

www.ingramcontent.com/pod-product-compliance
Ingram Content Group UK Ltd.
Pitfield, Milton Keynes, MK11 3LW, UK
UKHW061937120926
8588IPUK00008B/80